Genetic Analysis

There is a paradox lying at the heart of the study of heredity. To understand the ways in which features are passed on from one generation to the next, we have to dig deeper and deeper into the ultimate nature of things – from organisms, to genes, to molecules. And yet as we do this, increasingly we find we are out of focus with our subjects. What has any of this to do with the living, breathing organisms with which we started? Organisms are living. Molecules are not. How do we relate one to the other?

In *Genetic Analysis*, one of the most important empirical scientists in the field in the twentieth century attempts, through a study of history and drawing on his own vast experience as a practitioner, to face this paradox head-on. His book offers a deep and innovative understanding of our ways of thinking about heredity.

RAPHAEL FALK is Emeritus Professor, Department of Genetics and The Program for the History and Philosophy of Science at The Hebrew University of Jerusalem. His many works include *The Concept of the Gene in Development and Evolution: Historical and Epistemological Perspectives*, co-edited by Peter J. Beurton and Hans-Jörg Rheinberger (2000, 2008).

T0297268

CAMBRIDGE STUDIES IN
PHILOSOPHY AND BIOLOGY

General Editor
Michael Ruse, *Florida State University*

Advisory Board
Michael Donoghue, *Yale University*
Jean Gayon, *University of Paris*
Jonathan Hodge, *University of Leeds*
Jane Maienschein, *Arizona State University*
Jesús Mosterín, *Instituto de Filosofía (Spanish Research Council)*
Elliott Sober, *University of Wisconsin*

Recent Titles
Alfred I. Tauber *The Immune Self: Theory or Metaphor?*
Elliott Sober *From a Biological Point of View*
Robert Brandon *Concepts and Methods in Evolutionary Biology*
Peter Godfrey-Smith *Complexity and the Function of Mind in Nature*
William A. Rottschaefer *The Biology and Psychology of Moral Agency*
Sahotra Sarkar *Genetics and Reductionism*
Jean Gayon *Darwinism's Struggle for Survival*
Jane Maienschein and Michael Ruse (eds.) *Biology and the
Foundation of Ethics*
Jack Wilson *Biological Individuality*
Richard Creath and Jane Maienschein (eds.)
Biology and Epistemology
Jane Maienschein and Michael Ruse (eds.) *Biology and the
Foundation of Ethics*
Jack Wilson *Biological Individuality*
Richard Creath and Jane Maienschein (eds.) *Biology and Epistemology*
Alexander Rosenberg *Darwinism in Philosophy, Social Science,
and Policy*
Peter Beurton, Raphael Falk, and Hans-Jörg Rheinberger (eds.) *The
Concept of the Gene in Development and Evolution*
David Hull *Science and Selection*

James G. Lennox *Aristotle's Philosophy of Biology*

Marc Ereshefsky *The Poverty of the Linnaean Hierarchy*

Kim Sterelny *The Evolution of Agency and Other Essays*

William S. Cooper *The Evolution of Reason*

Peter McLaughlin *What Functions Explain*

Steven Hecht Orzack and Elliott Sober (eds.) *Adaptationism and Optimality*

Bryan G. Norton *Searching for Sustainability*

Sandra D. Mitchell *Biological Complexity and Integrative Pluralism*

Greg Cooper *The Science of the Struggle for Existence*

Joseph LaPorte *Natural Kinds and Conceptual Change*

Jason Scott Robert *Embryology, Epigenesis, and Evolution*

William F. Harms *Information and Meaning in Evolutionary Processes*

Marcel Weber *Philosophy of Experimental Biology*

Markku Oksanen and Juhani Pietorinen *Philosophy and Biodiversity*

Richard Burian *The Epistemology of Development, Evolution, and Genetics*

Ron Amundson *The Changing Role of the Embryo in Evolutionary Thought*

Sahotra Sarkar *Biodiversity and Environmental Philosophy*

Neven Sesardic *Making Sense of Heritability*

William Bechtel *Discovering Cell Mechanisms*

Giovanni Boniolo and Gabriele De Anna (eds.) *Evolutionary Ethics and Contemporary Biology*

Justin E. H. Smith (ed.) *The Problem of Animal Generation in Early Modern Philosophy*

Lindley Darden *Reasoning in Biological Discoveries*

Derek Turner *Making Prehistory*

Elizabeth A. Lloyd *Science, Politics and Evolution*

Manfred D. Laubichler and Jane Maienschein (eds.) *Form and Function in Developmental Evolution*

Genetic Analysis

A History of Genetic Thinking

RAPHAEL FALK

The Hebrew University of Jerusalem
Jerusalem, Israel

CAMBRIDGE
UNIVERSITY PRESS

CAMBRIDGE UNIVERSITY PRESS
Cambridge, New York, Melbourne, Madrid, Cape Town, Singapore,
São Paulo, Delhi, Dubai, Tokyo, Mexico City

Cambridge University Press
The Edinburgh Building, Cambridge CB2 8RU, UK

Published in the United States of America by Cambridge University Press, New York

www.cambridge.org
Information on this title: www.cambridge.org/9780521182812

First published 2009
First paperback edition 2010

A catalogue record for this publication is available from the British Library

Library of Congress Cataloguing in Publication data

Falk, Raphael.
  Genetic analysis : a history of genetic thinking / Raphael Falk.
    p. ; cm. – (Cambridge studies in philosophy and biology)
  Includes bibliographical references and index.
  ISBN 978-0-521-88418-1 (hardback)   1. Genetics–History.   I. Title.   II. Series:
  Cambridge studies in philosophy and biology.
  [DNLM: 1. Genetics.   2. Genetics–history.   3. History, 19th Century.   4. History,
  20th Century.   QU 450 F191g 2009]
  QH428.F35 2009
  576.5–dc22
              2009004699

ISBN 978-0-521-88418-1 Hardback
ISBN 978-0-521-18281-2 Paperback

Is the growth of science essentially so slow and so continuous that our attention is attracted only by the sudden showy change, which, like the bursting of a chrysalis, is merely the sequel to something of more importance which went before? Or, does a particular piece of work . . . have a value per se *which transcends the others completely? Probably both questions should have affirmative answers.*

(East, 1923, 227)

Contents

List of figures   *page* xi

Acknowledgments   xiii

Introduction   1

PART I   FROM REPRODUCTION AND GENERATION
          TO HEREDITY   11

1 The biologization of inheritance   14

2 Mendel: the design of an experiment   25

PART II   *FAKTOREN* IN SEARCH OF MEANING   39

3 From *Faktoren* to unit characters   44

4 The demise of the unit character   58

PART III   THE CHROMOSOME THEORY OF INHERITANCE   75

5 Chromosomes and Mendelian *Faktoren*   77

6 Mapping the chromosomes   94

7 Cytogenetic analysis of the chromosomes   108

PART IV   GENES AS THE ATOMS OF HEREDITY   125

8 Characterizing the gene   128

9 Analysis of the gene by mutations   141

10 From evolution to population genetics   158

Contents

PART V   INCREASING RESOLVING POWER   171

11 Recruiting bacteria and their viruses   178

12 Molecular "cytogenetics"   191

13 Recombination molecularized   202

PART VI   DEDUCING GENES FROM TRAITS, INDUCING TRAITS
           FROM GENES   209

14 How do genes do it?   211

15 The path from DNA to protein   220

16 Genes in the service of development   231

PART VII   WHAT IS TRUE FOR *E. COLI* IS NOT TRUE FOR
            THE ELEPHANT   245

17 Extending hybridization to molecules   249

18 Overcoming the dogma   259

19 Dominance   268

20 Populations evolve, organisms develop   274

Concluding comments   287
Bibliography   293
Index   321

Figures

2.1 Reconstruction of the design of Mendel's
monohybrid experiments, following Fisher (1936). *page* 31

5.1 Morgan's original interpretation and the formal
genetic interpretation of the experimental results
of white eye-color inheritance in *Drosophila*. 82

5.2 *Camera lucida* drawings of oogonial metaphase plates
of mitoses of a normal (XX) female (left), and a
non-disjunctional (XXY) female (right) (Bridges, 1916). 85

5.3 Scheme of production of primary non-disjunction in
matings of *Drosophila* and of their progeny, producing
secondary non-disjunction progeny, according to
Bridges (1916). 86

6.1 Mitotic crossing over between two of four chromatids
that may produce homozygosis for distal markers. 99

6.2 "Left end" of the polytenic chromosome-X of *Drosophila
melanogaster* with the aligned linkage map above it
(after Bridges, 1938). 102

8.1 The *ClB* mating scheme, screening for mutations on the
X-chromosome of *Drosophila melanogaster*. 137

8.2 The *Cy/Pm* mating scheme, screening for mutations on
chromosome II of *Drosophila melanogaster*. 138

9.1 A. Scheme of a multi-site gene or a nest of genes so
closely linked that no crossing over can be detected
between them. B. A mutation in each site of the
gene/nest of genes (Stadler, 1954). 151

9.2 Scheme of a multi-site gene, or a nest of tightly linked
genes. Sites may be deleted, which would make the

operational distinction between one gene or a nest
of genes irrelevant (Stadler, 1954). 152

11.1 Map of the *r*II locus of the T4 bacteriophage: two
cistrons, A and B, each contain many recons. Mutons
in recons differ in their specific mutation rates (after
Benzer, 1957). 189

12.1 Scheme of replication of a chromosome of *Vicia faba* in
the presence of a radioactively labeled medium. Each
line represents one subunit of the chromosome or
chromatid. Broken lines indicate labeled subunits; solid
lines indicate unlabeled subunits. C-metaphase indicates
metaphase in the presence of colchicine that prevents
separation of centromeres (Taylor *et al.*, 1957). 199

13.1 The Holliday model for enzymatically guided
recombination between two chromatids, each
composed of a double helix molecule of DNA
(Stahl, 1994). 207

14.1 The relation between eye-color and gene dosage or
activity at the *white* gene of *Drosophila*. w – white eye
allele; w^a – white-apricot allele; w^+ – wild-type allele
of the eye-color gene (after Muller, 1950a). 216

16.1 Schematic description of Morgan's conception of
"wild-type" and mutant wing production as reflected
in his mutation nomenclature. 232

16.2 "Yellow" (upper right) and "singed" (lower right) twin
spot of neighboring cells produced by somatic
recombination in a cell of a "wild-type" fly (left). 240

17.1 Minimum amount of DNA per cell in various
systematic classes (Britten and Davidson, 1969). 252

C.1 Image of the complex relationships of a cell at the
protein level. Insert shows in detail an (enlarged)
minute fraction for discerning and studying single
(inter)-action. 290

Acknowledgments

I am very fortunate to have had most inspiring and stimulating teachers. I have been especially influenced by Georg Haas in Zoology, and Elisabeth Goldschmidt in Genetics at the Hebrew University, and by Gert Bonnier at the Genetiska Institutet in Stockholms Högskola in Sweden. My insights were formulated further during the years of my post doctoral studies with Herman J. Muller, Curt Stern, and Paul Margolin. Jim Crow, Larry Sandler Dick Lewontin, and Ed Novitski helped me to examine my understanding and develop it further.

Interacting with my colleague Jacob Wahrman was an inexhaustible intellectual and scientific challenge to me. Yehuda Elkana and Yemima Ben-Menahem were most instrumental in assisting me to proceed, after thirty years at the laboratory bench, into the disciplines of the philosophy and history of science. I was most fortunate to spend a year in the Institute of Advanced Studies in Berlin where I tried to absorb as much as possible from Bob Cohen, Everett Mendelsohn, "Peter" Hempel, Tim Lenoir, Peter Bieri, Peter Weingart, and many other colleagues.

Over the years of writing this book I have been greatly helped by consulting many colleagues; more specifically, I was assisted by Moshe Feldman, Snait Gissis, Oren Harman, Eva Jablonka, Staffan Müller-Wille, Vítězslav Orel, Sam Schweber, Giora Simchen, Frank Stahl, Dudy Tzfati, and many others to whom I apologize for not naming them.

Miriam Greenfield turned my English into a respectable language. John Forder saved the text by his conscientious proof-reading. Yael Oren helped with the graphics. Michael Ruse, my editor, was a vital source of support and encouragement in the innumerous crises, from the planning of this volume to its publication.

My wife, Ruma Falk, has been an untiring support and an incredible source of inspiration and insight. Her intellectual devotion and thorough, uncompromising analysis has always been for me a beacon to strive to follow.

I wish to thank them all, and the many whom I did not mention by name, for their help in bringing to fruition this attempt to evaluate the role of genetic analysis in the intellectual context of contemporary life sciences.

Introduction

The subject of this book is genetic analysis. I have been involved in genetic analysis for over a half century, first in active experimental research and later doing research on the history and philosophy of genetics.

In 1965, the centenary of Mendel's presentation to the Natural History Society in Brno, two books were published with almost identical titles by two leading geneticists of that time: Alfred H. Sturtevant's *A History of Genetics* (1965) and Leslie C. Dunn's *A Short History of Genetics* (1965). Sturtevant's preface was very brief and succinct: "The publication of Mendel's paper of 1866 is the outstanding event in the history of genetics; but . . . the paper was overlooked until 1900, when it was found. Its importance was then at once widely recognized. These facts make the selection of topics for the early chapters of this book almost automatic" (Sturtevant, 1965, vii). I will discuss this notion at some length in later chapters. Dunn's approach was more reflective; he focused on the role and significance of the history of science. With respect to the history of genetics, Dunn noted:

> One of the interesting things about the history of genetics is that a few relatively simple ideas, stated clearly and tested by easily comprehended breeding experiments brought about a fundamental transformation of views about heredity, reproduction, evolution and the structure of living matter. It was chiefly the elucidation of the theory of the gene and its extension to the physical basis of heredity and to the causes of evolutionary changes in populations which gave genetics its unified character.
>
> Dunn (1965, vii)

Nonetheless, despite the magnitude of the achievement, Dunn observed that there was no interest in the history of genetics among historians of science because "[t]he events leading to its rise have been too recent to attract the interest of professional historians" (Dunn, 1965, ix). And the same was true of researchers who were practicing genetics. In the introduction to his book Dunn noted his surprise when a fellow geneticist explained why he was not familiar with the work of a predecessor: "if I read everyone else's paper, I wouldn't get my own written." Dunn noted that "an adequate perspective is an essential element in all historical research. [But f]or those who have participated in the development of genetics, the interest in the unfolding facts and theories and the opportunity to influence its surging progress have in general outweighed any temptation to stand aside long enough to reflect on the origin of its ideas and where they were leading." He agreed that "this on the whole is as it should be" (Dunn, 1965, ix), but commented that although "that attitude ... is not a useful view for science generally ... it is understandable in a field like genetics, where liberation from restrictions imposed by traditional ideas is sometimes a necessary condition for developing new views." And he stressed that "this aspect of genetics is especially marked today [1965]" when

> the attention of both the scientific and the lay public has for the past ten years been focused on the molecular basis of heredity and on the mode of transmission and transcription of a code of instructions which guides progeny in repeating the biological patterns of their ancestors. The discoveries in this field have been so rapid and exciting and so recent as to create an impression that genetics began in 1944 with O. T. Avery's discovery that the nucleic acid DNA is the vehicle of hereditary transmission.
>
> Dunn (1965, xii)

Dunn referred to the book of Alfred Barthelmess of 1952 that represented "the first attempt to trace the origin and path of development of the science of heredity."

> Whether one places the date of the birth of this branch of biology in the year 1900 or 1866 or even farther back, it nevertheless remains astonishing that until now no history of it has been written. The science of heredity has unfolded itself so precipitately and flowers today so vigorously that one could easily think, in seeking a reason for this lack, that there has been no time for reflection.
>
> Dunn (1965, xv)

The situation has changed radically since then. The history and the philosophy of genetics have attracted a great deal of attention by historians and philosophers of science (e.g., Harman, 2004; Keller, 2000;

Kohler, 1994; Moss, 2003; Olby, 1985; Orel, 1996), and to a more modest extent by scientists themselves (e.g., Carlson, 1966 /1989, 2004; Falk, 1986; Glass, 1963; Lederberg, 1990; Portin, 1993; Zuckerman and Lederberg, 1986). Special attention has been devoted to the history of genetics in the molecular era (e.g., Holmes, 2001; Judson, 1979; Kay, 2000; Morange, 1994, 1998; Olby, 1974; Rheinberger, 1997; Watson, 1968; Weiner, 1999). Many modern texts have claimed that a break in the continuity of genetic theories occurred in the 1950s with the introduction of the Watson–Crick model of DNA, the establishment of experimental research at the bacterial level, and the introduction of molecular methodologies to genetic analysis (see Olby, 1990, for a discussion). Thus philosopher Philip Kitcher has suggested: "There are two recent theories which have addressed the phenomena of heredity. One, *classical genetics*, stemming from the studies of T. H. Morgan, his colleagues and students, is the successful outgrowth of the Mendelian theory of heredity rediscovered at the beginning of this century. The other, *molecular genetics*, descends from the work of Watson and Crick" (Kitcher, 1984, 337). Of considerable influence has been Evelyn Fox Keller's thesis that the change from a linear mode of thinking to that of a cybernetic, informational mode changed the image of the gene from that of an *acting* agent to that of an *activated* agent (Keller, 1995, 2000, 2002). Moreover, Lenny Moss suggested that the gene concept should be dichotomized into a gene-*P* which is identified by a phenotypic marker and a gene-*D* which is defined by its molecular sequence (Moss, 2003).

I claim that it is wrong to conceive of the phenomena of heredity as involving two theories, classical genetics and molecular genetics. There are not two theories one of which (classical) should be reduced to the other (molecular). Indeed, philosophers of science have shown that formally such a reduction is futile (e.g., Kitcher, 1984; Schaffner, 1976. See also Sarkar, 1998). I propose that it is more meaningful historically and more helpful scientifically to view these not as two theories, but as one continuous theory that deals with the same array of problems at different levels of resolution. In the biological sciences, claims of regularity (and "lawfulness") are contingent on past events that happen to have taken place and were (nearly) fixed by natural selection and by the constraints of structure and function that have prevailed. In the physical sciences, foundational laws involving the nature of matter have been found to be essentially ahistoric – that is, time-translation-invariant over time scales close to the age of our universe. As Dobzhansky famously stated: "Nothing in biology makes sense except in the light of evolution"

(Dobzhansky, 1973), or, in the words of a philosopher of science: "the aim of biological theorizing is not, as it is in physical science, the identification of natural laws of successive generality, precision, and power, but the sharpening of tools for interacting with the biosphere" (Rosenberg, 1979, 254).

This book is an argument against a conceptual discontinuity between "classical" and "molecular" theories of genetics. In it I claim that molecular genetics is an organic extension of the so-called "classical" conceptions of genetic analysis, an evolution by refinement of methods, for example adopting biochemical and molecular markers (and eventually simply specific nucleotide bases, SNPs) to replace the traditional phenomenological markers such as wrinkled pea seeds or white eye-color of flies. Genetic analysis is the art of analyzing the phenomena of heredity by hybridization. Hybridization is a very ancient art, practiced primarily by breeders. The science of heredity is based on this ancient art: starting with Linnæus in the eighteenth century this art became a research tradition. Defined this way, the tradition is based on a methodology of interfering. Experimental examination of (preconceived) theories should be viewed as parallel to what I call the morphogenist tradition, which relies mainly on observations in the field and on the dissecting table. Although hybridization nowadays incorporates a wide array of techniques, including many at the level of DNA molecules, since 1865 the art has developed as an integral and consistent discipline on the foundation of Gregor Mendel's experiments with hybrids of garden peas. In the 1940s, the aggressive developments of what many view as a new research tradition of molecular biology began to increasingly affect not only the practical application of molecular methodologies to genetic problems, but also the conceptualization of the issues of genetics, to the extent that molecular genetics was claimed to comprise a discipline distinct from classical genetics.

Genetic Analysis presents the study of inheritance as a *conception directed by a methodology*. As such the book is organized as a historical study of the design of experimental evidence and its application to genetic theories.

As the art of analyzing the phenomena of heredity in the tradition of hybridization, genetic analysis is a discipline characterized by *methodological reductionism*, the assumption that empirically following single variables is the effective way to bridge realms. *Conceptual reductionism*, on the other hand, assumes that phenomena may be determined by a component or components from a more basic realm, and that the component or components individually or interactively bridge the phenomena to a higher realm. Methodological reductionism may be considered an epistemological statement, whereas conceptual reductionism is essentially an ontological one (see Sarkar, 1998, 19ff.). The distinction is one between explanation and resolution (see Falk, 2006, 219). Once we accept this, the problem of a formal, classic attempt to reduce one theory to the other – problematic as this by itself may be – becomes irrelevant to genetic analysis (see Fuerst, 1982).

In the introduction to his *Short History of Genetics* Dunn confessed that what interested him most in the history of science was "the relationship between ideas held at different times, couched in similar terms, yet obviously having different contents and meanings . . . What, if anything, does the second concept owe to the first? How, if not derived from the first, did the second arise?" (Dunn, 1965, xvii). Once we overcome the issue of the formal conceptual reduction of theories, we may, as Dunn suggested, trace the evolutionary change in the meaning of concepts. The understanding of this evolution of concepts is significant not only to the historian or the philosopher of science; it should also be of primary interest to the practicing geneticist.

Consider the concept of the gene: When practicing geneticists involved in deciphering the human genome at the turn of the millennium officially bet on the number of genes of the human genome, what were they referring to? Certainly not the concept formulated by Johannsen, in 1909 nor the dictum of "one gene – one enzyme" formulated by Beadle and Tatum in 1942. In 2003–4 at a workshop on "representing genes," organized by Karola Stotz and Paul Griffiths at the University of Pittsburgh, participants discussed roughly a dozen descriptions of generating transcripts and/or polypeptides that were considered to be genes. Why is the polypeptide translated on the ribosomes less of a phenotype than the vermilion eye-color of a Drosophila fly? Or for that matter, would it be wrong to refer to the transcribed RNA molecule (before splicing or afterwards) or even to the DNA sequence itself as phenotypes of the "something" that is conceived as the genotype? Aren't we actually reading off the genotype directly from the DNA sequence "this most basic of all phenotypes"? (Griffiths, Gelbart, Miller, and Lewontin, 1999, 576). A recent TV program claimed: "Tell me your genes and I'll tell you who you are." Having been trained as an experimental scientist I examine my claims empirically. The issue of whether the concepts of genetics have changed continuously or whether fundamentally different concepts have been generated at different periods is an issue that should be examined by juxtaposing the experiments done and quoting from the discussions of the researchers involved and the textbooks of the time. This I wish to do in the present book.

Mendel did not introduce a Kuhnian paradigm shift in biological research with his paper of 1866. Rather his work was profoundly integrated in the social, religious and scientific tradition of his Central European community. Acting within the hybridist research tradition, Mendel believed in a world constructed from the bottom up on the basis of God-directed lawfulness that had to be discovered and explicated. In that sense Mendel's ideas relied conceptually and therefore also methodologically on notions of the physical sciences using numerical analyses. His experiments were reductionist, bottom-up examinations of his theories based on his beliefs. This contrasted with the traditional top-down morphogenist research methods employed in comparative anatomy, embryology, or natural history, which viewed life as being a property that emerged *per se*, and was not (or not necessarily) reducible to simple phenomena that could be analyzed numerically in terms of physical science.

In 1900, Mendel's work was "rediscovered" only in the sense that researchers – foremost among them William Bateson and Hugo de Vries – had encountered difficulties with the evolutionary morphogenist tradition, whether in field observations or at the embryologist's and cytologist's laboratory bench, and had tried to overcome these by imposing the heuristics of the hybridist tradition onto their morphogenist conceptions. I suggest that when genetics was established as a discipline of the life sciences at the beginning of the twentieth century it was on the basis of an attempt to reconcile the two research traditions. However, the result was that genetics became a discipline of confrontation between material hypothetical constructs and instrumental intervening variables (MacCorquodale and Meehl, 1948) rather than a discipline of a reductionist research heuristics that formulated its regularities in lawful terms. A focal point of this confrontation was when R. A. Fisher (1936) challenged the experimental data in Mendel's paper, asking "Has Mendel's work been rediscovered?" Many years later Robert Olby would reformulate the question by asking "Mendel no Mendelian?" (Olby, 1979). For Mendel and for Wilhelm Johannsen – who introduced the genotype and gene conceptions – the hereditary factors were only *a priori* helpful instrumental variables, while for R. A. Fisher they were experimental material constructs. The "too good" fit of data and expectations led to suspicion of Mendel's findings or the actions of some of his associates rather than acceptance of his findings as evidence of a well-designed experiment of a preconceived theory (see, e.g., Sapp, 1990, chapter 5, 104–119).

With the adoption of the chromosome theory of inheritance by Thomas H. Morgan and his associates in the 1910s, genetics achieved its independence as a research discipline. It adopted the analytic reductionist research heuristics but maintained a dialectical conceptual confrontation between materialists and instrumentalists, or equivalently, between those who believed that they were dealing with hypothetical constructs and those who insisted that their entities were nothing but intervening variables. The evolution of the concept of the gene reflects this methodologically based conceptual tension as an ongoing dialectical confrontation between instrumental and material entities (Falk, 1986, 2000b, 2004).

Genetic analysis was inherently a phenomenological research discipline. Mendel used variables that were experimentally discernible and adequate for gathering considerable data to represent his *Faktoren*, irrespective of what their specific properties were. Once Johannsen overcame the identification of the Mendelian factors with "unit characters," the observable characteristics served only as "markers" of the genes. The chromosome theory of inheritance provided a firm cytological basis for the Mendelian analysis, and the analytic genetic linkage theory provided strong support for the cytological observations. The improvement in the sophistication of the phenomenological reductionist research methods turned the balance increasingly toward material "genocentrist" determinism, and genetic research increasingly introduced biochemical, even molecular, marker-variables instead of the classic phenomenological variables. Reductionist determinism triumphed with the evidence for DNA being the material basis of genetic claims and Watson and Crick's presentation of the model of the complementary double helix in 1953. Fungal and microbial screening methods increased the resolving power of genetic analysis by many orders of magnitude, and within a decade phenomenological genetics turned into molecular genetics. Reductionist genetic analysis reached a new peak with the acceptance of Crick's Central Dogma of genetics in the late 1950s: Genetic specificity is maintained by the sequence of bases in the DNA and expressed in the corresponding colinearity of the sequence of the amino acids of the polypeptides; DNA determines RNA which informs proteins. What was true for *E. coli* would be true for the elephant. Indeed, the triumph of methodological reduction was conceived as the victory of conceptual reduction (Monod, 1972), to the extent that some philosophically minded researchers believed that the science had exhausted itself, and no more fundamental principles of living organisms could be discovered (see Stent, 1969).

This picture started to change in the mid-1960s when inconsistencies arose within reductionist molecular genetic analyses. The more the reductionist heuristic of molecular analysis progressed, the more it became obvious that conceptual reductionism must be modified, and researchers returned to a conception of top-down systems. As it turned out, the simplistic reduction of genes to DNA sequences collapsed when it appeared that not all DNA was "genetic" – terms like "redundant" and even "junk" DNA prevailed. Even more traumatic was the increasing evidence that DNA sequences were not "simply" and unequivocally transcribed into messenger-RNA, which is straightforwardly translated into polypeptides. It became recognized that DNA sequences were also involved in "regulation" rather than merely in "coding," and it became increasingly clear that it was the cell (if not the organism) – rather than DNA, or even DNA transcribed into RNA that is translated to a poly-peptide – that was the critical *sub-system.* Conceptually, it was the perspective of the system that had to be clarified.

Even though researchers were aware that biological systems must be conceived as such, they were restricted by complexity because of limitations on human computational and cognitive powers, and there was often an irresistible temptation to continue to extend the efficient reductionist heuristics to reductionist conceptions. However, with the increasing computational power of modern computers and the parallel development of the computational sciences in capacities such as modeling and simulation, some of these human cognitive limitations were overcome. The triumph of the Human Genome Project at the turn of the millennium was proof of this expansion of technology and its power to affect theory. Once this conceptual top-down perspective was imposed on the bottom-up experimental heuristics, "genetic analysis" became less genetic. Biochemistry, cell biology, embryology and development, evolution, even comparative taxonomy, all became players in "system analysis," which transformed the life sciences. Today there is no longer a distinct science of genetics; neither neurobiologists nor medical doctors can avoid the involvement of genes in their research and practice. Yet, genetic analysis as a research method prevails, and now two DNA strands from organisms as distant as a mosquito fly and a Mangrove tree may be the ones that are hybridized *in vitro.*

Introduction

When I was an undergraduate, professor Georg Haas at the Department of Zoology of the Hebrew University used to complain in his comparative anatomy class that he was unable "to talk as an orchestra": He was reduced to linearly and sequentially presenting processes that occurred simultaneously and interactively. I too am restricted by this limitation and must present my evidence successively, but I hope to convey the reality of interactive integration by occasionally telling the same story from a different angle. As may have become clear, my belief in the intellectual continuity of genetic analysis makes my story rather "Whiggish" in spite of my attempts to stress the incessant emergence of new ideas and notions along a continuous road. I present in some detail not only experiments that I consider to be pivotal for genetic analysis but also some that serve to illuminate specific issues of genetic analysis, by giving both the rationale of the experiments and the methodology chosen to answer the challenge, often with quotations from the original sources. Admittedly, the presentation of the experimental evidence is heavily biased towards Drosophila, since this was the main object of my research work.

Each part of this book introduces a central idea of genetic analysis and comprises chapters that give the experimental and theoretical evidence for that central idea.

Part I "From Reproduction and Generation to Heredity" discusses the significance of Linnæus and his followers, who established a *science* of heredity. It recounts the role of Mendel in establishing the parameters of genetic analysis by the design of his experiments.

In Part II on "*Faktoren* in Search of Meaning" I discuss the intellectual circumstances surrounding the acceptance of the Mendelian principles, the constraints of evolutionary and cell biology and the establishment of the foundations of an independent discipline when these constraints were overcome.

Part III is devoted to "The Chromosome Theory of Inheritance," the development of new instruments of analysis, including the establishment of analytic cytogenetic research.

Part IV explains the concept of the gene. It describes the confrontation between the instrumental and the material conception and discusses the concept of the gene at the heart of genetics as a reductionist science.

After introducing the emerging genetic analysis research tradition in the earlier parts of the book, in the later parts I shift towards describing the expansion of this research tradition to the level of molecular research.

Part V, "Increasing Resolving Power," is devoted to the expansion of genetic analysis with the establishment of the details of the material basis of heredity. This increase in the resolving power of the analysis was enabled by a transition in study from eukaryotes to bacteria and from phenomenological markers to biochemical and eventually molecular markers. I also discuss the arguments for and against the conception of a molecular biology theory (or research program) comprising distinct theories of "classical" and "molecular" genetics.

Part VI discusses the experimental evidence of gene function and its dependence on the cellular system that turns the nucleotide sequence into one component of gene function rather than its determinant.

In Part VII I discuss the breakdown of the reductionist conception together with the elaboration of reductionist molecular methodologies, the return of the top-down systems analysis to genetics research and the realization that the elephant is not a large-scale *E. coli*, which culminated when genetic research expanded into all disciplines of the life sciences. Genetic analysis became an integral part of the new biology of the genomic age, and maintains its role in the study of the development of the individual organism and in the dynamics of evolution.

In the concluding remarks, I suggest that the triumph of genetics in the genomic (and post-genomic) era is precisely in its maintaining the dialectics of adopting bottom-up methods and heuristics in resolving top-down analyses of organisms as systems.

I

From reproduction and generation to heredity

> And Adam lived thirty and a hundred years, and begot a son
> in his own likeness, after his image.
>
> Genesis 5, 3

To beget a son in one's own image was considered an attribute of reproduction and generation. In biblical times, inheritance referred to the transmission of material commodities or land-ownership from one person to another:

> And Abraham said, Lord God, what wilt thou give me, seeing I go child-less, and the steward of my house is this Eliezer of Damascus? . . . And behold, the word of the Lord came unto him saying, this shall not be thine heir; but he that shall come forth out of thine own bowels shall be thine heir.
>
> Genesis 15, 2–5

Although inheritance also extended to the succession of titles and rights, it only rarely referred to the transmission of the natural traits of living creatures. Eventually, however, inheritance acquired more meta-phoric connotations: "What must I do to inherit eternal life" (Mark 10, 17; Luke 10, 25 and 18, 18).

In Greek philosophy biological continuity was acknowledged as early as the fifth century BC in the *Iliad*, where the metaphoric inheritance of the heroic qualities of the father by the son was taken for granted. And in Euripides' *Electra* the continuity of traits by descendants is alluded to when a servant, finding a lock of hair, attempts to identify Orestes by its resemblance to his sister's hair.

In the Roman Empire inheritance of property was encoded in a voluminous set of laws. The term *Hērēdĭtās* referred to the successor of the rights and liabilities of a deceased person, or to a successor to a throne. It was also used metaphorically, to describe transmission of characteristics such as glory, hatred, or eagerness. However, when the Romans related to breeding practices in agriculture they referred to transmission by reproduction (Sirks and Zirkle, 1964).

Also in the Middle Ages references to biological transmission were made in the context of reproduction and generation: St. Thomas in the thirteenth century alluded to "bodily defects" that are "transmitted by way of origin from parent to child," or "man generates a likeness to himself in kind" (Zirkle, 1945).

At the turn of the sixteenth century Shakespeare used "inherit" repeatedly in various metaphoric contexts:

> Youth, thou bear'st thy father's face; Frank nature, rather curious than in haste, Hath well compos'd thee. Thy father's moral parts Mayst thou inherit too.
>
> *All's Well that Ends Well*, I, ii, 20–22

Although physicians have used the metaphor of "hereditary" to refer to transmission of disease at least since the sixteenth century, it must be kept in mind that diseases were not considered "properties" of living organisms, but rather scourges that *inflicted* the organism.

The terms *reproduction* and *generation* were used to signify biological continuity and change.

The change in the role of the term "heredity" to imply a theory was closely linked to the increasing upheavals in social awareness that followed the discoveries of new continents in the sixteenth and seventeenth centuries. Stephen Toulmin (1972, 41) noted that Captain James Cook's arrival in Tahiti on H.M.S. *Discovery* in 1769 was the beginning of a new era in the recognition of natural diversity. Cook's mission had been to make some astronomical observations for the Royal Society, but on the *Discovery*'s return the voyage became the talk of Europe for quite other reasons. The customs of the Tahitians proved far more intriguing than the astronomical distances of the planets. The scene, however, changed not only with respect to cultural anthropology. The number of new, unknown and unexpected species of animals and plants found on every new voyage caused a profound revolution in the conception of the world of nature and the species that inhabited it.

Taxonomy became a central issue of science and philosophy in the seventeenth century. Two notions of taxonomy of natural systems emerged: One notion conceived of taxonomy as a human device to control nature's variability using morphological and anatomical observations as its methods. Starting with Buffon's pragmatic classification this morphogenist research tradition was able to perceive and conceive change, and this culminated in the theories of evolution of species of Lamarck and Darwin. The second notion was the diametrically opposing Linnæan taxonomy that conceived of each species as a distinct essential entity given by Nature.

Hybridization was the time-honored tool of animal and plant breeders to defy the given order of Nature and Linnæus and his followers adopted it as the analytic tool of their research tradition. As a breeders' device hybridization was often directed at the transmission of specific traits. This tradition reached its peak in the middle of the nineteenth century when the monk Johann Gregor Mendel combined his notions of the divine lawfulness of nature, the reductionist insights of his university education in physics and mathematics, and the experience he gained from his community of breeders to design the experiments that allowed him to formulate in analytic terms the laws of inheritance.

1

The biologization of inheritance

Similarity and variation among living creatures has long been a mystery that has stimulated the classification and organization of life into hierarchical systems, as well as the induction of the differential ranking of individuals in society. Variation and similarity also provided the material for one of the most basic ancient unfolding developments of human culture, animal and plant breeding. The first chapter of *Genesis* beautifully reflects the ancients' conception of the hierarchical, as well as anthropocentric, catalog of the universe:

> In the beginning God created the heavens and the earth . . . Then God said, "Let the land produce vegetation: seed-bearing plants and trees on the land that bear fruit with seed in it, according to their various kinds." And . . . the land produced vegetation: plants bearing seed according to their kinds and trees bearing fruit with seed in it according to their kinds. . . . And God said, "Let the land produce living creatures according to their kinds: livestock, creatures that move along the ground, and wild animals, each according to its kind." . . . Then God said, "Let us make man in our image, in our likeness, and let them rule over the fish of the sea and the birds of the air, over the livestock, over all the earth, and over all the creatures that move along the ground."
>
> Genesis 1, 1–26

Animals and plants were believed to have been generated by God as distinct kinds at the time of Creation, and it was reproduction that maintained the link of similarity in living creatures. Although creatures were generated "according to their various kinds," kinds were not necessarily conceived to be discontinuous types. A prevailing notion was that God in his goodness filled the world with a continuous, progressive presence of life, from the lowest to the highest, all distinct yet each adjacent to its next neighbor. This unique hierarchical notion of continuous-but-distinct was known as the *Scala Naturae*, or the Great Chain of Being (Lovejoy, 1936 [1950]). It was a completely static, ahistoric gradation of living forms that was maintained almost undisturbed until it was challenged in the eighteenth century by typological classifications, and finally toppled by nineteenth-century theories of evolution.

In such a static world, *reproduction* denoted repetition of conserved qualities in the processes of embryogenesis, whereas *generation* denoted the creation of new qualities (see Parnas, 2006). Passive conservation as opposed to active creation has been a *leitmotiv* in the history of what we would single out today as biological inheritance. Heredity and inheritance, which had been terms of social relations, were seldom used in biological contexts: in an impressive number of quotations on the beginning of plant hybridization (Zirkle, 1935a) and on the early history of inheritance (Zirkle, 1945), "heredity" and "inheritance" hardly appear to be mentioned as such.[1] But such terms were increasingly used metaphorically in biological contexts.

In the seventeenth century, Descartes' suggestion that living creatures be viewed "bottom up" as machines whose function could be reduced to that of their components, did not gain momentum. Life sciences maintained a distinct "top-down" perspective: the entities of reference were organisms as such, and the structure and function of their parts were the properties of the whole organism. Thus, it was mainly embryogenesis that provided the empirical foundations for the philosophies of reproduction and generation.

The term "development," and even more explicitly its German equivalent *Entwicklung* (and also its Hebrew equivalent התפתחות), denoted the centrality of theories of *preformation* that conceived of embryogenesis in terms of re-production, as a gradual unfolding of preexisting qualities.[2] These were, in essence, theories of *pangenesis* that conceived of the embryo as an unfolding of elements drawn from all parts of the body of the parent(s). Proponents of *epigenesis*, on the other hand, conceived of embryogenesis more in terms of generation. They relied on constraints imposed by specific (largely conserved) conditions in which embryos grew, a view that secured reproduction, but allowed for more flexible embryogenesis than that of the strict unfolding of preformed determinants. This dichotomy in theories to account for the development

---

[1] It must be noted, however, that I saw most of the quotations only in the English translation.
[2] *Entwicklung* was also translated as evolution.

of the individual organism – the input of nature *versus* the input of circumstances – was reflected in deliberations of the principles that determine the organization of living beings as taxonomic systems. Does taxonomy represent given structures of nature or does it reflect adaptive constraints of circumstances?

It was only in the first half of the nineteenth century that reductive notions of "bottom-up" explanations of physics and chemistry gained a foothold in the life sciences (see Lenoir, 1982). But by the beginning of the eighteenth century, Platonic *a priori* abstraction notions began to be accepted in the life sciences. Observed variability of individuals was considered to be noise that could blur but not deny the existence of the genuine type, just as the images on the walls of Plato's allegory of the cave were merely shadows of the "real thing." It was then that "heredity" started to acquire the status of an explanatory term *per se*, rather than merely a descriptive metaphor.

The introduction of heredity as an explanatory concept within the dichotomy of Nature *versus* Nurture, as eventually formulated by Francis Galton, explicated one of the elements, Nature, in terms that made it amenable to methods of experimental analysis. However crucial the methodology of "either/or" in the design of an experiment, its radiation back on the conceptual level of the dichotomy is a hurdle that has plagued thinking about heredity ever since.

The duality of heredity *versus* environment is closely related to the attempt to discern "biology" from "sociology" in human affairs. The socio-political upheavals at the end of the eighteenth century and during the French Revolution decisively effected a disjunction between biology and social culture. Biology was introduced as a distinct discipline at about 1800 (see McLaughlin, 2002). Whereas before the French Revolution the term "heredity" was not used in the sense of natural history, after 1830 the metaphor of "hérédité naturelle" became a widely used term (see also Pick, 1989, 133). In England "heredity" was relatively rare in biological texts before the end of the nineteenth century (although Darwin used the word "inheritance"). As noted by Galton in his memoir, "It seems hardly credible now that even the word heredity was then considered fanciful and unusual. I was chaffed by a cultured friend for adopting it from the French" (Galton, 1908, 288). In the first half of the nineteenth century the metaphor of biological inheritance ceased to be a self-evident observational fact of life; it assumed the role of a postulating abstraction, a force (Gayon, 2000) primarily involved in maintaining the types or the essences which singled out specific attributes of systems of

16

living creatures. The metaphor of inheritance of traits did not disappear from biology even after Wilhelm Johannsen introduced the discrimination of the phenotypic and the genotypic levels, and still is an impediment to the genetic analysis of systems (see Chapter 4).[3]

NATURAL TAXONOMIES

During the eighteenth century two notions of taxonomy of natural systems emerged. We may call these *nominalism* (or *instrumentalism*, or *conventionalism*) and *essentialism* (or *realism*), respectively (Amundson, 2006, 32ff.). For nominalists taxonomic categories did not necessarily represent objectively real relationships of natural systems; most were, according to Amundson, "cautious-reality nominalists." They considered taxonomy as primarily a system that represented the needs of researchers, and asserted that insights into the living systems would be gained preferably by morphological, anatomical, and physiological investigations, irrespective of species borders. Although new species were added to taxonomic lists according to need, attention was drawn to the common aspects shared by living systems; nominalists were, as a rule, rather open to notions of change and evolution of species. Essentialists, on the other hand, believed in the fixity of species, each species being a given, well-defined entity, with unique characters. Newly discovered species were merely species previously unknown. Since taxonomy emphasized the essential, specific properties that differentiated existing entities, hybridization – not individuals' characteristics – was the ultimate research tool of the essentialists that determined taxonomic status. I will call the research effort of the former proponents a *morphogenist-evolutionist tradition* and the latter proponents a *hybridist-typologist tradition*.

When Carlos Linnæus introduced in the eighteenth century an essentialist taxonomy based on the characteristics of plants' organs of

---

[3] In 1911, Johannsen opened his talk on "The genotype conception of heredity" by noting that "[b]iology has evidently borrowed the terms 'heredity' and 'inheritance' from every-day language . . . The *transmission* of properties . . . from parents to their children, or from more or less remote ancestors to their descendants, has been regarded as the essential point in the discussion of heredity, in biology as in jurisprudence. Here we have nothing to do with the latter. . . . The view of natural inheritance as realized by an act of transmission, viz., the transmission of the parent's (or ancestor's) *personal qualities* to the progeny, is the most naïve and oldest conception of heredity." (Johannsen, 1911, 129).

reproduction, it laid the ground for a notion of biological inheritance that examined the hybridability of organisms. Georges Louis Leclerc de Buffon, Linnæus' contemporary, suggested a nominalist taxonomy that considered the structures and functions of living creatures, including their utility for men. Although Linnæus himself hardly applied it in his experimental work, he suggested hybridization as the analytic tool for the study of what became inheritance of characteristics in plants and animals. This was in stark contrast to morphogenists, who maintained natural history, comparative anatomy and physiology as the principle research methods of similarity and variation and continued to think of similarity and variation within the framework of the reproduction and generation of organisms as such.

FROM *SYSTEMA NATURAE* TO HYBRIDIZATION

Linnæus endeavored to provide in his *Systema naturae* of 1735 a *methodus naturalis*, a comprehensive botanical system that expressed the "natural" relations of plants and animals. His *essentialist* or *typological* species concept is traceable to its metaphysical and methodological foundations in Aristotelian logic: plants belong to one and the same species inasmuch as their *form* is determined by their specific *essence*. Plants possess their species-specific form by virtue of its overall function, which consists in reproduction tending toward the preservation of the species' essence (Müller-Wille, 1998). For Linnæus the fructification systems of genera were the essences responsible for the existence of biological kinds. Species comprised plant individuals which were related by descent and distinguished from other plants of the genus by a complex of characters or "form" that remained constant no matter what external conditions those plants were subjected to (Ereshefsky, 2001, 200–208). Varieties, in contrast, were believed to result from the action of accidental external, physical factors in the environment. Species were *types* in a strictly rational Platonic hierarchical classification, and we can count as many species now as were created in the beginning. The dogma of the immutability of species met with general acceptance only in the late-eighteenth and early-nineteenth centuries (Zirkle, 1935b, 443). Yet, this strict essentialist typology had already been disturbed in the 1740s when Linnæus was forced to admit that new species could arise through hybridization after God's original creation (Müller-Wille and Orel, 2007, 179–182).

18

Hybridization – understood as cross-breeding – has its roots since antiquity in the practices of domestication of animal and plant breeders. Notwithstanding, hybrids, natural or induced, were embarrassments to good order. They had a completely separate status in the hierarchy of nature's taxonomy, and for a long time hybrids were rejected as a source of new species. Although mating of unlikes or hybridization had been a common practice of domestication already in pre-biblical times – witness, for example, the relief from the ninth century BC of Assyrian priests pollinating palm florescences (Gray, 1969, 58 and www.bible-origins.net/EzekielsCherubim.html) – hybridization was considered an act that conflicted with natural generation. The word "hybrid" stems probably from the Greek ὕβρις which means an insult or outrage, especially when the insult is offered to the gods (*hubris*) or the outrage is connected to sex (Zirkle, 1935a, 1). For the Hebrews the mating of unlike types seems to have been a form of bestiality:

> Ye shall keep my statutes. Thou shalt not let thy cattle gender with a diverse kind; thou shalt not sow thy field with mingled seed; neither shall a garment mingled in linen and woolen come upon thee.
>
> Leviticus 19, 19

For Linnæus, who believed that "there are as many species as there were different forms created by the Infinite Being at the beginning," hybrids seemed irrelevant to his taxonomy:

> I distinguish the species of the Almighty Creator which are true from the abnormal varieties of the Gardner: the former I reckon of the highest importance because of their author, the latter I reject because of their authors.
>
> Linnæus, quoted by Olby (1985, 32)

Although hybrids were rejected because of the *laws of generation*, this did not mean that it was impossible for one kind of organism to give birth to another kind of organism, or that two different kinds could give birth to a third, mixed kind of organism (Müller-Wille, 1998). Still, the only experiment that Linnæus performed with hybridization forced him to relinquish his confidence in the constancy of the number of species, though not in the typological conception of natural kinds.

Central to Linnæus' distinction between species and varieties was that plants belonging to one and the same species when brought under the regime of perfectly homogenous external conditions should be identical in all respects. Any trait difference appearing under such circumstances would therefore have to count as a specific difference. Individual plants, cultivated and propagated under stable, homogenous conditions that exhibit a difference with respect to any one trait, provoked Linnæus to speculate about hybridization as a source of "new" species. Thus, in 1751 in *Plantae hybridae* Linnæus addressed the formation of hybrids as a regular natural phenomenon amenable to scientific analysis (Müller-Wille and Orel, 2007, 10–14). In a contest essay submitted to the Academy of Sciences at St. Petersburg in 1759, Linnæus admitted that it was necessary to accept the origin of species by hybridization:

> There can be no doubt that these [three or four real plants, to whose origin I have been eyewitness] are all new species produced by hybrid generation. . . . it seems probable that many plants, which now appear different of the same *genus*, may in the beginning have been one plant, having arisen merely from hybrid generation. . . . I am, however, convinced this mode of multiplying plants does not interfere with the system or general scheme of nature.
>
> Linnæus, quoted by Zirkle (1935a, 195)

Consequently Linnæus concluded that when relegated to exceptional circumstances, such as those of domestication and cultivation, specific forms of living beings arose from hybridization and biological transmutation.

Linnæus' speculations about the role that hybridization played in speciation inspired Joseph Gottlieb Kölreuter, in whose hands hybridization became *a central research tool* for the elucidation of the nature of species and varieties (Müller-Wille and Orel, 2007). Although Kölreuter deemed Linnæus' speculation about the hybrid origin of what he called constant varieties "adventurous and going against all reason," Kölreuter's careful, large-scale experimental project of plant hybridizations established the hybridist research tradition. He did not doubt that plant hybrids *could* be produced, but was also sure that Nature had her own ways of preventing them from producing fresh species, so as to save nature from a "monstrous swarm of imperfections" that would have followed from Linnæus' theory (Müller-Wille and Orel, 2007; Olby, 1985).

Kölreuter's successor, Carl Friedrich von Gärtner, adhered avidly to the theories of inheritance of *types* and applied hybridization to study the permanence of species, the production of new species, and the question of whether species could be transformed by hybridization. Gärtner's "natural classification" of hybrids was analytic, exclusively based on the "composition and descent" of hybrids, and not on empirical generalizations. Attention was directed at the specific properties of organisms rather than at species *per se.* Even if a character may be predicated empirically *of* hybrids, that in no way defines them *as* hybrids (Müller-Wille and Orel, 2007, 186–191). Of almost 13,000 hybridizations that he performed, all of which he believed to be interspecific,[4] less than 10% produced "nearly complete fruits" and only 0.4% gave normal fruit and seed number, and even most of these were lost after some generations due to "successive reduction in fertility of the seeds"(Falk, 1991, 460). Significantly, Gärtner's wide-ranging hybridizations supported the study of inheritance "bottom-up," by paying increasing attention to specific traits rather than to the organism as such.

For those who adhered to the typological Linnæan species concept and adopted hybridization as the experimental tool for its verification, biological heredity was a self-evident concept. The hybridist tradition immanently focused on heredity as transmission. There appeared to be a clear-cut distinction not only at the methodological level but also between causal factors that might induce variability, transient or persistent, and hereditary elements that maintain the natural essence of the species as a type: Biological heredity was conceived as the transmission of characteristics that comprised the *essences* of the species as a type, whereas reproduction and generation were relegated to the *instantiation* of the type that was renewed in every generation.

With Linnæus *heredity* as a biological concept came of age, and with the *hybridist* research tradition the inheritance of *specific traits of individuals* became evident.

FROM *HISTOIRE NATURELLE* TO MORPHOGENETICS

In retrospect, the French biologist Georges Louis Leclerc de Buffon evidently presents a more dynamic, natural-history conception of biological similarity and variation than Linnæus, his contemporary. In his *Histoire naturelle* of 1749 Buffon did not accept the conception of an objective, given typology. His criterion of classifying living creatures by their usefulness to humans highlighted his notion of pragmatic rather than intrinsic values. Whereas Linnæus presented the notion that

---

[4] Though it is quite possible that what Gärtner described as fertile interspecific hybrids would be described in modern terms merely as intraspecific variety hybrids.

scientific work reflected subjugation to some transcendental ideology given to man "from above," to which we must be unconditionally committed, for Buffon science was a creation by man and for man, its sources coming from within, as integral parts of human life. Contrary to Linnæus, who promoted qualities of some ideal individual and its specific traits, Buffon's concept allowed for flexibility and even for change of kinds. The main tools of scholars who followed the morphogenic tradition were natural history, paleontology, comparative anatomy, and eventually, experimental embryology; though hybridization was also practiced. Thus, controversies like the famous vehement public discussions between Étienne Geoffroy Saint Hilaire and Georges Cuvier in the 1830s on the role of structural constraints *versus* functional dictates as the *determinants* of organisms' similarity and variation, should be conceived as morphogenist extensions of the concepts of preformation and epigenesis to organisms' reproduction, respectively. Jean Baptiste Lamarck and Charles Darwin, on the other hand, argued that the form and function of organisms were the *consequences* of evolutionary processes rather than the *determinants* of organisms' variation.

Toward the mid-nineteenth century a hybridist tradition that presented regularities as *deductions* from *a priori* principles was juxtaposed against a morphologist-evolutionist tradition that *discovered* the principles of natural history that caused similarity and variation. Mendel epitomizes the hybridist tradition; Darwin, the morphogenist tradition.

Of course, the distinction between the research traditions is rather contrived. Charles Naudin was nominally a typical follower of the hybridizationist tradition; he "came to the study of hybridization as a systematist who sought to use it to clarify the taxonomic relationships between the genera and species of the potato and cucumber families. . . . From this empirical observation and from his own *a priori* belief that nature abhors hybrids he hypothesized a process of segregation of the two species within the hybrid" (Olby, 1985, 48). Yet, although he conceived of segregation among the progeny of his hybrids and recognized the need for a representative sample, his thinking was in morphogenist terms of the species as a whole, unaware of the power of a rigorous statistical analysis, which practically could be carried out only by following simple, discrete traits (Olby, 1985).

I will discuss Mendel in Chapter 2. It should be kept in mind, however, that for Darwin too, species were distinct types and the properties that maintained species as distinct types came from within, they were inherent in the generative-ecological relationships of the members of the species in their environment (see Depew and Weber, 1995). As pointed out by Paul Farber (1976), the concept of type was used in the second half of the eighteenth century and in the nineteenth century in at least three distinct ways. Thus, Darwin's species type, unlike Linnæus' essentialist type, but similar to Buffon's type-conception as a model that would consolidate his materials, were not abstract universals of logic and taxonomy but rather systems of concrete relationships between creatures (Sloan, 1976, 372). And although Darwin too used hybridizations, his classifications were primarily based on natural history, comparative anatomy and paleontology. The *Origin of Species* was not based on deduction from *a priori* principles; it was based on induction from discoveries in nature. Rather than changing the conception of a species, Darwin changed the argument about the creative forces that (objectively) brought about the origin of species.

NUCLEAR CONTINUITY AND TRANSMISSION

An important shift in the morphogenist tradition occurred in the 1880s, when German cytologists and descriptive embryologists acknowledged that a *continuum of structure* must be preserved during the development of the individual and that it extended to inter-generation continuity. Much of the evidence that compelled them to offer theories of heredity that were associated with individual transmission was obtained thanks to new microscopic and dyeing techniques. The impact of the enormous achievements of cytology in the 1880s and of Virchow's slogan of *omni cellula e cellula* may best be appreciated when compared to the impact of today's achievements in genomics thanks to the new techniques of DNA sequencing and Crick's formulation of *The Central Dogma* of molecular biology.

August Weismann's and Moritz Nussbaum's concept of a continuity of a germ cell line was substantiated by Eduard Strasburger's findings that fertilization consisted of the copulation of a single male pronucleus of a germinal cell with a female gamete, with no possibility of participation of the cytoplasm of the pollen. Strasburger's observations led him to believe that if "the nuclear filament [the chromosomes] of every cell nucleus remain preserved in the reticulum of the nucleus, then ... the nuclear filaments of all the following nuclear generations will contain approximately similar pieces of the nuclear filaments from the father and the mother" (Churchill, 1987, 347–348, quoting Strasburger). Such

studies of fertilization and early cleavage also led Oscar Hertwig directly
to a theory of transmission:

> The maternal and paternal organization will be transmitted through the
> reproductive act to the offspring by means of substances, which are
> themselves organized . . . In the sequence of individuals there take place
> only changes in organization . . . We may consider the nuclei as the sites
> [*Anlagen*] of complicated molecular structure, which transmit the maternal
> and paternal characters . . . [They] never undergo a dissolution but only a
> transformation in their form to become the nucleus of the egg and sperm.
>
> Hertwig, quoted by Churchill (1987, 350)

Even more explicit in material-molecular terms was Albert Kölliker,
who claimed that the processes of gamete production, fertilization,
and early cleavage indicated the transmission of "morphologically
determined substances," thanks to a *Vererbungsstoff* in the nuclei of the
egg and spermatozoon, probably associated with the nuclein described
by Friedrich Miescher (Churchill, 1987, 351). Finally, Eduard van
Benden demonstrated that a similar nuclear structure participated in
both fertilization and normal mitotic divisions.

It was out of these developments in the morphogenist research tradition that Weismann and de Vries promoted theories of heredity
independent of theories of development. Against this background de
Vries and William Bateson conceived of the power of the hybridist
research tradition to provide the analytic tool for the examination of
these speculations. And it was on these analytic hybridist foundations
that the embryologist Thomas H. Morgan constructed the material
perspective of the hybridist research tradition of early genetics.

2

Mendel: the design of an experiment

Johann Gregor Mendel was born in 1822 to a peasant family in a small village in Moravia. Much attention was devoted to his education. As it turned out, Mendel's education was affected by the pansophy teachings of the devout seventeenth-century pedagogue and philosopher Jan Amos Comenius who, in Descartes' words, "too closely combined human science and theology," yet preached for the threefold virtue of "fullness, order and truth" (Orel, personal communication a). Mendel's interests were in natural history and agriculture. He pursued his studies at the Olomouc University in spite of severe economic difficulties. As he wrote in his curriculum vitae notes in 1850: "It was impossible for him to endure such exertion any further. Therefore, having finished his philo-sophical studies, he felt himself compelled to enter a station in life that would free him from the bitter struggle for existence. His circumstances decided his vocational choice. He requested and received in the year 1843 admission to the Augustinian monastery of St. Thomas in Brno" (Orel, 1996, 43–44). This was a fortunate choice, primarily because the head of the monastery at the time was the abbot František Cyril Napp, a "scientist, secret freethinker, and an expert in state affairs and eco-nomics" (Peaslee and Orel, 2007).

MENDEL IN CONTEXT

Moravia of the first half of the nineteenth century was a prosperous center of industry and agriculture as well as of scientific activity. In Moravia, Brno was of special significance as a center for theoretical and experimental scientific research on the role of hybridization in animal and plant breeding, thanks to a close cooperation of breeders and naturalists (Orel, unpubl. ms. b). Johann Karl Nestler, one of Mendel's professors at Olomouc, asserted that the most important problem for the improvement of methods in sheep breeding was the investigation of heredity as a separate issue from that of generation, the enigmatic process of reproduction and development (Orel, 2005, 100). The atmosphere at the meetings of members of the Brno Agricultural Society was described in 1842 by one of its members as one in which "differences of opinion, freely expressed in the meetings, stimulated experiments that open the way for new reflection, experimentation and progress which, according to natural laws, cannot be stopped" (Orel, 2005, 100). It was primarily with the blessing of Abbot Napp that Mendel pursued his studies in 1851–3, mainly in mathematics, physics, and plant physiology at the university in Vienna. As Mendel's biographer Vítězslav Orel notes, it is no wonder Mendel "returned to Brno with the idea of plant hybridization experiments in his mind" (Orel, 2005, 101).

Mendel's intellectual world seems to have been shaped by three major interests: theology, science, and breeding. He believed in the laws of nature, was geographically and culturally deeply committed to breeders' efforts to improve specific traits by hybridization, and was acquainted with mathematical analysis and theories of modern physics. Mendel combined all three interests into an experimental method of testing nature's laws of inheritance of specific traits. However, contrary to reductionist explanations in physiology, or even those in "developmental mechanics" as a causal study of *form* that strived to reduce all phenomena to "the more recent concepts of physics and chemistry" (Roux, 1894), Mendel offered an explanation that was purely phenomenological and not committed to any particular theory (though he must have had assumptions about the determinate relationship between *Anlagen* and characters, see Müller-Wille and Orel, 2007).

Christian Carl André, who held Moravia as a center of intensive interactions of animal and plant breeders and scholars, emphasized the importance of developing both basic and applied research in natural science (Wood and Orel, 2005). He speculated on the scientific basis of breeding practices in his publication of empirical *genetic laws* (*genetische Gesetze*), such as "traits of grandparents may disappear and then reappear in later generation" or "the precondition for successful application of inbreeding is scrupulous selection of stock animals" (Orel and Wood, 1998, 81). When he was forced to leave Brno, his former colleagues Nestler and Franz Diebl continued the tradition of developing scientific animal and plant breeding. Abbot Napp stressed that the proper question to be asked should be *"What is inherited and how?"* (Orel and Hartl, 1994).

Nestler was convinced that nature produces species with unequivocal constancy, and that man can imitate nature and control the reproductive process by the formation of modified organisms. He juxtaposed the crossings of merino rams with local sheep to those of Kölreuter's back-crossing experiments with *Nicotiana*. Nestler's formulation of the enigma of heredity of wool quantity and quality anticipated to a large extent the experimental approach that was eventually taken up by Mendel in his study of the laws of inheritance in peas. Diebl's textbook which described artificial fertilization as the most important method for creating new varieties also influenced Mendel's thinking.

After joining the Augustinian monastery in 1843, Mendel studied theology and also attended lectures on agriculture. In 1851 Abbot Napp sent him to study physics with Christian Doppler in Vienna. Mendel also paid great attention to mathematics and to plant physiology, newly introduced by the biogeographer Franz Unger. Unger drew on two sources: the Humboldian plant geography research tradition of quantifying correlations between organic phenomena and physical variables, and the German idealist morphology that endeavored to uncover laws of developmental (*Entwicklung*) change. The realization that plant distribution had to result, at least in part, from historical developments kindled Unger's interest in evolution. Eventually he reasoned that the appearance of new species was not caused by spontaneous generation but rather by common descent, and he assumed that the process would follow a deterministic developmental law. In this he followed his teacher, Lorenz Oken, who viewed the organic world as a manifestation of abstract, numerical relationships (Gliboff, 1999). Sander Gliboff calls the group of plant physiologists who had accepted such a view of evolution before Darwin the "Austro-Ungerians" (Gliboff, 1999, 223). Unger believed that hybridizations would provide a means to "penetrate that labyrinth into which no human eye has yet reached," which might explain the origin of new plant species. From Unger Mendel acquired the theoretical background and the skill to perform experiments and to undertake independent research (Olby, 1985, 95–99). And it was Unger's studies in physics that convinced Mendel that nature might be reduced to basic elements whose properties are governed by laws written in the language of mathematics (Gliboff, 1999).

In 1861, Mendel was one of the founders of the reorganized Natural Science Society in Brno (*naturforschenden Verein zu Brünn*) (Orel,

2005), and it was to the monthly meeting of this society that he read in February and March 1865 his *Versuche über Pflanzen-Hybriden* (Experiments on plant hybrids), published one year later in the fourth volume of the *Proceedings of the Brünn Society* (Orel, 1996, 93–95).

Although Mendel worked within the hybridist typological tradition, he dissociated himself from Kölreuter's and Gärtner's conclusions that hybridization did not play a role in the formation of new species. But, as pointed out by Müller-Wille and Orel (2007, 174–175, and 177), Mendel adopted the definition of species in the "strictest sense" of Linnæus, that only those individuals that display identical traits under identical conditions belong to a species. Thus, no two of his distinct varieties of peas "could be counted as one and the same species" (Mendel, in Stern and Sherwood, 1966, 5). Experiments with *inter*specific hybrids were conceived to delve into the problems of the theory of the evolution of species. However, as Mendel emphasized, "no generally applicable law of the formation and development of hybrids has yet been successfully formulated." Therefore, "[i]t requires a good deal of courage indeed to undertake such a far-reaching task; however, this seems to be the one correct way to finally reaching the solution to a question whose significance for the evolutionary history [*Entwicklungs-Geschichte*] of organic forms must not be underestimated" (Mendel, in Stern and Sherwood, 1966, 2). The problem with the hybridization experiments of previous investigators like Gärtner and Wichura, he stated, was that "not one has been carried out to an extent and in a manner that would make it possible to determine the number of different forms in which hybrid progeny appear, permit classification of these forms in each generation with certainty, and ascertain their numerical inter-relationships" (Mendel, in Stern and Sherwood, 1966, 2). And this was exactly what Mendel, the student of the reductionist methodology of experimental design in the physical sciences, set out to do. As suggested by Gliboff, a point may be made that the motive for Mendel's entire project was an Ungerian one of elucidating a mathematical law of evolution (Gliboff, 1999, 219).

A WELL-DESIGNED EXPERIMENT

In 1936 R. A. Fisher published a paper asking "Has Mendel's work been rediscovered?" In it he pointed out the improbability of getting numerical results that fit expectations to the extent that Mendel presented in the published version of his talks: Mendel's results were too good to be true (Fisher, 1936). Hypotheses of foul play by Mendel or by one of his helpers abounded (see, e.g., Sapp, 1990, chapter 5). However, as had already been suggested by Leslie C. Dunn, "the excessive goodness of fit to a theory that runs through his data certainly indicates that he had a theory in mind when the data as reported were tallied" (Dunn, 1965). Accumulating evidence supports that, indeed, Mendel had a theory in mind and that he carefully designed his experiments, considering the given constraints.

Mendel succeeded in integrating his notions of the experimental biological tradition of hybridization with the elements of reductionist methodology of the physical sciences and formulating them as the given laws of nature. Two operational principles turned Mendel's work into a powerful analytic research methodology:

1. Itemization of characters, viewing characteristics as *phenomena per se*, waiving their relations with other characteristics.
2. Analyzing hybridization results in terms of variables expressed by numerical data.

Mendel carefully chose the pea as the object of his experiments, being a domesticated plant with many varieties available on which he could easily perform large-scale hybridization experiments. Of the thirty-four initial varieties of peas he procured, after two years of preliminary experimentation he chose twenty-two that showed persistent differences in seven traits, apparently in different combinations, each coming in two alternative forms, according to which the data might be convincingly classified into qualitative alternative groups. Two of the seven traits were properties that could be determined and classified on the very plants in which the hybridizations (cross pollinations) were performed (shape of the ripe seeds, and color of seed albumen); for the remaining five traits (color of seed coat, shape of the ripe pod, color of the unripe pod, position of flowers, and stem length) it was necessary to examine the plants grown from the seeds, which demanded planting a large number of progeny to facilitate statistical analysis of segregation. The patch of land at Mendel's disposal in the garden of his monastery was very small indeed (35m × 7m; plus some more space in Abbot Napp's greenhouse measuring 22.7m × 4.5m. See Orel, 1996, 96). All this dictated meticulous planning for his experiments from which he obtained results from many thousands of plants within seven years (1856–63). There was absolutely no room for wasteful trial-and-error experiments.

Mendel's experiments culminated in the law of segregation and the law of independent assortment: alternative *Faktoren* for specific traits may exist in the cells of hybrid plants, irrespective of the characteristic the traits stand for. Segregation occurs at the production of gametes, such that only one of the alternatives is transmitted with equal probability to each progeny.[1] *Faktoren* for different traits segregate independently of each other.[2] He also pointed out that as a rule, one of the two alternative factors dominates and the other is recessive, such that the hybrid being heterozygous for that trait cannot be discerned from the homozygote without further breeding tests.

In order to determine whether the expected proportion of 2:1 of heterozygotes to homozygotes among hybrid seeds showed the dominating pattern of the first two characteristics it was enough to grow plants from some 500 seeds for each trait, respectively, allow the plants to self-fertilize, and then check the seeds of these plants. But for the remaining five traits, six experiments (one trait had to be tested twice) and an additional plant generation were needed. In each of these experiments 10 progeny seeds were planted from each of 100 plants in order to determine how many of those hundred plants were heterozygous and how

[1] In a scathing criticism of Mendel's work Bishop asserts that Mendel "deliberately chose characters exhibiting a most unusual pattern of inheritance . . . because he wanted to demonstrate stasis, formulated a highly improbable theory, and then extrapolated to all other modes of inheritance" (Bishop, 1996, 212). Bishop discusses Mendel's wrinkled pea, a trait that had been found later, at the level of the molecular analysis, to be due to the absence of a relevant enzyme activity. Furthermore, he claims that since this inactivity is due to an insertion into an intron, all coding sequences for the production of non-wrinkled seeds are actually present in the so-called wrinkled seed variety. This reveals Bishop's basic misunderstanding of the Mendelian analysis which had been elucidated already in 1909 by Johannsen: Mendel's traits are only the "markers" of the *Faktoren*. One of the major achievements of Mendel has been that he overcame the details of the factors, the nature of the elements being completely irrelevant to the patterns of their transmission.

[2] Mendel's choice of peas as his experimental material turned out inadvertently to be fortunate. Peas have only seven chromosome pairs, and later studies showed that of Mendel's seven factors, three were on chromosome 4, two on chromosome 1, and one on each of chromosomes 5 and 7 (Blixt, 1975). Recombination frequencies in peas are, however, quite high and in all but one combination the factors would have segregated independently even when located on the same chromosome (being syntenic). Only the pair shape of the ripe pod and stem length would have been expected to be conspicuously linked, giving on average only 10 percent recombination. It is possible that Mendel did not examine the segregation of this specific pair of traits; but it is also possible that the shape of the ripe pod was controlled by a gene located on chromosome 6, rather than on chromosome 4, and accordingly segregated independently of stem length (Mendel's experimental stocks are no longer available for examination) (Novitski and Blixt, 1978).

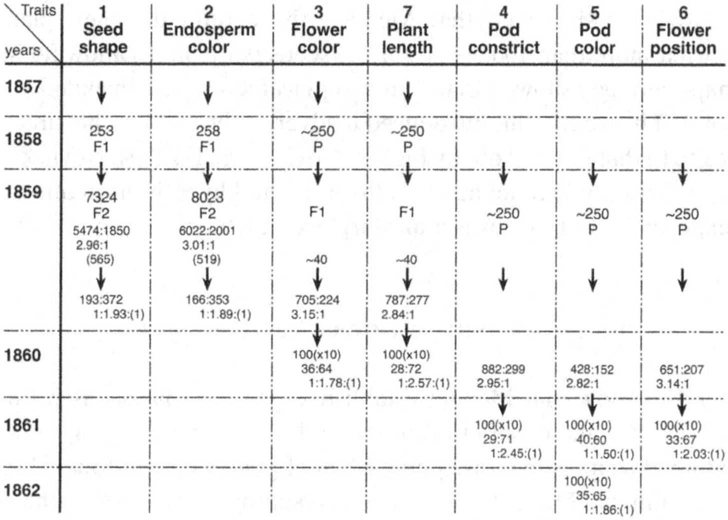

Traits / years	1 Seed shape	2 Endosperm color	3 Flower color	7 Plant length	4 Pod constrict	5 Pod color	6 Flower position
1857	↓	↓	↓	↓			
1858	253 F1	258 F1	~250 P	~250 P			
1859	7324 F2 5474:1850 2.96:1 (565) ↓ 193:372 1:1.93:(1)	8023 F2 6022:2001 3.01:1 (519) ↓ 166:353 1:1.89:(1)	F1 ~40 ↓ 705:224 3.15:1	F1 ~40 ↓ 787:277 2.84:1	~250 P ↓	~250 P ↓	~250 P ↓
1860			100(x10) 36:64 1:1.78:(1)	100(x10) 28:72 1:2.57:(1)	882:299 2.95:1	428:152 2.82:1	651:207 3.14:1
1861					100(x10) 29:71 1:2.45:(1)	100(x10) 40:60 1:1.50:(1)	100(x10) 33:67 1:2.03:(1)
1862						100(x10) 35:65 1:1.86:(1)	

Figure 2.1. Reconstruction of the design of Mendel's monohybrid experiments, following Fisher (1936). Top: The sequence of traits as listed by Mendel (p. 29).

many homozygous for the relevant trait (see Figure 2.1). If at least one of the ten progeny-plants revealed the recessive pattern the parent-plant was deemed heterozygous for that trait. If none of the ten revealed the recessive trait the parent-plant was assumed to be homozygous for that trait. This demanded that an extraordinarily complex experimental design had to be carried out in the limited space of the monastery's garden. Mendel tested two traits one year, the remaining three the following year (and the repeat of one experiment, the year thereafter) together with a trihybrid cross. Since there is a calculable probability that a heterozygous parent-plant might not reveal the recessive trait among ten progeny, an excess of those deemed "homozygotes" should be expected among the hundred tested plants, but Mendel's published results fit the expectations to an extent that was "too good to be true" (Fisher, 1936). Various attempts were made to suggest how such results could be obtained when Mendel designed his experiments so as to examine his preconceived hypothesis (see, e.g., Di Trocchio, 1991). Recently Edward and Charles Novitski pointed out that it is hardly feasible that ten seeds would grow into discernible plants for all 6×100 tested parent-plants. At least in some cases fewer than ten non-recessive progeny plants grew, and in such cases it was not possible to classify the parent-plant as either homozygous or heterozygous for the relevant trait.

The Novitskis suggested that Mendel, the careful designer, planted somewhat more than 100 groups of ten seeds per plant to allow for such mishaps and they showed that if such replacements were included in the samples of hundreds, this introduced inadvertently a bias in the direction opposite to that pointed out by Fisher (Novitski, C. E., 2004; Novitski, E., 2004). Obviously Mendel had a well-formulated hypothesis in mind and his experiment to test it was a masterpiece of design.

THE ESTABLISHMENT OF GENETIC ANALYSIS

I intend to show that Mendel established genetic analysis as we have developed it in the twentieth century, and, to the extent that genetics is maintained as a distinct discipline, Mendel guides it even today. Moreover, I claim that Mendel's experiments resulting in his paper on the laws of heredity was not a Kuhnian paradigm shift in biological research: his work was embedded in the research tradition of his time and place. Furthermore, the thirty-five years when his work was overlooked is not the time it takes for a paradigm shift to simmer. Being a hybridist rather than a natural historian, he did not conceive of hybrids as abnormal productions, neither did he make a distinction between species' hybrids and varieties' mongrels (bastards) as Darwin (and Kölreuter) did. His was not a research of mounting "anomalies" resisting the paradigmatic treatment that scientists were trained to follow (Fuller, 2000). Rather, the hiatus in recognizing Mendel's contribution was the consequence of the domination of two research traditions. True, Mendel expressed his intention to formulate the "law of the formation and development of hybrids," not because of a need that had evaded his predecessors, but rather because he believed that his predecessors had not applied the proper research heuristics.

An "Austro-Ungerian" reading of his hybridization experiments might even present Mendel as one who was addressing an evolutionary question (Gliboff, 1999, 225). Like Darwin he believed that the same laws govern the behavior of both naturally occurring species and the products of domestication (Campbell, 1982, 39), and claimed that it is legitimate to make deductions from the one to the other: "No one would seriously want to maintain that plant development in the wild and in garden beds was governed by different laws" (Mendel, in Stern and Sherwood, 1966, 37). Yet, working very much in the heart of the hybridist research tradition, his notions were rather tangential, not perpendicular, to those of Darwin's *Origin*. Like Darwin, Mendel was convinced that the discipline to which he directed his attention was subject to law. However, unlike Darwin, for whom the laws were discoverable by following natural history, paleontology, and embryology, Mendel maintained the belief that the laws were deducible from first principles and that hybridization experiments could only uphold them (Falk, 2001a, 2006). Finally, Mendel's education in physics helped him to overcome the resistance of many biologists to accept laboratory results as natural, and to *experimentally* reduce a problem of nature to one manageable in controlled conditions, even without adopting a reductionist *conception*.

It must, however, be kept in mind that at the time Mendel undertook his studies in Vienna, cell theory became one of the greatest unifying principles of biology. There is no doubt that in formulating his *Faktoren* hypothesis Mendel was aware of the cellular theory of reproduction: "According to the opinion of famous physiologists, propagation in phanerogams is initiated by the union of one germinal and one pollen cell into one single cell, which is able to develop into an independent organism through incorporation of matter and the formation of new cells" (Mendel, in Stern and Sherwood, 1966, 41–42). Margaret Campbell claimed that cell theory, and especially Unger's insistence that division rather than formation *de novo* of cells took place, were crucial to Mendel's conception of his units or elements "as stable and self-replicating" (Campbell, 1982, 44). However, in a perfect argument of genetic analysis, Mendel made a fundamental distinction between cells and "elements": in a long footnote he commented that "It is presumably beyond doubt that in *Pisum* a complete union of elements from both fertilized cells has to take place for the formation of a new embryo" although "experiments have in no way confirmed this up to now." He then went on, reasoning that presumably, "When a germinal cell is successfully combined with a *dissimilar* pollen cell we have to assume that some compromise [*Ausgleichung*] takes place between those elements of both cells that cause their difference." But such a "compromise" – or equalization, compensation – would not do for those hybrids whose offspring are *variable*, because of the finding "that this balance between the antagonistic elements is only temporary and does not extend beyond the lifetime of the hybrid plant" (Mendel, in Stern and Sherwood, 1966, 41–43). Contrary to the cytological *observations* according to which "we have to assume" the loss of the individuality of the contribution of the two parents, *deduction* from hybridization analysis proves otherwise, pointing to the existence of "elements" that maintain their individuality: "In the formation of these cells all elements present participate in completely free and uniform fashion, and only those that differ separate from each other. In this manner the production of as many kinds of germinal and pollen cells would be possible as there are combinations of potentially formative elements" (Mendel, in Stern and Sherwood, 1966, 43). Mendel granted that empirically "[w]hether variable hybrids of other plant species show complete agreement in behavior ... remains to be decided experimentally" but on the basis of the given lawfulness of nature, "one might assume, however, that no basic difference could exist in important matters since *unity* in the plan of development of organic life is beyond doubt" (Mendel, in Stern and Sherwood, 1966, 43).

Mendel the hybridist, directing his attention to inheritance as transmission, made the notion of reproduction irrelevant. It was *genetic analysis*, rather than fertilization redefined in the context of *cell theory*, that induced Mendel to speculate on the role of independent elements for different variable traits and to subject this speculation to experimental test. The conception of *Faktoren* did not necessarily involve any material basis, and as Campbell concluded: "It seems then that Mendel's theory was plausible to him because he cast his thought in a metaphysical framework which supposed that Forms could not be known directly. ... The non-empirical nature of the hereditary elements ... posed no difficulties for him." Cell theory provided a context within which the view that single cells were involved in the fertilization process was plausible, but it was "the mathematical theory concerning combination series" that made the structure of Mendel's theory plausible to him (Campbell, 1982, 52–53).

Following Mendel, genetic analysis became a research logic that employed the notion of hybridization as its basic experimental tool, and the concept of heuristic reduction of biological transmission between generations to basic elements that serve as devices in the effort to establish the immanent lawfulness of phenomena.

MENDEL IN PERSPECTIVE

In retrospect, as noted by Curt Stern in the introduction to *Mendel Source Book*, "Mendel's triumph was a lonely one," neither his audience nor the readers of his reprints appreciated the significance of his achievement (Stern and Sherwood, 1966, v). Yet:

The observations and their analysis which Mendel supplied have indeed led far. Without a break in intellectual continuity they became the Mendelism of the early years of this century, joined with the study of chromosomes to expand into the cytogenetics of the next decades, took on new meaning when the biochemical and developmental activities of Mendel's "cell elements" came into the foreground of investigations, and remain conceptually one of the bases of the most advanced contemporary molecular genetics.

Stern and Sherwood (1966, vi)

Others, however, questioned the extent to which the founder of modern genetics and the breeder-monk were the same, or as phrased by Robert Olby: "Mendel no Mendelian?" (Olby, 1979). I maintain that Mendel's 1866 paper was the founding treatise of the science of heredity *as we know it today* in that it (a) established the methodology of genetic analysis by inferring hereditary *Faktoren* from the segregation of discrete traits, and (b) did not commit his *Faktoren* theory to the nature of these entities (Falk, 2006). Mendel's contribution was not a revolutionary *deus ex machina*, but rather an evolutionary process that was intimately grounded in the conceptual framework of his intellectual and practical environment (Campbell, 1982; Gliboff, 1999; Orel, 1996). But he did turn the hybridist experimental tradition into a powerful analytic instrument of reducing the phenomena of living systems, whether of the *shape of the ripe seed*, or the *nucleotide sequence of a transcription factor*, into conceptually amenable entities.

Allegations that Mendel's objective was merely practical, namely to find the empirical design that would describe the formation of hybrids and the development of their offspring over several generations rather than uncovering universal laws of nature (Monaghan and Corcos, 1990), must be rejected (Falk and Sarkar, 1991). The claim that Mendel fell short of expectation at the "level of theory formation in which he proposed and tested a theory of hybridization based upon a simple mechanism of gamete formation and their union in fertilization" (Monaghan and Corcos, 1990, 279) may also be rejected. Mendel carefully designed his experiments and limited his attention to those traits for which there was obvious correspondence between the potential for a trait and the trait proper (Falk and Sarkar, 1991, 449). To the extent that one can reconstruct Mendel's intentions, I suggest that he was looking for the plan of the Creator for eternal transcendental realities, and like Kepler, two hundred and fifty years earlier, he believed that this plan of reality is formulated in simple mathematical relationships (Falk, 2001a). Mendel did invoke *Faktoren* in trying to account for the constancy of traits (Mendel, in Stern and Sherwood, 1966, 24), though, significantly, he made no claims about the *Faktoren* being particulate. Sadly, because of experiments with *Hieracium* (the hawkweed),[3] and his growing duties as abbot of the monastery, Mendel did not pursue his hybridization work (Orel, 1996).

That it was the estrangement of the hybridist and morphogenist traditions in the decades immediately following his presentation to the *naturforschenden Verein* in 1865 that kept Mendel's work from being combined with developments in microscopic techniques, significant to Mendel's acceptance in the 1900s, finds support in quotes such as that of the plant physiologist Julius Sachs. In 1875 Sachs wrote the history of botany up to 1860, disparaging the kind of reductive reasoning that accepts a hypothesis merely because the "geometry happened to come out right." Sachs claimed that this was "an undesirable holdover from German idealism and *Naturphilosophie*" (Gliboff, 1999, 229). Gliboff takes Carl Nägeli's criticism of Mendel's work, claiming that his formulae were "empirical" and "would not be demonstrably rational" to reflect a claim that Mendel's elements and mechanisms were beyond the range of assumptions considered under the prevailing Kantian scheme of "rational" inductivism: rational induction from observed empirical regularities to general laws and explanations are allowable only with the aid of a limited set of *a priori* assumptions (such as time, space, causality, and the cells) (Gliboff, 1999, 230).

Also, Mendel's disassociation of heredity and developmental processes was inherent to the hybridist research tradition, rather than to an "eccentric definition of the problem of heredity" (Sandler and Sandler, 1985, 69). Such confounding of heredity and development was true for the prevailing morphogenist research tradition. The watershed in morphogenists' thinking of heredity as a distinct discipline occurred

---

[3] *Hieracium* was the main research object of Carl Nägeli, with whom Mendel corresponded for a number of years (see Gregor Mendel's "Letters to Carl Nägeli, 1866–1873," in Stern and Sherwood, 1966, 56–102). The topic of *Hieracium* experiments was discussed in Brno by members of the Natural Science Society before Mendel came in contact with Nägeli (Orel, personal communication). Obviously Nägeli did not follow Mendel's train of thought. More significantly, as it turned out years later, many of the seeds of *Hieracium*, especially those of hybrids between two different varieties, are apomictic: they arise directly from the maternal tissue, without the fertilization of gametes, i.e., they are not hybrids! Experiments with *Hieracium* led the Abbot Mendel to the idea that their hybrids were "constant hybrids," though later he accepted the polymorphic heredity of *Hieracium* hybrids (see Orel, 1996, 183–188).

eventually in the 1880s (see Churchill, 1987, and Chapter 1). As I claim in the next chapters, it demanded a re-evaluation of the morphogenists' notion of Darwinian evolution in order to notice and appreciate Mendel's paper.

Finally, the impact of Mendel's logical commitment to ideas of evolution, implied in his numerical presentation of the results, is highlighted by Bishop's juxtaposition of Mendel and Darwin: by *selecting* properties that may be classified into two alternative patterns, without intermediates, Mendel significantly biased his experimental results in favor of his hypothesis of evolutionary discontinuity; Darwin, by concentrating on continuously varying traits, biased his data in favor of his hypothesis (see Bishop, 1996). It took many years to overcome this presumably inconsistent antagonism between these two great concepts of nineteenth-century biological research.

II

Faktoren in search of meaning

On March 26, 1900, a notice, *Sur la loi de disjunction de hybrids* [Concerning the law of segregation of hybrids] appeared in the *Compes Rendus de l'Academie de Sciences*. Its author, the well-known Dutch botanist Hugo de Vries, stated:

> According to the principles which I have expressed elsewhere (*Intracelluläre Pangenesis*, 1889) the specific characters of organisms are composed of separate units. One is able to study, experimentally, these units either by the phenomena of variability and mutability or by the production of hybrids.
>
> de Vries (1950 [1900], 30)

In this short notice, de Vries reminded his readers of his hypothesis of reproduction and generation, according to which an organism may be reduced to preformed units in the gametes that will unfold into mature organs at embryogenesis. His thesis, however, suggests that, besides the morphogenists' tradition of the study of these units "by the phenomena of variability and mutability," an effective research method is offered by the hybridist tradition, namely, hybridization. This notice presents the segregation principle of the progeny of monohybrids, of "75 per 100N and 25 per 100B" of the dominant character to that of the recessive character.

A more detailed report, submitted twelve days earlier, on March 14, 1900, as a preliminary communication to the *Berichte der deutschen botanischen Gesellschaft*, was published after the short notice:

1. *Of the two antagonistic characteristics, the hybrid carries only one*, and that in complete development. Thus in this respect the hybrid is indistinguishable from one of the two parents. There are no transitional forms.

39

2. *In the formation of pollen and ovules the two antagonistic characteristics separate*, following for the most part simple laws of probability.

> These two statements, in their most essential points, were drawn up long ago by Mendel for a special case (peas). These formulations have been forgotten and their significance misunderstood. As my experiments show, they possess generalized validity for true hybrids.
>
> de Vries, in Stern and Sherwood (1966, 110)

As soon as Carl Correns read de Vries' early communication he submitted, on April 6, 1900 his note, *G. Mendel's Regel über das Verhalten der Nachkommenshaft der Rassenbastarde* [G. Mendel's law concerning the behavior of progeny of varietal hybrids]:

> In my hybridization experiments with varieties of maize and peas, I have come to the same results as de Vries. . . . When I discovered the regularity of the phenomenon, and the explanation thereof . . . the same thing happened to me which now seems to be happening to de Vries: I thought that I had found something new. But then I convinced myself that the Abbot Gregor Mendel in Brünn, had . . . given exactly the same explanation, as far as that was possible in 1866. . . . At the time I did not consider it necessary to establish my priority for this "re-discovery" by a preliminary note.
>
> Correns, in Stern and Sherwood (1966, 119–120)

Whereas de Vries presented the "rediscovery" of Mendel's paper as a significant contribution of the hybridist research tradition to his own conception of evolution and development, Correns presented it, in the morphogenist context, as a case that corresponded to the assumptions that cytologists had formed in previous decades, and accordingly he did not consider it to be "something new."

In the following issue of the *Berichte der deutschen botanischen Gesellschaft*, Erich Tschermak published a note *Über künstliche Kreuzung bei Pisum sativum* [Concerning artificial crossing in *Pisum sativum*]:

> Stimulated by the experiments of Darwin on the effects of cross- and self-fertilization in the plant kingdom, I began, in the year 1898 to make hybridization experiments with *Pisum sativum*.
>
> Tschermak (1950 [1900], 42)

Tschermak found that in hybrids "quite regularly" the dominating character developed exclusively. However, when the hybrid is fertilized by the parental type with the recessive character, the number of bearers of the recessive character are increased over that of self-fertilization of the hybrid. The influence of the character "yellow" in the seeds in the hybrid was in this case reduced by 57 per cent, while that of the character "green" was reduced by 43.5 percent.

Tschermak (1950 [1900], 46)

This and similar notes in Tschermak's paper convinced Curt Stern that "Tschermak's papers of 1900 not only lack fundamental analysis of his breeding results but clearly show that he had not . . . been a rediscoverer of Mendelism but only an experimenter whose understanding – to use a phrase coined by Bateson in another context – has 'fallen short of the essential discovery'" (Stern and Sherwood, 1966, xi–xii). Olby (1985) came to similar conclusions. This is not to diminish Tschermak's considerable "influence on the recognition of Mendelian genetics by plant breeders" (Stern and Sherwood, 1966, xi).

There can be no doubt that in the spring of 1900 Mendel's work moved to center-stage, but I doubt whether it was "rediscovered." As noted in the previous chapter, Mendel of 1900 was presented in a different context than Mendel of 1866: Mendel's disassociation of heredity and developmental processes was immanent in the hybridist research tradition. Dominance as a phenomenon of development, central to the morphogenist theory of de Vries, would never have been considered by Mendel as a law of inheritance. On the other hand, advances in microscopy that resolved much of the confounding of heredity and development, apparently allowed Correns to contribute to the "rediscovery" by diverting attention away from Mendel's laws of inheritance and toward the correspondence of his findings with the morphogenists' mechanics of cellular process.

Operationally, the person who, upon reading de Vries' paper on May 9, 1900, really established Mendelism was William Bateson (Olby, 1987). Riding the train to give a talk at the London Meeting of the Royal Agricultural Society, he realized the significance of Mendel's paper and at once introduced changes in the text of his London talk. Whereas de Vries' interest in Mendel abated, Bateson's enthusiasm only increased with time. Bateson, a student of comparative embryology who turned to the study of systematics and natural history, was interested in the role of ecology in the evolution of species. He was an ardent supporter of stepwise evolution of species by *homoeosis*, discontinuous repeats of organs. As to Darwin's theory of evolution by continuous modification, he had pointed out:

The fact that in certain cases there are forms transitional between groups which are sufficiently different to have been thought to be distinct, is a very important fact which must not be lost sight of; but . . . it remains none the less true that at a given point of time, the forms of living things may be arranged in Specific Groups, and that between the immense majority of these there are no transitional forms . . . The existence, then, of Specific Differences is one of the characteristics of the forms of living things.

. . . specific diversity of form is consequent upon diversity of environment, and diversity of environment is thus the ultimate measure of diversity of specific form. Here then we meet the difficulty that diverse environments often shade into each other insensibly and form a continuous series, whereas the Specific Forms of life which are subject to them on the whole form a Discontinuous Series.

Bateson (1894, 2–5)

By adopting Mendel's hypothesis of heredity as the transmission of discrete factors, Bateson brought the issue of discontinuous evolution to the forefront of biological discourse. But alongside the issue of evolution that Bateson audaciously fought for, he put the universality of Mendelian *Faktoren* at the center of the discourse on biological heredity. In adopting Mendel's research method of experimental hybridizations he established the new discipline that in 1906 he called "genetics."

Mendel, the researcher of transmission, by choosing discrete traits each of which could be classified (more or less) unequivocally into two alternative discernible classes was spared the commitment to the nature of the factors. Although he indicated that cellular processes were involved in the process of reproduction, he treated his *Faktoren* as entities determined by the laws of nature. In 1900, de Vries placed the particulate, preformed pangenes as determinants of Mendelian *unit characters* at the center of the picture. This notion of a direct relationship between the segregating Mendelian *Faktoren* and the observed unit characters prevailed throughout the first decade of the twentieth century. Correns consistently insisted on giving the *Faktoren* more of a cell-physiological interpretation, whereas Bateson described the *Faktoren* in terms of the mechanics of unit characters, to the extent that he interpreted morphological, physiological, or even behavioral characteristics as unit characters as long as they exhibited Mendelian segregation in hybridization experiments.

Thus, although the notion of the unit character had its source in the morphogenists' tradition, it was now gallantly defended in terms of genetic analysis acquired from the hybridists' tradition. It took the work

42

of Wilhelm Johannsen, arguably the last of the Linnæan taxonomic essentialist hybridists, to demote the unit character (Falk, 2008). Johannsen introduced the differentiation of the genotype and the phenotype by – contrary to Mendel – directing his attention to continuous variables that can *not* be classified as discrete alternative characteristics.

3

From *Faktoren* to unit characters

Heredity is, of course, at the heart of any theory of evolution, and more specifically at the heart of the Darwinian hypothesis of evolution by natural selection: it is inherited variability that provides the raw material upon which natural selection can operate. In 1868 Darwin presented a "provisional hypothesis of pangenesis," in *The Variation of Animals and Plants under Domestication*. Pangenes, representing all organs, were assembled in the gametes and merged in the zygote to be farmed out to the respective newly formed organs (Darwin, 1868). Although Darwin conceived of "[t]wo distinct elements [that] are included under the term 'inheritance,'" namely the transmission and the development of characters" (Darwin, 1981 [1871], 279), his hypothesis was essentially a top-down hypothesis of reproduction. As a hypothesis of inheritance it was unacceptable on both theoretical and empirical grounds. Already in 1867, Fleeming Jenkins had shown that in Darwin's blending model of inheritance all variation would be "swamped out" long before it could be established by natural selection (see Hull, 1973, 302–350). Others, like Thomas H. Huxley and, later, Francis Galton, were just as worried about the empirical evidence of the nonheredity of (most of) the intraspecific variation (Falk, 1995, 226).

PANGENES AND *FAKTOREN*

Contrary to common lore, Mendel's paper was not unknown in the thirty-four years between its publication and its acceptance as significant (see letters to Roberts, in Stern and Sherwood, 1966; Jahn, 1957/58; and Weinstein, 1977).

Hugo de Vries' theory of *Intracelluläre Pangenesis* (1889) was a theory of reproduction and generation intended as a criticism of the

Darwinian hypothesis of continuous, gradual evolutionary change. Although de Vries repeatedly declared that his theory was constructed on Darwin's notions of heredity (Stamhuis, Meijer, and Zevenhuizen, 1999, 240), Onno Meijer pointed out not only that de Vries turned Darwin's hypothesis upside down, but also that "A more non-Mendelian view is hard to imagine." Still, de Vries was convinced that Mendel's principles "would make *Pangenesis* stronger" (Meijer, 1985, 206).

Taking notice of the inroads into cellular processes of gamete formation and fertilization, de Vries rejected notions of force or chemical character in heredity and insisted that the carriers of heredity were particles of a specific order: the characters of a species are independent of each other and their transmission calls for highly independent hereditary particles, the *pangenes* (Stamhuis *et al.*, 1999, 241). These pangenes are transmitted in cell nuclei from one generation to another through the gametes. The zygote inherits a set of maternal and a set of paternal pangenes that are transmitted indiscriminately to all cells of the organism, where some of them are moved to the cytoplasm of their respective cells, thus directing the development and differentiation of these cells. Contrary to Darwin's theory, in de Vries' theory there is no way for the pangenes of the cells of the soma to affect the pangenes of the gametes, although conditions, notably nutrition, could affect the quantitative relations of the pangenes in the cells, causing the commonly observed "fluctuations" in organisms of a given species (Meijer, 1985, 207–210). In this preformationist conception of organisms, pangenes were held to be *active* or *latent* in specific cells of the organisms, according to the cell's role in development and differentiation (de Vries, 1889[1910]). Originally, de Vries did not limit the number of pangenes for a unit character per cell. Only after 1900, following Mendel, did he modify his theory to that of two pangenes per unit character per cell that may assume two alternative states. Results obtained by de Vries in 1899 with *Lychnis*, showing a segregation into three types, were later reinterpreted by him as segregating into only two alternative patterns at a ratio of 73:27 (Kottler, 1979).

As noted by Bishop (1996, 211), the question of the occurrence of intermediate types is very significant for a theory of evolution, for it may categorically rule out the possibility of establishing a definite ratio between only two types resembling the parent form. de Vries was aware of this and emphasized that "the lack of transitional forms between any two simple antagonistic characters in the hybrid is perhaps the best proof that such characters are well delimited units" (de Vries, in Stern and Sherwood, 1966, 110). Pangenes provided for a theory of evolution also by being assigned the property of mutability, of obtaining new properties discontinuous with the original ones (de Vries, 1902–3). A *retrogressive* mutation changes a pangene that was active in specific cells to become latent, whereas a *degressive* mutation changes a latent pangene into an active one. A third, *progressive* class of mutations changes a pangene to become involved in new unit characters. These, according to de Vries, were the main contributors to a saltatory, discontinuous evolution of new species by natural selection (Meijer, 1985; Stamhuis *et al.*, 1999; Theunissen, 1994; Zevenhuizen, 1998; see also Falk, 2001b, 288–289).

de Vries thus adapted Mendel's construct to his preformationist theory of development, of inheritance *and* of generation. Reproduction of *unit characters* provided for de Vries not only a hypothesis of development but also a mechanism for discontinuous evolution, intended to replace the Darwinian model. Regarding the process of species formation, Mendel's analysis was for him merely a theory of inheritance (Brannigan, 1979). He ignored Mendel's painstaking methodological design, such as the selection of seven traits that could be gainfully analyzed by his experiments, and imputed to Mendel the confirmation of his viewpoint that *"the concept of species recedes into the background in favor of the consideration of a species as a composite of independent factors"* (de Vries, in Stern and Sherwood, 1966, 108).

de Vries elaborated on his conception of evolution by discontinuous mutations in the voluminous tome *Die Mutationstheorie* (de Vries, 1902–3; for the English version, see de Vries, 1912, *c.* 1904). Variation between organisms, to the extent that it is not due to "fluctuating" environmental factors, is discontinuous: "No slow, gradual changes can have taken place" (de Vries, 1912, *c.* 1904, 40). Volume I, *Die Entstehung der Arten durch Mutation* [The origin of species by mutation], was already in name an explicit challenge to Darwin's notion. Although it was published after his "rediscovery" of Mendel, in it de Vries paid little attention to Mendel's hypothesis or to his experimental design.

de Vries was not interested in Mendel's work beyond its support of his theory of pangenesis. When Bateson adopted Mendelism as his research project, de Vries implored Bateson:

> I prayed you last time, please don't stop at Mendel. I am now writing the
> second part of my book which treats of crossing, and it becomes more and
> more clear to me that Mendelism is an exception to the general rule of

crossing. It is in no way *the* rule! It seems to hold good only in derivative cases, such as real variety-characters.

de Vries, in Provine (1971, 68)

Still, the value of de Vries' particulate approach to the study of inheritance cannot be overestimated: the 1900 superposing hybridist *methodology* on his morphogenist *conceptions* directed Mendelism and genetic analysis, for better or worse, to encompass evolution and development, rather than merely the mechanics of heredity. William Bateson and colleagues accepted de Vries' notion that organisms are "built up of distinct units," and it was the inheritance and development of such unit characters and their evolution that Mendel's laws were supposed to explain (Falk, 2003, 88).

Like de Vries, Bateson contended for a stepwise discontinuous evolutionary progress. Based on his field observations he suggested already in 1894 evolution by homoeosis or "heteromorphosis," i.e., the alteration of one organ of a segmental or homologous series from its characteristic form to that of some other member of a series. According to his "Theory of Repetition of Parts," organisms are constructed as discontinuous repeats with variations on the same theme. Different organisms are systemic variations of such repeats (Bateson, 1894). Although he was well acquainted with horticulturalists' hybridization methods, Bateson needed an experimentally testable theory of inheritance of discontinuous variability. Upon reading de Vries' "rediscovery" paper Bateson immediately adopted Mendel's hypothesis of segregation of unit characters in the gametes of hybrids and enthusiastically proceeded to "Mendelize," i.e., to test by hybridization the inheritance of alternative unit characters. Unlike de Vries, Bateson conceived of Mendel's work as a breakthrough in transmission biology rather than merely support for the theories of *Intracelluläre Pangenesis* and the *Mutationstheorie*.

Bateson did not heed de Vries' plea to abandon Mendelism in favor of the theory of mutations. Instead he adopted the reductionist, preformationist concepts of de Vries of the inheritance of unit character. He applied to unit characters the Mendelian laws of independent segregation, and although he got deeply involved in their implications for the theory of evolution, it was their transmission properties that became the focus of his genetic analysis. As a matter of fact Bateson soon *defined* unit characters as those traits that "Mendelize," i.e., segregate as Mendelian entities: upon encountering cases in which the structural or physiological unit characters did not correspond to the unit characters of inheritance, Bateson and coworkers resorted to *ad hoc* helping hypotheses. To the extent that the morphologist's or physiologist's unit character did not agree with those of Mendelian segregation units, the former were not really "unit characters," but rather complex characters. Mendelian segregation became for Bateson a device to assess spurious morphological unit characteristics (Schwartz, 1998, 2002). Such was the case of the "compound character" of fowls' combs, which appeared in four "antagonistic" forms ("walnut," "pea," "rose," and "single") that were inherited according to Mendel's rule of two independently segregating unit characters (Bateson, 1905 [1928]). Bateson suggested a resolution of the "outward" or zygotic unit character of the comb morphology, "being represented in the gametes by more than one factor" (Hurst, 1906): a zygotic entity breaking up in the gametes into two unit characters, each segregating strictly according to the Mendelian rule, and synthesized again into a compound entity once a zygote was produced (maintaining the material identity between factor and character, rather than assuming the unit character was a functional interaction between distinct factors). When Bateson encountered cases of two or more Mendelian unit characters in which "the proportions do not accord with Mendel's assumption of random segregation," i.e., the characters proved to be "coupled" or "repulsed," he interpreted these according to his 1891 Theory of Repetition of Parts, namely that "reduplication or proliferation" of gametes may take place, and that "some factors are distributed according to one of the duplicated series and other factors according to the normal Mendelian system" (Bateson and Punnett, 1911).

When Bateson suggested that the two alternatives of a unit character (which he called allelomorphs, later shortened to alleles) represented the respective presence and absence of the unit character (Schwartz, 2002), he actually abandoned the de Vriesian claim on the role of pangenes in evolution, according to which there were old pangenes which were dominant over new ones (see below):

> It will be noted that the view of many naturalists that the phylogenetically older character is prepotent, or more correctly, dominant, is by no means universal. In poultry, for instance, both pea and rose combs are dominant against single, though the latter is almost certainly ancestral; the poly-dactyle foot is dominant against the normal, though a palpable sport.
>
> Bateson and Saunders (1902)

Both genetic analysis and observations in nature intimate that the rule of dominance being ancestral does not hold. However, when Bateson tried to accommodate the Presence and Absence Hypothesis to his own ideas about evolution he too ran into trouble (see Swinburne, 1962). "Recognition of the distinction between dominant and recessive characters has, it must be conceded, created a very serious obstacle in the way of any rational and concrete theory of evolution" (Bateson, 1913 [1979], 93), at least if one considered mutations as being (structural or functional) losses. The confounding of transmission with reproduction and generation remained problematic (see also Chapter 19).

In the critical years at the beginning of the twentieth century it was Bateson more than anybody else who assumed the role of Mendel's "bulldog," as Thomas Huxley did for Darwin (Falk, 2007). Bateson's Mendelism, like that of de Vries however, confounded essential concrete material preformed unit characters with Mendel's commitment to methodological reduction to factors that follow numerical regularities. To a great extent, the disentanglement of this bias shaped the future of genetic research.

THE STRUGGLE WITH THE CONCEPTION
OF GENETIC ANALYSIS

Bateson's enthusiasm for Mendelism as a way to resolve the difficulty of Darwinian continuous evolution was bound to collide with the ideas of the positivist and author of the influential treatise *Grammar of Science*, Karl Pearson (Pearson, 1900). Contrary to Bateson's conclusions that relied on the *a priori* theory of "Mendel's Law of Hybridization" concerning the organization of biological material, Pearson argued for a position that was based on statistical observations ostensibly without prior theoretical assumptions. This, he claimed, was represented in the "Law of Ancestral Heredity."

Pearson was drawn to the problems of inheritance by Francis Galton's *Natural Inheritance*. Galton had shown him how with statistical exposition one could study nature without the need to commit oneself to causative interpretations: "It was Galton who first freed me from the prejudice that sound mathematics could only be applied to natural phenomena under the category of causation" (see Provine, 1971, 51). As noted by Jean Gayon, Karl Pearson's treatment of ancestral heredity was exemplary for the epistemological approach that conceived it as a purely descriptive matter (Gayon, 2000, 74–75). Thus, whereas Bateson confronted Pearson with a hypothesis *within the theory of evolution*, Pearson confronted Bateson with a statement *on the grammar of science*. Whereas Pearson developed statistical methods for the analysis of data on variation, presumably without being committed to any hypothetical speculation, Bateson's starting point was the development of empirical methods for the examination of a hypothesis of particulate inheritance, or genetics.

Galton put forward the "stirp" theory in which the correlation coefficient between parents and offspring became for him and his followers the typical measure of the hereditary force (Galton, 1875). Pearson further developed Galton's "stirp" theory of inheritance, according to which the traits of the individuals in one generation are the sum of the contribution of those of their ancestors, one half of each parent, a quarter of each grandparent, an eighth of each great-grandparent, etc. On Mendel's particulate hypothesis ancestors were helpful only to analyze the specific particulate status of an individual so as to allow the prediction of that individual's genetic combination and the possible results of hybridization in generations to come; but specific progeny did not *necessarily* carry specific ancestral factors.

Bateson's confrontation with Pearson and with his former friend and colleague W. F. Raphael Weldon was highly emotional and involved science politics at least as much as basic and empiric scientific deliberations. However, G. Udny Yule had already shown in 1902 that analytically the two approaches converge. As formulated by Yule, the statisticians' notion of "ancestral heredity" was the question: to what extent will knowledge of the grandparent's character enable one to increase the accuracy of the estimate of the character of the grandchildren above that obtained from the knowledge of the character of the parent? "If the answer to the question be in the affirmative ... then there is what may be termed a *partial* heredity from grandparent as well as from parent" (Yule, 1902, 201). This was the case. Accordingly, a law may be deduced "that *the mean character of the offspring can be calculated with the more exactness, the more extensive our knowledge of the corresponding characters of the ancestry*," and this was Pearson's Law of Ancestral Heredity (Yule, 1902, 202).

Yule regarded the dispute between Pearson the biometrician and Bateson the Mendelian not only as a dispute on the epistemology of science as such, but more as a dispute that involved the legitimacy of a top-down phenomenological descriptive analysis of the variability of a race (= breeding population) against that of a bottom-up particulate reductive analytical interpretation of individual (or pure line) character differences in hybridization experiments (Yule, 1902; see also Tabery, 2004).

There has always been a good deal of misunderstanding between biologists ... due in great part, I believe, to the fact that [they] use such terms as *heredity, variation, variable, variability*, in precisely the same signification. The employment of quantitative methods necessarily leads to the use of such expressions in a more precise signification ...

Quite generally, the statistician speaks of a character as *inherited* whenever the number or "constant" B [in the equation Y=A+B.X] is greater than zero ...

The distinctions between continuity and discontinuity of variation, between inheritance of attributes and of variables do not seem to me to be of *necessary* importance for the theory of heredity ... The real and important distinction seems to lie between the phenomena of *heredity* within the race, and the phenomena of *hybridization* that occurs on crossing two races admittedly distinct.

Yule (1902, 195–199)

Although mathematically one could reduce the Law of Ancestral Heredity to that of Mendel's Law of Hybridization, epistemologically, one should discern the difference between heredity and hybridization: Mendelism is concerned with hybridizations. *Hybridization* is the method for the study of the hereditary variation of specific characteristics between *individuals. Heredity*, on the other hand, deals with the *population*-aspect of inheritance (Tabery, 2004). The Law of Ancestral Heredity is concerned with *heredity*, it is a law that regards the correlation of variance in one generation of the population with that of another; today we would say that it is a law in population genetics. "The statistical theory of heredity, as developed in the work of Galton and Pearson, concerns itself with aggregates or groups of the population and not with single individuals" (Yule, 1903). As for Pearson's claim for no hypothesis, Yule pointed out that "It is difficult to suppose that the weight attached to pedigree is based on nothing but illusion" (Yule, 1902, 202) – it begs for a hypothesis, and this was provided by Mendel's theory of hybridizations.

Whereas Pearson and Galton deduce (describe) backwards, to ancestral generations, presumably with no hypothesis implied, Mendel's laws, *ex hypothesis* deduce forward, to future generations (Gayon, 2000).

The value of the work of Mendel and his successors lies not in discovering a phenomenon inconsistent with that law [of Ancestral Heredity], but in shewing that a process, consistent with it, though neither suggested nor postulated by it, might actually occur.

Yule (1902, 227)

Sixteen years later R. A. Fisher too showed that mathematically the Law of Ancestral Inheritance can be reduced to that of Mendelian Inheritance (Fisher, 1918). Fisher is generally regarded as "the first to successfully put forth a theory of the relationship between biometry, Darwinian evolution, and Mendelian inheritance," and his paper "is considered to be a direct descendent of Yule's earlier suggestion" (Tabery, 2004, 82). Yule and Fisher, however, had different agendas.

Contrary to Yule's demographic top-down perspective of inherited variation in populations for which Mendelism provides a satisfactory theoretical explanation, Fisher was an ardent Mendelian reductionist who endeavored to explain inherited variance of populations bottom-up. As put by Tabery, "For Fisher the ancestral law was a special case of the Mendelian principle. For Yule the Mendelian principles were a special case of the ancestral law" (Tabery, 2004, 90–91). Yule claimed that the Mendelian hypothesis of inheritance was a legitimate explanation if one observed the correlations in traits between progeny and ancestors. Fisher claimed that "if one *first* supposes Mendelian inheritance, *then* one can derive the correlation between relatives, resulting in the ancestral law of heredity" (Tabery, 2004, 83, emphasis in the original). In other words, whereas Yule provided a hypothetical framework for the observed variability, Fisher provided an explanation in observable terms for the hypothetical Mendelian factors. Fisher took the Mendelian principles of inheritance as the explanatory base and then derived the biometric law of ancestral heredity to show that the statistical law was just a special case of the physiological law (Sarkar, 1998, 106; see also Chapter 10).

THE PRESENCE AND ABSENCE HYPOTHESIS

Significant to de Vries' approach to Mendel's work was his treatment of the phenomena of dominance and recessivity. Mendel was very explicit in stating that he deliberately *selected* for his experiments such traits that could be discerned qualitatively, so that he might classify his material into binary categories. Quite early in his paper he called attention to the important methodological point:

> Experiments on ornamental plants undertaken in previous years had proven that, as a rule, hybrids do not represent the form exactly intermediate between the parental strains. Although the intermediate form of some of the more striking traits . . . is indeed nearly always seen, in other

cases one of the two parental traits is so preponderant that it is difficult, or quite impossible, to detect the other in the hybrid ... This is of great importance to the definition and classification of the forms in which the offspring of hybrids appear. In the following discussion those traits that pass into hybrid association entirely or almost entirely unchanged, thus themselves representing the traits of the hybrid, are termed *dominating*, and those that become latent in the association, *recessive*.

Mendel, in Stern and Sherwood (1966, 9)

Mendel did not treat the phenomenon of dominance as a law of heredity. Throughout his paper he describes a *phenomenon*, not a property, using the verb *to dominate* [*dominirende*] rather than the adjective *dominant*. When de Vries took notice of Mendel's paper he adapted it to his theory of elementary types, accepting that as a rule only two factors represented each unit character, and that these segregated regularly. Reluctantly de Vries traded his *active* and *latent* pangenes for Mendel's terms *dominating* and *recessive* factors, respectively. As such, in the role of pangenes, dominance became an intrinsic, fundamental property of the *Faktoren* (see Falk, 2001b; de Vries, in Stern and Sherwood, 1966, 108–110).

Contrary to Mendel the hybridist, who methodologically took notice of the phenomenon and *used it* for a more clear-cut analysis of qualitative differences, de Vries the morphogenist conceived of dominance as an immanent property of unit characters, and *imputed to it* the status of a (Mendelian) law.

Dominance and recessivity as reflected in hybrids indicated to de Vries not only the *developmental* state of the *Anlagen*, but also their *evolutionary* history. The older the pangenes, the more dominant they were:

> ...a hybrid bears the active character of its parent-species and not the inactive character of the variety chosen for the cross.
>
> We may put this assertion in a briefer form, stating that the active character prevails in the hybrid over its dormant antagonist. Or as it is equally often put, the one dominates and the other is recessive. In this terminology the character of the species is dominant in the hybrid while that of the variety is recessive. Hence it follows that in the hybrid the latent or dormant unit is recessive ...

de Vries (1912 [*c.* 1904], 280)

Carl Correns was not committed like de Vries to preformationism. His interest was rather in the relationship between the hereditary *Anlage* and the development of character. Not surprisingly, although he made notes from Mendel's Pisum paper as early as 1896 (Rheinberger, 2006, 75–113), he did not see a need to hurry and publish his findings and claim priority of a "rediscovery." For him there was nothing to be rediscovered in Mendel's paper since Mendelian segregation and assortment were nothing but what occurred in the cell nucleus during the process of reduction division. He confirmed that indeed "in *many* pairs one trait, or rather the *Anlage* thereof, is much stronger than the other trait, or its *Anlage*, that the former alone appears in the hybrid plant," but he could not understand "why de Vries assumes that in *all* pairs of traits which differentiate two strains, one member must always dominate" (Correns, in Stern and Sherwood, 1966, 121, emphasis in the original). Two years later Correns emphasized that in spite of dominance being a common phenomenon "the rule that the trait 'colorless' – the lack of pigment – is recessive towards the 'colored' cannot anymore be maintained as a law" (Correns, 1903 [1924a], 1903 [1924b], 346). By then he already stressed the distinction between the concepts of pairs-of-traits (*Merkmals pare*) and pairs-of-*Anlagen* (*Anlagen pare*) – the trait being only the marker of the *Anlagen* – and realized that dominance was a "vegetative" phenomenon of development, rather than one of hereditary factors.

Because for Correns the transmission of *Faktoren* was a consequence of cellular processes, and as he was not biased by an anti-Darwinian prejudiced hypothesis of preformationist intracellular pangenes, his problem was essentially one of "how do they do this?" Correns insisted that the presentation of the traits of segregating allelomorphic *Anlagen* as two opposing (qualitative) alternatives, a dominant one and a recessive one, was wrong and we should view these as only two extremes of a continuous (quantitative) variable (Falk, 2001b). By conceiving that the developing traits were the physiological consequences of the action of the *Anlagen* he actually anticipated the distinction of genotype and phenotype. As Gliboff (1999, 230) commented, Correns' role, more than "rediscovering" Mendel, was to shift attention from the *law* to the *mechanism*, and this may have been crucial in allowing Mendel's paper to become the founding document of modern genetics.

Bateson, unlike Correns, adhered to de Vries' qualitative notion of unit character that might acquire one of two (or more) alternative states, and in 1906 expressed this by suggesting that the dominance–recessivity relationship reflected the presence and absence, respectively, of a *single unit character* (Carlson, 1966 [1989], 58). Bateson's Presence and Absence Hypothesis deviated, however, from de Vries' notion of different kinds of pangenes also in that it did not refer to structures but rather to functions: Bateson emphasized vibration, motion, and waves rather than static material units (Coleman, 1970) as the articles of the hereditary transmission of unit characters, and did not consider dominance an "essential" part of Mendel's theory (Falk, 2001b):

> Mendel's hypothesis [of independent segregation of traits] is so ingenious and remarkable, and possibly of such far-reaching importance, that one can understand Mr. Bateson speaking of it as the "essential part" of his discovery, to the complete exclusion of the law of dominance and the various laws of numerical proportions which summarized the *facts* observed. These last two laws obviously enough do not hold in many cases.
>
> Yule (1902, 224)

For a while the Presence and Absence Hypothesis for antagonistic unit characters worked. When Bateson came across "compound unit characters," like the above-mentioned birds with "rose" comb and "pea" comb that produced in the first generation (F_1) hybrids a new character, the "walnut" comb, and in the next generation (F_2) *four* antagonistic characters, which appeared to segregate according to Mendel's rule of two independent traits, he first suggested – as noted – "the breaking up of the compound allelomorphs of the original parents" into "gametic synthetic" unit characters (Schwartz, 1998). However, after analyzing and resolving some "preconceptions that the types of comb were definite entities alternative to each other," Bateson and his coworkers concluded that "the *presence* of a given modification must be regarded as alle-lomorphic to the *absence* of the same modification" (Bateson, 1905 [1928], 136–137, emphasis in the original). As they explain:

> The critical point of difference between the two views lies in the way in which single [the comb obtained as the double recessive in the F_2 of mating between fowls with "pea" combs and fowls with "rose" combs] is regarded. While formerly we treated it as a positive condition segregating from the other, on what we may call the presence-and-absence hypothesis it is to be recognized as the original state into which the factor – or pangen, if we use de Vries' term – for rose, pea, etc., has been introduced.
>
> Bateson (1905 [1928], 137)

Bateson's coworkers asked: "Is it not possible, for instance, that some of the original Mendelian characters in peas may be due to more than one gametic factor?" (Hurst, 1906, 119). To detect a hybridists' unit character it must vary, but is this unit necessarily also the morphogenists' unit character? Mendel's pea varieties were not *yellow* versus *green*, but rather *yellow* and *non-yellow*: "the contrasting pair, yellow and green, might be regarded as presence and absence of yellow on a basis of green" (Hurst, 1906, 119). "Allelomorphism may be represented as ... the *presence* of something constituting the dominant character which is *absent* from the recessive gametes" (Bateson, 1907, 653). However, both Bateson and Hurst became aware that this confounds notions of heredity and those of development. Bateson mentioned, for example, barley crosses, "where the absence of female organs in the lateral florets ... was found to be dominant over the presence of fully developed flowers."

> Doubtless the statement could be inverted, and it could be suggested that the absence of florets, etc. was due to the *presence* of some element which prevented their growth, but that would be to abandon all judgment based on the actual appearance, and the terms would become meaningless.
>
> Bateson (1926 [(1928], 138)

Yet, as late as 1909, George Shull declared that, "notwithstanding ... difficulties... there can be no question that most of the phenomena of Mendelian inheritance are more simply stated in terms of presence and absence than in any other way" (Shull, 1909, 412). Also Bateson's colleague and coworker Reginald C. Punnett insisted on the confounded notion of the Presence and Absence Hypothesis of each unit character of inheritance, adding or not adding a property to the existing background of the organism:

> [It] offers a ready explanation of one of the most widespread phenomena of heredity – the existence of characters in alternative pairs. For there are but two relations into which the unsplittable unit-character can enter with the individual. It may be present or it may be absent, and no third relation can be conceived. From this we are led to ask, whether the hypothesis can be brought into any simple relation with the phenomenon of dominance. Is dominance the outcome of the presence of the given factor, and recessiveness the condition implied by its absence? At present we can only say that such a point of view is not in variance with the great majority of the cases hitherto worked out...
>
> Punnett (1909, 38–39)

Several years later, when the distinction between genotype and phenotype was established, Punnett turned once more to Mendel, who carefully phrased his *Faktoren* within the boundaries of his experimental methodology:

> Mendel considered that in the gamete there was either a definite some-
> thing corresponding to the dominant character or a definite something
> corresponding to the recessive character, and that these somethings
> whatever they were could not coexist in any single gamete. For these
> somethings we shall in the future use the term **factor**. The factor, then, is
> what corresponds in the gamete to the **unit-character** that appears in some
> shape or other in the development of the zygote.
>
> Punnett (1911, 27–28)

In retrospect it is amazing to realize how close Correns and even Bateson
were to resolving the preformationist one-to-one link between unit
character and Mendelian *Faktoren*. Being captive in their preconceived
images and notions they could not overcome the difficulty. It took
Wilhelm Johannsen, a hybridist, who from the outset was deliberately
interested in quantitative variability and thus *not* trapped in the confu-
sion that de Vries introduced by turning Mendel's methodology of
studying easily discernible discrete characters into a conception of
the nature of organisms being composed of unit characters, to cut the
Gordian knot.

Notwithstanding, although the unit character disappeared from
genetic analytic discourse, its impact was compelling and it cast a long
shadow on genetic thinking. For many years to come dominance was
considered to be an inherent character of the specific factor; in Drosophila genetics nomenclature dominant mutations of wild-type alleles
were denoted with a capital letter, and recessives with a lower-case letter
(see, e.g., Lindsley and Grell, 1968; see also Chapter 19).

4

The demise of the unit character

de Vries and Bateson introduced the hybridist analytic methodology to challenge morphogenists' thinking about Darwinian evolution. Asking to what extent could ecologically related continuous variations of local populations become characteristics of distinct taxonomic strains and eventually species, they adopted not only a reductionist, particulate conception of organisms but also a rather determinist notion of the inherent nature of these unit characters. They established Mendelian genetics on determinist foundations in which *Faktoren* served as preformed unit characters.

Deterministic notions such as *Faktoren* becoming unit characters were anathema to both natural historians in the field and embryologists in the laboratory. Richard Woltereck's objective was to provide a rejoinder to what he called the Mendelian teaching following Weismann's and de Vries's conception of the origin of species (Woltereck, 1909). In his "investigations on the change of species, with emphasis on quantitative species-specific difference" he showed that when the conditions under which strains of *Daphnia* species from different lakes in Germany existed were modified, their morphology changed to simulate that of another acknowledged species. Furthermore, each strain had its specific inherent *Norm of Reaction*, that is, its characteristic morphological response pattern to variation in environmental conditions. Such specific norms of reaction made it meaningless to predict the properties (the form characteristics) of one strain under one set of conditions from those under another set of conditions or from those of another strain under the same conditions (Woltereck, 1909; see also Falk, 2000a). Mendelism as it was espoused in the footsteps of de Vries insisted on the lawful constancy of unit characters (excluding irrelevant environmental "fluctuations") unless some uncontrollable discontinuous "mutation" occurred. The concept of

58

the norm of reaction of morphological patterns, especially those considered essential taxonomic type-characteristics, undermined the Mendelian conception *as long as its* Faktoren *were unfolded unit characters.*

Contrary to August Weismann's germ plasm theory and Wilhelm Roux's analysis of mosaic development, which provided the foundations upon which de Vries proposed his *Intracelluläre Pangenesis* and the notion of unit characters, other embryologists conceived of embryogenesis as progressive cell–cell interactions. Most explicit was Hans Driesch, who showed experimentally the regulative developmental capacity of the blastomeres of the sea urchin embryo. Driesch eventually came to believe that development could not be reduced to explanations in physico-chemical terms and invoked a metaphysical vital force, *entelechy*. Driesch's friend and colleague, Thomas H. Morgan, who came "to look upon the problem of heredity as identical with the problem of development," also rejected the notion of the unit character, "for we now realize that it is not characters that are transmitted to the child from the body of the parent, but that the parent carries over the material common to both parent and offspring" (Morgan, 1910a, 449). He ignored entirely "the possibility that characters first acquired by the body are transmitted to the germ," not only on empiric considerations but also for analytic reasons:

> Were there sufficient evidence to establish this view, our problem would be affected in so far as that we should not only have to account for the way in which the fertilized egg produces the characters of the adult, but also for the way in which the characters of the adult modify the germ-cells.
>
> Morgan (1910a, 449)

A particulate theory "is the more picturesque or artistic conception of the developmental process" that "seems better to satisfy a class or type of mind that asks for a finalistic solution, even though the solution be purely formal," but for an experimentalist like Morgan it seemed "less stimulating for further research" (Morgan, 1910a, 451).

But the Mendelian unit characters encountered difficulties also from within the hybridist research program. William E. Castle explicated what was understood by unit character:

> (1) any visible character of an organism which behaves as an indivisible unit of Mendelian inheritance and (2) by implication, that thing in the germ-cell which produces the visible character.
>
> Castle (1919c, 127)

Thus, hybridization experiments followed by generations of selection of a unit character were obvious tests for the expected stability of the units. Castle's experiments with rats, however, showed unexpected variation in fur color of the progeny in selection experiments following hybridization of "hooded" rats with wild-type rats (Castle, 1906). The "hooded" unit character did not segregate cleanly from its alternative allelomorph; in the hybrids the alternatives rather "contaminated" each other. Castle wondered: "Can Mendelian unit-characters be modified by selection?" (Castle, 1913a, 106–127). The instability of the Mendelian unit character appeared to be a blatant refutation of Mendel's Law of the stability of the *Faktoren*. Also Darbishire's hybridization experiments with mice revealed variation in the trait fur color in F_1 (first generation hybrids) (Darbishire, 1902). This violated the preformationists' principle of dominance that predicted uniformity of hybrids, challenging the expected reproductive autonomy of the Mendelian unit characters (Ankeny, 2000, 328). The *ad hoc* explanations were that these were not really unit characters but rather complex, multi-unit characters (as noted, Bateson actually *re-defined* unit characters as those segregating according to Mendelian laws). Another alternative would have been to abandon the preformationist relationships between the Mendelian factors and the unit character. This needed a fundamental shake-up of the young science of heredity, namely the formal dissociation of inheritance from development.

This was done by Wilhelm Johannsen, arguably the last of the Linnæan typologists (Churchill, 1974; Roll-Hansen, 1978). Johannsen, like Mendel, was acting within the hybridist tradition of experimental analysis; however, unlike Mendel and the early Mendelians, he was interested in the transmission of quantitatively varying characteristics rather than in well-defined binary unit characters of plants. He was deeply committed to mathematical analysis of data, and was particularly influenced by Galton's rule of ancestral inheritance, as revealed by the effect of selection in normally distributed populations. Empirically, the mean of the progeny of selected parents regressed toward the mean of the population, rather than corresponding to that of the selected parents. Galton's regression coefficient corresponded to Darwin's expectation of evolution by slow, gradual, and continuous change of a population that accumulated over the generations. Johannsen, however, was also occupied with breeding procedures, in particular the programs operating in the Svalöf Experimental Station in nearby Sweden. The station's head, Nils Hjalmar Nilsson, had already juxtaposed in the early 1890s the two prevalent strategies of selection for the breeding of crops: the painstaking step-by-step mass selection program of the best plots year after year, as indicated by the Darwinian notion of evolution, and that of picking up outstanding individual plants, independent of ancestral inheritance – "sports" – and breeding from them. The latter practice was based on the pedigree method, elaborated by the French breeder Louis Vilmorin, and on de Vries's notion of evolution by mutational jumps. Nilsson, who believed in the constancy of biological types, a principle according to which "continued one-sided selection of variants does *not* lead to a gradual replacement of the type," voted for the latter strategy (Roll-Hansen, 1978, 204–205). Consequently, Johannsen was caught between Nilsson's belief in the Linnæan research tradition of essential types, which may be changed by discontinuous saltations, and Galton's morphogenist law of ancestral inheritance, which suggested that progeny inherit half their characteristics from each parent, a quarter from each grandparent, an eighth from each great-grandparent, and so on.

Johannsen's solution was to conceive of two levels of variation in populations: the observable, empiric level and the deduced, conceptual level. Continuous changes at the superficial level did not refute the notion of the constancy of types at the deeper level.

TYPES OR TRAITS?

Like Mendel, Johannsen was interested in a quantitative, numerical analysis of the problem of inheritance, and both persons relied on breeders' experience. But unlike Mendel, who was concerned with the inheritance of specific, individual characters and for which "the study breaks up into just as many experiments as there are constantly differing traits in the experimental plants" (Mendel, in Stern and Sherwood, 1966, 5), Johannsen was interested in the general aspect of inheritance of the species as a type. He maintained that the varieties and subspecies of the Linnæan species, which show typical characters from generation to generation, were "systematic units," or "constant form-types." He rejected the Darwinian conception of species as continuously changing entities and "reconciled evolution and unchangeable (stable) species by letting the elementary species change discontinuously. Through such sudden changes, mutations, new elementary species appear spontaneously" (Roll-Hansen, 1978, 222). Unlike Mendel, whose experimental design was based on the logic of physical "bottom-up" reduction, Johannsen's approach was the "top-down" conception of a typologist who considered the organism to be the fundamental entity. Thus, whereas Mendel judiciously *selected* individual unit characters that had proven in preliminary experiments to provide two distinct qualitative appearances – yellow versus green, smooth versus wrinkled, tall versus small – Johannsen judiciously *analyzed* quantitative characters that did not assume discrete unit characters. It was this very difference in the experimental procedure, inadvertently allowing Mendel to obviate the need to distinguish between the unit character and its *Faktor*, which forced Johannsen to discern hereditary and non-hereditary inputs of observed variability.

Johannsen started his critical experiments in 1900 with the purchase of eight kilograms of bean seeds. He followed the efficiency of selection of two quantitative characters: seed weight and seed circumference. Beans reproduce by self-fertilization; consequently, any changes achieved via selection for a character in successive generations would indicate the breeding potential of the examined sample, rather than the outcome of some possible hybridization with foreign types. To start with, both characters showed normal distribution about a mean. When he selected the twenty-five heaviest and the twenty-five lightest beans as seeds for the next generation, their progeny in 1901 showed, as expected, partial regression to the mean of the population. The progeny of nineteen of the lightest seeds (and in a parallel manner, the progeny of the heaviest seeds) were individually tracked further in the following years. After two generations of self-fertilization there was no more regression to the mean of the original population, instead there was full regression to the mean of the individually selected seeds: "pure lines" were thus established, each being characterized by its mean, which was maintained, in spite of continued selection, with complete regression to the mean of the line.

These experimental results convinced Johannsen that the Linnaean species concept should now be conceived more specifically, that is, in terms of a "geno-species" (*Antkægsart*). A geno-species, being a type that includes all individuals with the same hereditary make-up, is a *geno-type*. Ordinarily, such types could not be easily identified or maintained because organisms interbreed. They were, however, identified by inbreeding among self-fertilizing beans. The establishment of pure lines provided Johannsen with the experimental evidence for his notion of essential types. He concluded that the empirical, statistical mean of a *population* was not necessarily identical with the notion of the biological *type* (Johannsen, 1903). The observed mean is a superficial *appearance statistic*, a **phenotypic** variable that must be conceived as distinct from *something inherent* in the biological type, or the **genotype** (Johannsen, 1909, 113–128).

This was a conceptual breakthrough. In one fell swoop Johannsen severed the Gordian knot tying the character to its hereditary factor (Falk, 2000b, 320). However, by deriving the term *gene* from his genotype, Johannsen actually provided new legitimization to heredity of particulate entities. By thus dissociating transmission from development, not only did the unit character become superfluous. Johannsen went further and as a direct consequence of his organismic conception he rejected any material interpretation of the notion of the genotype, such as the chromosomal theory of inheritance or genes as the material units of inheritance. To the end of his career Johannsen continued to talk of genotypical (and phenotypical) variation and remained reserved about the meaning of the concept of the gene (Johannsen, 1923, 136–137; see also Falk, 1986, 135–141).

Garland Allen suggested that there was continuity between Johannsen's distinction between genotype and phenotype and August Weismann's hypothesis of distinct germ plasm and somatoplasm, which emphasized the disjunction of the impact of heredity from the impact of environment, thus placing "one more nail . . . in the coffin of neo-Lamarckism" (Allen, 1979, 205). Jan Sapp, on the other hand, believed that "the distinction between the genotype and the phenotype . . . served as a polemic against descriptive, speculative, and morphological approaches to the study of heredity – categories within which Weismann's theory itself proliferated" (Sapp, 1983, 326). Johannsen was quite explicit in rejecting any attempt to relate his conceptions to those of Galton and Weismann. The conception of the genotype, even if initiated by Galton and Weismann, was completely revised, and "of all the Weismannian armory of notions and categories it may use nothing" (Johannsen, 1911, 132). A decade later Johannsen was yet more explicit in distancing himself from attempts to reduce his top-down notion of the genotype into particles, though he admitted that originally he was "somewhat possessed with the antiquated morphological spirit in Galton's, Weismann's and Mendel's viewpoints" (Johannsen, 1923, 136).

By dividing a population's variance into a continuous and a discontinuous component, Johannsen indicated that the genotype is the cause of the phenotype of individual organisms (see also Churchill, 1974). "The genotype is a theoretical entity somewhat like the ideal

Aristotelian form, belonging to the organism as a whole" (Roll-Hansen, 1978, 224).

Fred Churchill referred to the conceptual distinction between phenotypes and genotypes fashioned by Johannsen as "one of the major accomplishments of the history of biology" and quoted the geneticist Leslie C. Dunn: "Johannsen's place in the history of biology may come to be seen as a bridge over which nineteenth-century ideas of heredity and evolution passed to be incorporated, after critical purging, into modern genetics and evolutionary biology" (Churchill, 1974, 6). According to Dunn it was the failure to appreciate that the phenotype depends on the interaction of many genes with each other and with the environment that "had led to the retention of the notion of unit characters which plagued genetics for two decades and doubtless delayed the clarification of some of its basic concepts" (Dunn, 1965, 99). Nonetheless, eventually, major aspects of the science of genetics were profoundly overhauled, such as the stability of Mendelian factors, Darwinian continuous evolution, and Weismann's preformationism (see Falk, 2008).

STABILITY OF MENDELIAN FACTORS

In his experiments with selection of the unit character Castle observed that characters "were unmistakably *changed* by crosses" and so he "for many years advocated the view that the gametes are not pure" and "that unit-characters are 'inconstant' in varying degrees" (Castle, 1919b, 126). In 1914 Castle responded to Muller's claim that such a view was not "in harmony with the results of Johannsen and other investigators," that he did not understand how "the experiments of Johannsen have any direct bearing on the case since no single *Mendelizing unit-factor* was demonstrated in that connection." As Castle saw it, Muller "might with propriety cite the bean work as bearing on the interpretation of the inheritance of body size in animals ... But neither of these cases has any direct bearing on the question of unit-character constancy, since in neither case has a unit-character, either constant or inconstant been shown to exist" (Castle, 1915, 37). Eventually Castle's colleague Edward M. East convinced him that Johannsen's insight that administered the final blow to the concept of unit character was relevant to his findings (East, 1912). Unit character is an efficient hereditary unit as long as it is not modifiable. But once we accept that not only must environmental conditions be standardized and kept constant in order for the unit character to be discretely definable but also all internal conditions and effects of other traits must be kept constant, the notion of the unit character as defining a hereditary factor loses its meaning. East wrote:

> I believe that we may describe our results simply and accurately by holding that unit factors produce identical ontogenetic expressions under identical or similar conditions. If under identical conditions the expression *is* different, then a new standard, a new unit, must be assumed . . . To be sure there are numerous changes of expression of characters when external and internal conditions are not so uniform . . . these changes can all be described adequately and simply by ascribing them to modifying conditions both external and internal. When external we recognize their usual effect in what we called non-inherited fluctuations, when internal we recognize their cause in other gametic factors inherited independently of the primary factor but modifying its reaction during development.
>
> East (1912, 648–649)

From his top-down perspective East pointed out that "This is a physiological conception of heredity, as it recognizes the great cooperation between factors during development" (East, 1912, 649). Castle's conclusions with respect to hooded (and piebald) rats made sense only as long as he regarded the traits from a bottom-up perspective, as immanent unit characters. But because they existed not only in an external environment but also in the internal environment of the organism, the unit character became meaningless beyond its instrumental level.

East summarized his analysis of Castle's argument in an italicized section:

> *Taking into consideration all the facts, no one can deny that they are well described by terminology which requires hypothetical descriptive segregating units as represented by the term factors. What then is the object of having the units vary at will? There is then no value to the unit, the unit itself being only an assumption. It is the expressed character that is seen to vary; and if one can describe these facts by the use of hypothetical units theoretically fixed but influenced by environment and by other units, simplicity of description is gained. If, however, one creates a hypothetical unit by which to describe phenomena and this unit varies, he really has no basis for description.*
>
> East (1912, 651)

East advanced a strictly instrumental approach to the concept of the Mendelian factors: "a factor, not being a biological reality but a descriptive term, must be fixed and unchangeable" (East, 1912, 634). Johannsen made a clean disjunction: "Personal qualities are the reactions of the gametes joining to form a zygote; but the nature of the gametes is not determined by the personal qualities of the parents or ancestors in question" (East, 1912, 644).

DARWINIAN CONTINUOUS EVOLUTION

According to Garland Allen, "[t]he most significant effect of Johannsen's 1903 work was to reinforce what the Mendelians and other critics of Darwin had long claimed, that natural selection was powerless to produce new species" (Allen, 1979, 197; see also Chapter 3). In spite of his indebtedness to Galton, Johannsen soon found himself criticizing the biometric school precisely because he fully assimilated the statistical meaning of their conceptions. Whereas Weldon stated that "[i]t cannot be too strongly urged that the problem of animal evolution is essentially a statistical problem," and Pearson insisted that the solutions to these problems were essentially statistical (Provine, 1971, 31 and 51), Johannsen stated at the outset of his *Elemente der Exakten Erblichkeitslehre* that "we must pursue the science of heredity *with* but not *as* mathematics" (Johannsen, 1909, 2; see also Churchill, 1974, 8).

Johannsen accepted Darwin's historical role in stimulating general interest in heredity and variation, in spite of the details of Darwin's ideas of heredity, but he wanted to liberate the theory of biological heredity from the damnation of "Darwinism" (Roll-Hansen, 1978, 208). By the conceptual dissociation of phenotypic variability from genotypic variability he did exactly this. Selection worked on phenotypic variation, heredity on genotypic variation. The question of whether selection acted on continuous or discontinuous variation was misguided. Conceptually the variation that Pearson was interested in was not what the Mendelians were interested in (Allen, 1979). Although the phenotypic variation was the only variation that could be measured empirically the conceptually significant genotypic variation was just one of its components. Variation of type did occur even in pure lines owing to environmentally induced "fluctuations" as well as inadequacies in measuring techniques. These fluctuations, indeed, follow the Quetelet–Galton law of binomial distribution. Mutations, on the other hand, are defined as "suddenly occurring larger or smaller deviations" from the parent type and are not predictable by the rules of binomial distribution

(Roll-Hansen, 1978, 210). In *pure lines* all variation is fluctuating, and regression to the mean is complete. In contrast, variation about the mean of a *population* is due also to variability of types, or genotypes. It is only this latter variation that is inherited and amenable to evolutionarily significant selection.

Johannsen realized that reproduction by self-fertilization provided a very special case to establish the "individual" rather than the species as the essential type – the *Formtypus* – over successive generations. Thus, Galton's law of regression may have correctly described the behavior of populations because they contained a mixture of hereditary types (Roll-Hansen, 1978, 211–212). In 1905 Johannsen introduced the Danish terms *Livs-type* (life-type) as distinct from *Tal-type* (number-type), which in 1909 he translated into *genotype* and *phenotype*:

> The "type" in Quetelet's sense is a superficial phenomenon which can be deceptive ... Therefore I have designated a statistical, i.e., purely descriptively established type, as an "appearance type" [*Erscheinungstypus*], a phenotype ... Through the term phenotype the necessary reservation is made, that the appearance itself permits no further conclusion to be drawn. A given phenotype may be the expression of a biological unit, but it does not need to be.
>
> Johannsen (1909, 123, as translated by Dunn, 1965, 91–92)

The stable type of the pure line provided a stable equilibrium. Johannsen did not consider contributions to variation due to segregation of factors besides those of mutations. Only the rare sporadic "sports" due to new mutations in existing factors provided raw material for evolution; in the early 1900s Johannsen ignored possible contributions of existing variation of factors in a population to its evolution (Roll-Hansen, 1978, 216).

THE "GENOTYPE CONCEPTION"

Arguably, the most profound aspect of Johannsen's insight was the physiological and embryological one. Johannsen's segregation of genotype and phenotype made the distinction between preformationists and epigenesists irrelevant for genetics. Heredity could now be conceived as a process, a *production*: it was competent to investigate the causes of development and function, unbiased by the notions of transmission of factors that unfold to become traits.

Like Johannsen, Morgan was skeptical with regard to the meaning of the Mendelian factors. Morgan felt a "distinct disinclination to reduce the problem of development to the action of specific particles in the chromosome," at least in the sense of the Roux–Weismann assumption, which argued for "nicely separating at each division the different kinds of materials of which the chromosomes are composed" (Morgan, 1910a):

> [T]hose not engaged in the immediate work itself have, I believe, often been misled in regard to the meaning attached to the term factor, and by the assumed relation between a factor and a unit character. The confusion is due to a tendency . . . to speak of a unit character as the product of a particular unit factor acting alone.
>
> Morgan (1913a, 5)

With the introduction of Johannsen's distinction between the pheno-type and the genotype, Bateson's interpretation of the dominance and recessivity of Mendelian characters as the presence and absence of the corresponding preformed determinants, was also considered by Morgan to be merely a "*system of nomenclature*" that was useful "when properly interpreted." Once Johannsen's distinction was adopted, it became "unwise to commit ourselves any longer to a view that a recessive character is necessarily the result of a loss from the germ-cell" (Morgan, 1913a, 11).

"SOMETHING" IN THE GAMETES

> *Bloß die einfache Vorstellung soll Ausdruck finden, daß durch "etwas" in den Gameten eine Eigenschaft des sich entwickelnden Organismus bedingt oder mitbestimmt wird oder werden kan.* [Merely the simple idea should be expressed that a property of the developing organism is or may be con-ditioned or co-determined, through "something" in the gametes.]
>
> Johannsen (1909, 124)

Although Johannsen's original analysis was a statistical one of the components of population variance of traits, his interest was primarily in the physiology of the plant as an organism. Realizing that in the pure lines that he had established the Galtonian regression was "complete, quite up to the type line" (Johannsen, 1903; see also Johannsen, 1903 [1955], 206), he argued for the need to *conceptually* discern two levels of variation: phenotypic and genotypic *variability*. This would resolve the dispute between proponents of continuous and discontinuous Darwinian evolution in terms of selection *of* continuous (phenotypic) variation and selection *for* discontinuous (genotypic) variation (Sober, 1984; Falk,

1995, 239). However, Johannsen's attention was primarily directed at the developmental connotations of his distinction between the "phenomenon of superficial nature" [*Erscheinung oberflächlicher Natur*], or phenotype, and the "something" in the gametes that "determines, or at least influences very considerably, the character of the organism that was established from the gametes at fertilization" (Johannsen, 1909, 123; see also Churchill, 1974).

The emphasis was on the genotype, but although in 1909 Johannsen asserted that "the gene for the trait" is rather a *shorthand expression* for a much more complicated relation, he could not detach himself from the notion of the organism being composed of unit characters: "Every trait at the basis of which there is a specific gene (a gene of a specific kind) may be designated a '*unit character*' as may be shown by hybridization experiments" (Johannsen, 1909, 125). Two years later he was, however, very explicit about the futility of the unit character:

> Mendelism . . . overthrows totally the idea of "*organs*" as being represented by the unities of [Galton's] "stirp," pointing out that the personal qualities of *the organism in toto* are the results of the reactions of the genotypical constitution. The segregation of one sort of "gene" may have influence upon the whole organization. Hence the talk of "genes for any particular character" ought to be omitted . . . So, as to the classical cases of peas, it is not correct to speak of the gene – or genes – for "yellow" in the cotyledons or for their "wrinkles" – yellow color and wrinkled shape being only reactions for factors that may have many other effects in the pea plants.
>
> Johannsen (1911, 147)

Observable characteristics of individuals, or the phenotype, are the responses to the genotype which the individuals inherited. The phenotypes of specific traits that segregated as Mendelian units were *markers* of units of "something" that he called "genes." However, "no hypothesis on the existence of this 'something' should therewith be construed or supported" (Johannsen, 1909, 124). Johannsen never committed himself to the nature of these genes. The term "genes for" was a highly misleading relic of de Vriesian *Intracelluläre Pangenesis* at the beginning of the twentieth century as much as it is at the beginning of the twenty-first century. A useful distinction is that between *hypothetical constructs* and *intervening variables* (see Falk, 1986). Intervening variables are purely "summarizing" characteristics: they "neglect certain features of experience and group phenomena by a restricted set of properties into classes" (MacCorquodale and Meehl, 1948, 96). An intervening variable is simply "a quantity obtained by a specified manipulation of the values of empirical variables: it will involve no hypotheses as to the existence of the observed entities or the occurrence of unobserved processes." Hypothetical constructs, on the other hand, encompass "words . . . which are not explicitly defined by the empirical relations." In hypothetical concepts something is added to the empirical data; we add to the coordinated grouping of data "certain existence propositions, i.e., propositions that do *more* than define them" (MacCorquodale and Meehl, 1948, 96–103).

For Johannsen, who extended the top-down typological notion to the essences of individual organisms, the genotype was an abstraction of even metaphysical dimensions. Genes remained intervening variables, of instrumental value to the practicing experimentalist:

> My term *"gene"* was introduced and generally accepted as a short and unprejudiced word for unit-factors. . . . From a physiological or chemico-biological standpoint . . . *there are no unit characters at all!* . . . We may in some way "dissect" the organism descriptively, using all the tricks of terminology as we please. But that is not allowed in genetical explanation. Here, in the present state of research, we have especially to do with such genotypical units as are separable, be it independently or in a more or less mutual linkage.
>
> Johannsen (1923, 136–137)

In the third edition of his *Elemente* Johannsen is explicit: "The phenotypic characters of an organism are always reactions of the *entire-genotype* in interplay with the factors of life-position." And further on:

> Our formulas, as used here for not directly observable genotypic factors – genes, as we used to say – are and remain *computational-formulas*, placement-devices that should facilitate our overview. It is precisely therefore that the little word 'gene' is in place; no imagination of the nature of this 'construction' is prejudiced by it, rather the different possibilities remain open from case to case.
>
> Johannsen (1926, 434)

Johannsen's attempt to identify the genotype as the fundamental entity of systematics beyond that of the species was, in a way, an attempt to preserve the Aristotelian typological notion of the organism in a world of Darwinian variation. However, by considering the immanent properties of the organism – its genotype – versus the ephemeral appearance of the individual organism – its phenotype – Johannsen laid the foundations for overcoming the dissonance between the particulate and the organismic approach. By suggesting that the (organisms') hereditary properties might be *described* as being composed of many intervening variables, the genes, he provided the way out of the Pearson–Bateson dispute by indicating that whereas the first one was referring to phenotypic variability, the second was referring to genotypic variability. The full impact of this effort came to light when Johannsen met with Morgan at the Meeting of the American Society of Naturalists in December 1910. By then Johannsen had extended his statistical, population conception of the genotype and phenotype to the process of development, providing for the dissociation of notions of reproduction from those of inheritance. For Morgan, who dismissed any concept that smacked of preformationism, the idea of material particles that uniquely determine traits of the organism transferred through secluded germ lines – as he understood Weismann's theory, or at least Edmund B. Wilson's version of it – was anathema to his organismic view (Falk, 2003, 88). Johannsen's severance of the conceptual link between development and heredity was crucial.

Wilson, Morgan's teacher and mentor, had been occupied for years with establishing the cell at the center of the theory of inheritance (Wilson, 1896, and later editions). Morgan was skeptical:

> The *modern* theory of particulate inheritance goes back no further than the discovery that the sperm transmits equally with the egg the characters of the race; and with the discovery that the most conspicuous thing that the sperm brings into the egg is the nucleus of the male cell or more specifically its chromatin. Around these simple statements a whole edifice has been erected. We owe to Weismann more than to any other biologist, the peculiar trend that this speculation has followed.
>
> Morgan (1910a, 452)

Johannsen's genotype conception allowed Morgan to maintain his organismic perspective and yet accept a particulate theory of inheritance, maintaining the notion of many-to-many relationships between genes and traits (Falk, 2003; Falk and Schwartz, 1993). When Morgan adopted and elaborated the chromosomal theory of inheritance, he resisted the tendency of his coworkers, notably Muller, to conceive of genes as its material atomic components (see Falk, 1986). As a matter of fact, Johannsen warned Morgan against the temptation of the "morphological spirit" to conceive of genes as particles with a certain structure (Johannsen, 1923; Roll-Hansen, 1978, 226). In *The Physical Basis of Heredity* (1919) Morgan maintained an instrumental conception of the "representative particles":

> The attempt to explain biological phenomena by means of *representative particles* has often been made in the past. The superficial resemblance of the theory of the gene to some older theories ... has furnished the opponents of the Mendelian theory of heredity an opportunity to injure the latter ... It need not be denied, however, that there is an historical connection between the mediæval theory of preformation and the particulate theory of heredity. ... Weismann, also, the most prominent modern adherent of preformation, held that the whole germs, ids, are present in the germ-plasm, each standing for a whole organism – each (or most or one?) becoming unravelled as the embryonic development proceeded. In fact, Weismann's entire theory was invented primarily to explain embryonic development rather than genetics. *Its connection with the modern idea of germ-plasm is little more than an analogy...*
>
> Morgan (1919, 234, italics added)

However, unlike Morgan, who accommodated the genes as "representative particles" into his chromosomal theory of inheritance, while remaining skeptical as to the reality of discrete material genes (see Part IV), for Johannsen "a new rival to the genotype theory was gathering momentum, the chromosome theory" (Roll-Hansen, 1978, 221). The question of chromosomes as the presumed "bearers of hereditary qualities" seemed to Johannsen to be an idle one: "I am unable to see any reason for localizing 'the factors of heredity' (i.e., the genotypical constitution) in the nuclei. The organism is in its totality penetrated and stamped by its genotype constitution. All living parts of the individual are potentially equivalent as to genotype-constitution" (Johannsen, 1911, 154).

Only in the third edition of his book (Johannsen, 1926), after reviewing some of the arguments against regarding the chromosomes as primary structures, did Johannsen reluctantly confess that "the behavior of the chromosomes ... shows in many cases such an astonishing parallel with certain hereditary phenomena after hybridization, that one is well persuaded to ascribe to them a wholly special significance" (Johannsen, 1926, 537ff; see also Churchill, 1974).

By the 1920s Johannsen's genie was out of the bottle. The chromosomal theory of inheritance overcame preformationism (Morgan, 1910a, 452–453). Likewise, evolution was conceived as "inheritance *of* variation" rather than "inheritance *and* variation," amenable to Fisher's Fundamental Theorem of Natural Selection of variation of discrete, discontinuous entities as the motor of Darwinian evolution (see Fisher, 1918, 1930; Muller, 1922). Finally, the constant stability of hereditary factors rather than that of their products was formulated in Beadle and

Tatum's "one gene – one enzyme" hypothesis (Beadle and Tatum, 1941b). The dialectic of the nature of the genes, however, whether instrumental intervening variables, or material hypothetical constructs, appeared to have ended only in 1953, with the presentation of Watson and Crick's chemico-physical model of DNA as the material basis of inheritance (Falk, 1986; and see Parts V and VI).

It erupted anew, however, with the developments in molecular biology towards the end of the twentieth century (Falk, 2000b; Griffiths and Stotz, 2006).

It is a major paradox that by eliminating the unit character and introducing the genotype as the "something" of inheritance behind phenotypic appearance, Johannsen provided a framework for claims of genetics *contraposing* environment: thus he inadvertently upheld the notion of Nature *versus* Nurture rather than Nature *cum* Nurture, and consequently spawned "genocentricity." In methodological terms this reductionism was most rewarding. Yet in conceptual terms this genocentric construct was bound to falter. By the end of the twentieth century, the results of the Human Genome Project, with its detailed mapping of genomes and their corresponding phenotypic proteomes, once more overcame the notion of autonomous discrete structural "atoms of heredity," leading us back to the nineteenth-century notion of epigenetic forces of the genotype as components of a system rather than that of particulate genes (see, e.g., Pearson, 2006).

III

The chromosome theory of inheritance

Cytology . . . merely established the zygote to be double with respect to the gametes. It was, however, discoveries like that of Mendel and his followers, which show that the gamete's simple nature corresponds also with respect to heredity.

Johannsen (1926, 432)

Already in 1900 Correns suggested that Mendel's "numerical ratio 1:1 strongly suggests that the separation occurs during a *nuclear division*, the reduction division of Weismann" (Correns, in Stern and Sherwood, 1966, 127). It was, however, only after the establishment of Mendel's laws that cytologists came to share a set of common assumptions which led them to agree on what they saw under the microscope and eventually to accept the link between chromosomes and Mendelian factors. Chromosome continuity and integrity and their specificity were derived from Boveri's experiments; chromosomes' simulation of multi-factorial Mendelian segregation further supported their role in inheritance, as Wilson and his associates indicated in studies of sex-determination in various insect species. But the behavior of chromosomes could not be simply *observed*; it had to be *interpreted*: were these just certain other structure-properties that *obeyed* the Mendelian laws of segregation or were these elements that causally *determined* the observed Mendelian laws of segregation? The establishment of the distinction between genotype and phenotype, overcoming the preformationist notion of unit character, was crucial for such an examination.

Morgan's discovery of a correlation between the pattern of inheritance of a Mendelian factor of Drosophila and that of its sex-chromosome suggested a causal relationship, but only the finding that the inheritance of several independent Mendelian factors was similarly correlated with sex-chromosomes allowed the hypothesis that the

75

chromosome was the causal mediator of inheritance. Morgan adopted Janssens' chiasmatype speculation of the cytological mechanics of chromosomes at meiosis, which also accounted for several exceptions from Mendel's rules, notably those of "coupling" and "repulsion" deviations from independent segregation of unit characters.

Three papers of Morgan's students that established the foundation of genetic analysis were based on experimental work with *Drosophila melanogaster*, which became the leading organism of genetic analysis in the next half-century: Bridges' proof of the chromosome theory of inheritance (Bridges, 1916), Muller's analysis of the mechanism of crossing over (Muller, 1916), and Sturtevant's linear mapping of the chromosome by recombination frequencies (Sturtevant, 1913b).

In the 1910s Morgan and his students initiated in the "Fly-Room" at Columbia University in New York an intensive analysis of the mechanisms of inheritance (Kohler, 1994). They amalgamated genetics into a research discipline of cytogenetics on the foundations of both the morphogenist tradition in cell biology and the hybridist tradition in experimental analysis. Once the causal role that the chromosomes' choreography played in Mendelian lawfulness was accepted, chromosomes' behavior and their exceptions became a major object for extending genetic analysis.

The following two decades were characterized by intensive efforts to unravel the mechanics of the chromosomes by genetic analysis, mainly by exploiting virtual or observable aberrations in chromosome numbers and structures, and the construction of genetic chromosome maps. Many of these studies involved the cytology of meiosis in hybrids, especially in plants that had conspicuously large chromosomes, turning cytology *per se* into an important instrument of genetic analysis. The induction of aberrations by X-rays as introduced by Muller in 1927 offered a new tool to the study of chromosome mechanics, to the location of genes on chromosomes, and to the understanding of the cytogenetic properties of the chromosomes. Finally, the unraveling of the giant polytenic chromosomes of dipterans at the beginning of the 1930s provided Drosophila geneticists with an almost "macroscopic" tool to study chromosome structure and mechanics. Solid foundations were laid for the extension in the 1950s of the analysis of the material basis of inheritance to the molecular level even though many of the notions of the mechanics of the chromosomes were rather re-interpreted in molecular and enzymatic terms.

5

Chromosomes and Mendelian *Faktoren*

The involvement of chromosomes as bearers of hereditary continuity had been proposed as early as in the 1880s, when cytologists and embryologists unraveled the processes of fertilization and cell division at mitosis and meiosis. "Only with the recognition that a continuum of structure must be preserved during the development of the individual did heredity assume its modern and more narrow meaning of *Vererbung*, or transmission" (Churchill, 1987, 364). The dispute over the facts was bitter, but that over the speculations and theories on the meaning of these processes was even more acrimonious. Most daring in his speculations was August Weismann. Weismann's concern with the theoretical problem of deriving totipotent germ cells from a mature organism with differentiated cells led him to argue that germinal cells are transmitted undisturbed from cell to cell in the germ line. He assumed that in the soma cells the chromosomes break down transversely into numerous particles or biophores, and different assortments are farmed out to the cells, which differentiate according to the biophores allocated to them. In the cells of sexually reproducing species chromosomes may often be discerned in a species-specific pattern – there are two of each, one maternal and one paternal, they are *diploid*. Since the chromosome number in the germ line of all individuals of a species is identical, Weismann further speculated that at the "maturation division" (meiosis) of germ cells, the number of chromosomes is halved – they are *haploid* (Churchill, 1987).

Once it was accepted that there is differentiation of chromosomes during development and continuity of chromosomes in transmission concerning different aspects of cell biology, inheritance and development were bound to separate (Churchill, 1987, 359–360; and also Chapter 1). But this did not happen; rather heredity maintained pre-formationist notions, as highlighted by the "unit character" of the early

Mendelians (see Chapter 3). Edmund B. Wilson, in *The Cell in Development and Inheritance* (Wilson, 1896) developed the theory of the cell as the fundamental unit of living organisms, and interpreted Weismann's hypotheses of the central role of the cellular nucleus, and especially of the chromosomes, in terms both of heredity and of development and differentiation (Allen, 1966, 1978; Griesemer, 2000). Wilson expected that "in some measure, reconciliation between the extremes of both the rival theories," preformation and epigenesis, should be found (Wilson, 1893 [1986], 78). The correlation between sex and the pattern of the chromosomes, as they are revealed at cell division in many animal species, made sex differentiation a favorable subject for the analysis of the role of chromosomes in heredity *and* development. Contrary to the constancy in the chromosome pattern in the germ line of all individuals of the species, in many species females carry a pair of sex-chromosomes in their diploid cells, whereas only one such sex-chromosome is present in males.[1]

The need for a full integral complement of chromosomes for normal development was demonstrated by Theodor Boveri. He fertilized eggs of sea urchins (*Paracentrotus lividus*) with two spermatozoa. Upon initiation of cleavage divisions, the three haploid nuclei that fused to form the zygote performed a *three*-pole mitosis. Consequently four embryo cells with excessive and deficient diploid numbers of chromosomes were produced. The embryos that developed from these soon showed major developmental aberrations and died. Only embryos that happened to contain a fully balanced complement of chromosomes survived (Boveri, 1902). The individuality of chromosomes and their causal involvement in differentiation was further indicated by the specific abnormalities of progeny with specific excessive or missing chromosomes. Twenty years later Blakeslee (1922) succeeded in hybridization experiments with the Jimson weed *Datura stramonium*, the diploid chromosome number of which is $2n = 24$, to produce all possible 12 trisomics – otherwise diploid plants each with one of the chromosomes represented three times instead of twice $(2n + 1)$. He correlated each different fruiting body phenotype to a specific excessive chromosome. Such correlations further supported the claims of a

---

[1] In some groups, like birds and Lepidoptera, it is the other way round: the females are the ones that carry the single X-like chromosome, and the males carry two such chromosomes. This was considered irrelevant to the universal principle, just as whether traffic is right-handed or left-handed is irrelevant to the fact that there are global traffic rules.

causative role of chromosomes in development, and more specifically, of X-chromosomes in sex determination.

The property of chromosomes as Mendelian *Faktoren* was suggested by Walter S. Sutton, one of Wilson's students. He showed that all chromosomes of the grasshopper *Brachystola magna* could be individually identified and that at meiosis the segregating chromosomes behaved *as if* each pair were alleles of a Mendelian unit character (Sutton, 1903). Sutton pointed out that *if* each pair segregated independently of the others, different combinations of the chromosomes (12 pairs in females) would produce a very large number of patterns ($2^{12} = 4,096$), as expected of the Mendelian theory of variation. However, as Mendel demonstrated, for confirmation of the independent segregation at least two alleles at two factors are needed. This was obtained for chromosome pairs in meioses of a male *B. magna* grasshopper that possessed, besides the single X-chromosome, a pair of chromosomes that could be individuated by the difference in their size. As shown by Carothers, in 146 cells the smaller element of the unequal pair segregated to the same pole as the X-chromosome in meiotic anaphase, whereas in 154 cells the larger element segregated with the X-chromosome (Carothers, 1913). Were chromosomes just another cell-morphology characteristic that segregates as Mendelian unit characters do, or did chromosome patterns at cell division prove that they had a causal role in heredity and development?

Thomas H. Morgan conducted from 1906 to 1909 a cytological study of several, mostly parthenogenetic, species of phylloxeran aphids in order to determine the role of chromosomes in sex determination, as suggested by Nettie Stevens and Edmund B. Wilson. Eggs destined to become females developed differently from those destined to become males. Also in male meiosis, two types of sperm were produced: one type with six chromosomes survived and eventually fertilized the eggs, resulting in females, the other type with only four chromosomes degenerated. Morgan considered such observations to be strong support that "the sex determinant was whatever cytoplasmic factor moved the chromosomes" (Falk, 1991, 469; Gilbert, 1978, 342). He too saw the *methodological* advantages of a heuristic that follows the development of parts of the organism as such:

> We no longer look for an actual embryo præformed but we look for samples of each part, which samples by increasing in size and joining suitably to other parts make the embryo. This is modern præformation.
>
> Morgan (1910a, 452)

However, Morgan saw two major difficulties with correlating chromosomes and Mendelism:

1. The constancy of chromosomal content, and their individuality, were inexplicable in terms of cellular differentiation during development. "If Mendelian characters are due to the presence or absence of a specific chromosome, as Sutton's hypothesis assumes, how can we account for the fact that the tissues and organs of an animal differ from each other when they all contain the same chromosome complex"? (Morgan, 1910a, 477).
2. The particulate nature of the Mendelizing characters was problematic. Sutton pointed out that if each character that Mendelizes is carried by a particular chromosome the mechanism of meiosis gives an explanation of the Mendelian rules.

> If we analyze the facts further we find that the hypothesis requires in order that pure gametes are to be formed by the hybrid that each particular character, or whatever it is that produces the character, be confined to a single chromosome . . . Since the number of chromosomes is relatively small and the characters of the individual are very numerous, it follows on the theory that many characters must be contained in the same chromosome. Consequently many characters must Mendelize together. Do the facts conform to this requisite of the hypothesis? It seems to me that they do not.
>
> Morgan (1910a, 466–467)

Even when he adopted the reductionist methodology, being convinced of the heuristic advantages of the *particulate theory of development*, Morgan repeatedly emphasized the need for a many-to-many relationship of genes and functions in organisms as developing wholes (see Falk and Schwartz, 1993, and discussion in Chapter 16). In fact, quite early on he drew an explicitly modern picture of "the gradual elaboration and differentiation of the various regions of the embryo" by suggesting that "[t]he initial difference in the protoplasmic regions may be supposed to affect the activity of genes. The genes will then in turn affect the protoplasm, which will start a new series of reciprocal reactions" (Morgan, 1934a, 10).

As for the issue of the number of chromosomes being too small for the many unit characters, or genes, he resorted to the phenomenon of partial coupling or repulsion of unit characters, discussed by Bateson (see below). Morgan resolved Bateson's difficulty of compound and coupled unit characters that "do not accord with Mendel's assumption of random segregation" by adopting the chromosome theory of inheritance.

Bateson, according to Morgan, confounded the *mechanics of the inheritance* of the Mendelian factors and their *physiological effects on development*. Coupling and repulsion were simple mechanical results of the location of the materials in the chromosomes, and the proportions that result are the expression of the relative location of the factors in the chromosomes.

> *Instead of random segregation in Mendel's sense we find "association of factors" that are located near together in the chromosomes. Cytology furnishes the mechanism that the experimental evidence demands.*
>
> Morgan (1911, italics in original)

THE ANALYSIS OF A *WHITE-EYED* FLY

In the spring of 1910 Morgan noted that in a culture of *Drosophila melanogaster* flies "which had been running . . . through a considerable number of generations, a male appeared with white eyes. The normal flies have brilliant red eyes" (Morgan, 1910b, 120). Flies from the culture of this "sport" white-eyed male and its progeny were inter-crossed for several generations and the number of female and male progeny with red and white eyes was scored. A "criss-cross" pattern, in which white-eyed females mated to red-eyed (wild-type) males produced red-eyed daughters and white-eyed sons, simulated the transmission pattern of the sex-chromosome. This pattern was complementary to a case worked out by Punnett and Raynor in the Magpie moth *Abraxas*. In this case pale female "sports" appeared. Mating of wild-type females with males of this culture gave the inverted "criss-cross" progeny pattern (in the moth the females are the ones that carry only one X chromosome). Morgan put forward *An Hypothesis to Account for the Results*. "Assume that all of the spermatozoa of the white-eyed male carry the 'factor' for white eyes 'W'; that half of the spermatozoa carry a sex factor 'X' the other half lack it, *i.e.*, the male is heterozygous for sex" (Morgan, 1910b, 120). In other words, Morgan suggested the *coordinated* inheritance of two factors, the sex-chromosome and eye color, according to the Mendelian rule of segregation, "in the sense that there are three reds to one white. But it is also apparent that all of the whites are confined to the male sex" (Morgan, 1910b, 121), contrary to the Mendelian rule of independent segregation. A hypothesis that the X chromosomes are the material correlate for the genetic factor in Drosophila and the Magpie moth was at hand (see Figure 5.1).

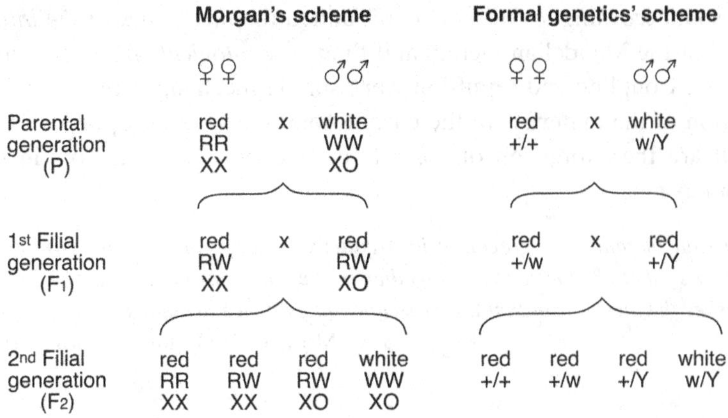

	Morgan's scheme		**Formal genetics' scheme**	
	♀♀　　　　♂♂		♀♀　　　　♂♂	
Parental generation (P)	red RR XX	x	white WW XO	red +/+	x	white w/Y
1st Filial generation (F₁)	red RW XX	x	red RW XO	red +/w	x	red +/Y
2nd Filial generation (F₂)	red RR XX	red RW XX	red RW XO	white WW XO	red +/+	red +/w	red +/Y	white w/Y

Figure 5.1. Morgan's original interpretation and the formal genetic interpretation of the experimental results of white eye-color inheritance in *Drosophila*.

An increasing number of cases came to light in which Mendelian factors did not follow Mendel's rule of independent segregation. "The most notable cases of this sort are found in sex-limited inheritance [later renamed sex-linked inheritance] in *Abraxas* and *Drosophila*, and in several breeds of poultry, in which a coupling between the factors for femaleness and one other factor must be assumed to take place" (Morgan, 1911).

The working hypothesis of the Morgan school was that chromosomes play a causal function in the transmission and development of traits, "cause here in the sense in which science always used this expression, namely, to mean that a particular system differs from another system only in one special factor" (Morgan, Sturtevant, Muller, and Bridges, 1915, 209). Morgan, following Mendel, implied a phenomenological analysis, with no commitment to a rigorous mechanical-causal relationship of cause and effect. This was always made in the wider context of biological research from embryogenesis to evolution. He seemed even to become more generous toward Weismann and his theory of a germ line as distinct from the soma line, conceiving it as a methodological device rather than a concept of preformation:

> The *most* important fact that we know about living matter is its inordinate power of increasing itself. . . . The simplest possible device is to divide. This makes dispersal possible with an increased chance of finding food, and of escaping annihilation, and at the same time reducing the mass permits of a more ready escape of the by-products of the living machine. . . . But there is another method of division that is almost universal and is utilized by high

and by low forms alike: individual *cells*, as eggs, are set free from the rest of the body. Since they represent so small a part of the body, an immense number of them may be produced on the chance that a few will escape the dangers of the long road leading to maturity. . . . The first step, then, in the evolution of sex was taken when colonies of many cells appeared. We find a division of labor in these many celled organisms . . . To-day we are only beginning to appreciate the far-reaching significance of this separation into the immortal germ-cells and the mortal body, for there emerges the possibility of endless relations between the body on the one hand and the germ-cells on the other. . . . Few biological questions have been more combated than this attempt to isolate the germ-tract from the influence of the body. Nussbaum was amongst the first, if not the first, to draw attention to this distinction, but the credit of pointing out its importance is generally given to Weismann, whose fascinating speculations start from this idea. For Weismann, the germ-cells are immortal – the soma alone has the stigma of death upon it.

Morgan (1913b, 2–16)

NON-DISJUNCTION AS PROOF OF THE CHROMOSOME THEORY OF HEREDITY

The journal *Genetics* was inaugurated in 1916 with Calvin B. Bridges' extensive paper on "Non-disjunction as proof of the chromosome theory of heredity" (Bridges, 1916). It was heralded earlier by two shorter publications (Bridges, 1913; Bridges, 1914). Rare exceptions to the "criss-cross" inheritance of the characteristic sex-linked markers of *Drosophila* flies were observed: instead of the expected wild-type daughters and vermilion sons from a cross of *vermilion* (v) females to wild-type males ($+$), an occasional vermilion daughter (or a wild-type sterile son) was obtained.[2] Upon crossing the rare exceptional vermilion daughter to regular wild-type males 3–4 percent secondary exceptional progeny (compared to the "criss-cross" expectations) were produced; such exceptional progeny could be maintained by mating the secondary exceptional females to males from a regular stock. Bridges hypothesized that the primary exceptional flies were due to "chromosomal non-disjunction,"

---

[2] Mutants are designated by italicized letters, like w for the white-eyed mutant, or v for the vermilion eye-color mutant. The "wild type" allele of all mutants is designated by $+$ (or w^+ and v^+, respectively). Two (recessive) mutants of independent origin are "complementing" when the compound heterozygote a/b is phenotypically non-mutant or "wild type." Two mutants, the phenotype of the compound heterozygote of which is not wild-type – they do not complement – are alleles assigned to the same gene and designated a_1/a_2.

that is, to an occasional meiosis in an oocyte in which the two X-chromosomes moved to the same pole rather than segregating to opposite poles (producing XX and no-X eggs, instead of normal eggs each with a single X; all other chromosomes – the autosomes – segregated normally). When such an XX-oocyte was fertilized by a sperm carrying a Y-chromosome a primary exceptional matroclinous daughter carrying XXY chromosomes was produced, and when a complementary oocyte missing any sex-chromosome was fertilized by an X-chromosome-carrying sperm, a sterile primary exceptional patroclinous XO male was born (the Y-chromosome was concluded to be "empty" except for male-fertility factors). Bridges showed that his *genetic* analysis of the chromosomal patterns of the flies was consistent with the *cytological* analysis of the chromosome patterns of the flies (see Figure 5.2).

Although Bridges' results strongly supported the chromosome theory of Mendelian inheritance, correlation does not prove causation. It could still be maintained that chromosomes behave like Mendelian factors, rather than that chromosome mechanics at meiosis cause Mendelian ratios. Additional mutant stocks, besides the *white* sport demonstrated the "criss-cross" inheritance pattern, in coordination with the transmission of X-chromosomes. Complementation tests confirmed that these markers were due to different gene functions rather than to different alleles of the same gene. This strongly supported the hypothesis that the cytologically observed maneuvers of chromosomes are the physical cause for the segregation pattern of these genes, and that the "sport" or mutant traits of these stocks were due to mutations at *different sites* of the X-chromosome. They comprised different *loci* of "sex-linked" genes. Other mutants of Drosophila could be grouped into two major non-sex-linked or "autosomal" linkage groups, presumably related respectively to each of the two pairs of larger chromosomes of the fly.

An important support for the chromosomal theory was provided by Muller's discovery that besides the three linkage groups of Drosophila that had already been explored, a fourth small linkage group existed. This was expected on the basis of Drosophila's chromosome pattern, which, with a pair of sex-chromosomes and the two pairs of major autosomes, carries also a pair of minute chromosomes ($2n = 4$). This finding filled "the chief gap yet remaining in the series of genetic phenomena that form a parallel to the known cytological facts in *Drosophila ampelophila*" (Muller, 1914b).[3]

---

[3] The name *Drosophila ampelophila* was changed to *D. melanogaster*.

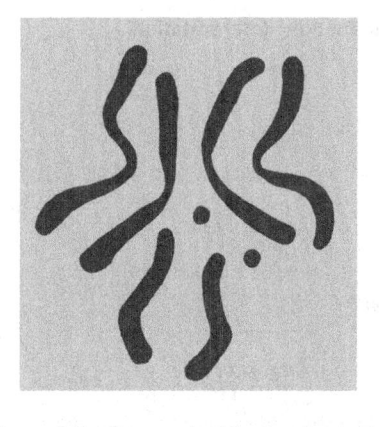 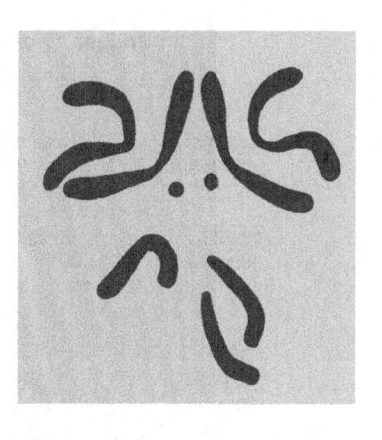

Figure 5.2. *Camera lucida* drawings of oogonial metaphase plates of mitoses of a normal (XX) female (left), and a non-disjunctional (XXY) female (right) (Bridges, 1916).

Bridges' analysis not only allowed him to establish the causal relations of chromosome segregation to Mendelian genetics, but also gave him insight into the mechanics that make meiotic division such an effective process. Cytologists showed that homologous chromosome pairing anticipated their segregation in meiosis. Bridges noted that if there were no preference as to which two of the three sex-chromosomes in XXY females were paired at meiosis – and consequently were directed to segregate, the third unpaired chromosome segregating at random – a maximum frequency of 33.3 percent secondary non-disjunction should be observed. The experimental results of a much lower frequency of secondary non-disjunction indicated that the two Xs pair (and segregate) preferentially with each other rather than either of the Xs pairing with the single Y. However, his experimental tools did not allow him to challenge this assumption. In 1948 Kenneth Cooper showed that in stocks in which one of the two Xs is grossly rearranged (and therefore the two Xs can hardly pair at meiotic prophase), as much as 90 percent secondary non-disjunction is obtained. This could be avoided if a one-armed Y-chromosome rather than the normal, two-armed Y-chromosome was used. Cooper hypothesized that a triad, of a Y-chromosome with each X-chromosome paired to (and segregating from) a different arm of the two-armed Y-chromosome was produced in XXY oocytes, rather than two chromosomes pairing and the third moving at random. Such triads may give up to 100 percent secondary non-disjunction since both Xs are directed to segregate to the same pole, away from the Y-chromosome

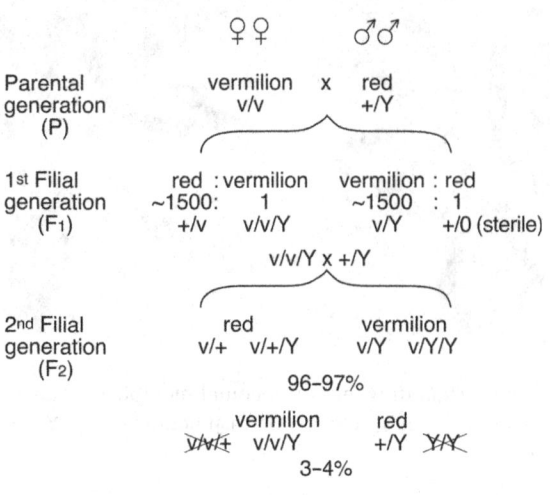

Bridges' "non-disjunction" interpretation

Figure 5.3. Scheme of production of primary non-disjunction in matings of *Drosophila* and of their progeny, producing secondary non-disjunction progeny, according to Bridges (1916).

(Cooper, 1948; see also Falk, 1955). No indications of the physical and/or chemical forces that cause specific pairing and segregation of the chromosomes were found. These were open to wide and often rather wild speculation.

Bridges' paper in *Genetics* was in many senses the flagship of genetic analysis and of its cytogenetic correlates (Figure 5.3). As we shall see, it also contributed significantly to the understanding of the deviations from independent segregation of Mendelian factors.

CHALLENGING THE INDEPENDENCE OF *FAKTOREN*

Exceptions to Mendel's law of independent segregation of factors had been reported early on, when researchers probed the universality of Mendel's rules. Were these exceptions due to some functional or developmental dependence of different traits, demanding new definitions of some unit characters, or were they due to an association of the factors themselves? In more modern terms: were these cases of developmental or physiological pleiotropy, in which the same factor is involved in the phenotype of more than one trait or characteristic, or were proper Mendelian factors sometimes linked rather than independently segregated?[4]

Correns was probably the first to take notice of deviations from the rule of independent segregation of different traits, and to put forward a physiological explanation in terms of differential lethality of gametes (Correns, 1902 [1924]). Such an explanation proved, however, problematic in another mating of his maize stocks:

> An inbred stock of blue maize (var. *cyanea* Kcke.) with a strong tendency for [male] sterility was crossed with an inbred stock of fertile white maize (var. *alba* Alef.). The marker pair white aleurone – blue aleurone followed approximately Mendel's rule. However, when I re-isolated from the hybrids the two parental-stocks, the blue stock was to my surprise as sterile as before. This sexual marker [sterility] was therefore coupled with a vegetative one (blue aleurone).
>
> Correns (1902 [1924], 296–297)

Attempts to explain this association of aleurone color with male sterility as a case of viability differences failed, and Correns concluded: "Of importance appears to me to be, to the best of my knowledge, the first objective evidence among plants of coupling of a *vegetative* character, such as that of the chemical constitution of the reserve-material in the endosperm, with a *sexual* character, being the reduced chances for the realization of fertilization. It appears that there exists a real 'correlation between a vegetative and a sexual marker'" (Correns, 1902 [1924], 299). Obviously Correns ascribed the coupling to a "real" common physiological cause. But it was Bateson and Punnett who elaborated on the phenomenon and proposed a theory of "coupling" and "repulsion."

Bateson and his associates repeatedly encountered cases in which the structural or physiological unit character did not correspond to the unit character of inheritance. Such a case was, for example, the "compound characters" of birds' combs (see Chapter 3). For technical reasons Bateson and his colleagues used to inbreed their first generation (F_1) hybrids, rather than backcrossing to one of the parental stocks. This was conducive to a distinct two-layer conception of gametic and zygotic characteristics. A 1:1 "gametic ratio" in segregation of a hybrid would produce in F_2 a 3:1 ratio of progeny or "zygotic ratio," given full

---

[4] Note that the very use of the term "pleiotropy" (see Rieger, Michaelis, and Green, 1976), like "penetrance" and "expressivity" (see Falk, 2000a; Sarkar, 1999), which were invented in the 1920s, was primarily a semantic technique to neutralize "disturbing" deviations from the genocentric dogma.

dominance of one alternative. For dihybrids a "gametic ratio" of 1:1:1:1 would thus produce a 9:3:3:1 "zygotic ratio." Other cases of "outward or zygotic characters" – what we would call today adult phenotypes – were interpreted as gametically compound, "being represented in the gametes by more than one factor." Red flower color in sweet peas, the red color of 'Fireball' tomatoes, the yellow-gray coat color of the Belgian Hare, "all are really compound characters, each being represented in the gametes by more than one factor" (Hurst, 1906, 114). Backcrossing the first generation hybrids to the parental stock with the recessive alternative, as was the routine in Morgan's lab with Drosophila flies, gave similar zygotic and gametic ratios. It was not accidental that such crosses were called "test crosses."

Matters became more complicated when compound characteristics of unit characters turned out to be only *partially* compound: the characteristics were "coupled" or "repulsed" only at specific proportions. Experimental coupling or repulsion was interpreted by Bateson and his colleagues as a phenomenon that took place between the *gametic* components of the unit characters, at specific proportions that led to deviations from the expected Mendelian ratios at the *zygotic* level. When Bateson detected cases of two or more unit characters in which "the proportions do not accord with Mendel's assumption of random segregation," he did not turn to analogies derived from cytological observations, but interpreted these along his 1891 theory of repetition of parts (Morgan, 1911).

For Bateson and Punnett "coupling" and "repulsion" – terms borrowed from electro-magnetic theories – indicated that some gametes, but not others, were "plainly a consequence of some geometrically ordered series of divisions" (Bateson and Punnett, 1911), along Bateson's theory of repetition of parts in segmentation in organisms (Bateson, 1894; see also Chapter 3). He explained the theory in a letter to his sister: "You see, an eight-petalled form stands to a four-petalled form as a note does to the lower octave" (Hutchinson and Rachootin, 1979, xi). Such a transcendental physico-chemical morphology principle that had been the central thesis of Bateson's work was now applied to the non-random segregation of unit characters. As he did with the metameric phenomena he had been dealing with earlier, Bateson interpreted the deviations from Mendel's rule in terms of field theories (Coleman, 1970; see also Falk, 2003; and Gilbert, Opitz, and Raff, 1996 for the establishment of the biological field theory).

In the context of this conception Bateson and his colleagues identified "series of numbers" in the gametes and consequently also in the zygotes, which, to their mind, were "plainly a consequence of some geometrically ordered series of divisions" (Bateson and Punnett, 1911, 298). They suggested that after segregation of factors at gametogenesis "some factors are distributed according to one of the reduplicated series and other factors according to the normal Mendelian system" (Bateson and Punnett, 1911, 301).[5] This model of series of differential numerical reduplication of some gametic combinations demanded increasingly Baroque-style combinations. Thus Punnett proposed to resolve the triple coupling of B, E, and L in peas by assuming a 13:3:3:13 ratio between factors B and L and also between factors L and E, whereas between factors B and E a ratio of 63:1:1:63 was suggested (Punnett, 1913, 82). An experimental difficulty of testing this model was that in F_2 of dihybrid crosses, the expected proportion of some zygotic types (such as the double recessive) is relatively rare, and may not be recovered at all in small samples. Thus, upon mating the *Be* × *bE* dihybrid of garden peas (*B-b* and *E-e* being two unit characters), Punnett (1913, 80) reported that he obtained only three types of progeny at the ratio of 2969:1379:1441 in the F_2, asserting: "it is not incompatible with the view that the [gametic] series is 1*BE*: 127*Be*: 127*bE*: 1 *be*. But as one hooded red [*be* zygote] is to be looked for in 65,536 plants it is not proposed to investigate this particular case any further." However, although not yet ready to accept a chromosomal theory of inheritance, in 1913 Punnett started to show signs of "retreat" from the gametic reduplication interpretation. He ended his paper on "Reduplication series in sweet peas" stating: "Where so many points remain doubtful, as at present, it is difficult to suggest any scheme by which this result could be brought about, and the problem must at present be left in the hope that fresh data may eventually lead to its solution" (Punnett, 1913, 94–95).

THE CHIASMATYPE THEORY

It is instructive to compare Morgan's chromosomally minded interpretation of the coupling and repulsion deviations from Mendel's rule of independence of factors with that of Bateson and Punnett, who

---

[5] At that time there were some indications that coupling and repulsion might not be two faces of the same phenomenon. For example, "Baur's work with *Antirrhinum* (snapdragon) suggesting that in this genus coupling may result from the cross AB × ab, while for the same pair of characters the mating Ab × aB may give an ordinary Mendelian result" (Punnett, 1913, 78).

completely ignored any chromosomal association in their interpretation. Many years later at the end of a talk on the "Early days of genetics," Punnett responded to the question of how he and Bateson managed to miss the tie-up of linkage phenomena with chromosomes: "The answer is Boveri. We were deeply impressed by his paper 'On the Individuality of the Chromosomes' and felt that any tempering with them by way of breakage and recombination was forbidden. For to break the chromosome would be to break the rules" (Punnett, 1950, 10). It appears that Punnett and Bateson's adherence to the particulate preformed unit character conception of Mendelism blocked them from conceiving of the chromosome as a string of functionally independent factors. Morgan, on the other hand, found that "[t]he discovery that there occurs in the formation of the germ-cells a process that supplies the machinery by means of which segregation might take place has aroused [his] interest in the application of the observations of cytology to the conclusions in regard to Mendelian segregation" (Morgan, 1910a, 465).

> I venture to suggest a comparatively simple explanation based on results of inheritance of eye color, body color, wing mutations and the sex factor for femaleness in *Drosophila*. If the materials that represent these factors are contained in the chromosomes, and if those factors that "couple" be near together in a linear series, then when the parental pairs (in the heterozygote) conjugate like regions will stand opposed. . . . [W]hen the chromosomes separate (split) . . . the original material will, for short distances, be more likely to fall on the same side of the split, while remoter regions will be as likely to fall on the same side as the last, as on the opposite side.
>
> Morgan (1911)

A chromosome theory of inheritance needed to explain genetic coupling and repulsion. It was necessary to spell out why cytologically coupling and repulsion were *not* absolute, or, why all factors linked to a given chromosome should not segregate as a unit. Morgan had to add to the chromosomal theory the notion of crossing over, which he borrowed from Frans Alfons Janssens' theory of *chiasmatypie*, based on his observations of meiosis at spermatogenesis in the slender salamander *Batrachoseps attenuatis* (Janssens, 1909). Microscopic observations suggested that during meiosis chromosomes pair (each has already been replicated to form two chromatids) and are increasingly coiled and twisted one about the other up to the pachytene stage of meiotic prophase. At diplotene, the following stage of meiosis, the tight pairing between the homologs appears to be relieved and the chromosomes start to segregate, *except at some sites*, at which sister chromatids appear to have crossed over. These *chiasmata* are resolved when chromosomes are further pulled apart to opposite poles at anaphase. Janssens speculated that the mechanical twisting at pachytene led to breaks in the paired chromosomes, allowing some relaxation of the tension by local untwisting, which was followed by repair of the broken "sticky" ends. The consequence of at least some such repair was the exchange of segments between the paired homologous chromosomes. This mechanism of exchange provided for the partial linkage that Morgan had to explain.

> We may make a general statement or hypothesis that covers cases like these, and in fact all cases where linkage occurs: viz. that when factors lie in different chromosomes they freely assort and give the Mendelian expectation; but when factors lie in the same chromosome, they may be said to be linked and they give departures from the Mendelian ratios. The extent to which they depart from expectation will vary with different factors. I have suggested that the departures may be interpreted as the distance between the factors in question.
>
> Morgan (1913b, 92–93)

Despite the increasing circumstantial evidence for the chromosomal theory of inheritance, and especially for the material basis of linkage maps, it was repeatedly pointed out that cytologically observed exchanges need not be the result of physical exchange, as predicted by the chiasmatype theory.

A competing "classical theory" suggested that paired chromosomes may just exchange pairing-partners that *appeared* as recombinations but were merely mispairings of sister-chromatids that are resolved as soon as chromosomes are pulled apart at anaphase: instead of two maternal chromatids and two paternal chromatids pairing all along, at chiasmata a maternal and a paternal chromatid change pairing partners. This interpretation of chiasmata could be examined when homologous chromosomes that differed in appearance at one end (such as different length or presence of a "knob" on one of the homologs) were examined. According to such an interpretation the ends of the homologs in bivalents with an unequal number of chiasmata would look identical (one long/"knobbed" chromatid and one short/ "knobless" chromatid on each homolog). This classic theory interpretation was refuted since in all such bivalents one homolog was marked with both the long/"knobbed" chromatids, the other homolog with none (see, e.g., Whitehouse, 1965, 114–120). As may be noted, these observations also indicated that – contrary to earlier hypotheses – it was the *first* meiotic division that was reductional, the second being equational.

A working hypothesis alternative to the mechanical torsion and break model of exchange for segmental interchange between homologous chromosomes in flowering plants was suggested by John Belling. According to his "copy choice" model, the new chromatids were synthesized during the meiotic prophase pairing of the twisted homologs. Such a linear copying synthesis of the new chromatids along the originally coiled homologs would lead to the copy running once along one homolog, once along the other homolog (Belling, 1928). This model was refuted, however, by recombination experiments in which multiple crossovers involved three and even all four of the chromatids rather than only the two new ones, as expected by the "copy choice" model. The model was further refuted in the 1950s, when it was shown that chromosomes were not only replicated in the interphase preceding each mitotic nuclear division (Howard and Pelc, 1951), but also at meiosis when one replication of DNA took place in the interphase preceding both meiotic divisions (Swift, 1950; Taylor, 1953).

Genetic analysis also provided other alternative interpretations to the chiasmatype theory. Richard Goldschmidt had noted already in 1917 that Morgan had merely proved that some forces were involved and that their proportional effects "may be represented geometrically as distances on a straight line" (Goldschmidt, 1917, 83). Goldschmidt suggested that exchange of markers between chromosomes might be related to chromosome replication, and the process of crossing over could be used to unravel "the real nature of these forces," rather than the location of the genes along the physical map. He provided a model where "individual genes are assembled to the chromosome after the manner of antigen-antibody fixation, with a variable force providing for the numerical rules" (Goldschmidt, 1917, 24). Hans Winkler elaborated another theory of direct change of individual alleles into each other, by some kind of mutual "gene conversion" (Winkler, 1930). Muller, however, continued to point out that "when the various conditions which have to be fulfilled at segregation are taken into consideration, any other explanation for these peculiarly linear linkage findings [other] than an arrangement of the genes in the spatial, physical line proves to be hazardously fanciful" (Muller, 1920, 101). He provided evidence against such physiological point-to-point interactions between the conjugated homologs at meiosis: genetic recombination was the correlate of the exchange of whole *segments* of the virtual linkage map. He experimentally studied more directly "the mechanism of crossing over" by following interactions between recombination events in multiply marked Drosophila stocks. He found that recombination at a given frequency between two markers affected also all other markers "down stream" (or "up stream") on the genetic linkage map. Furthermore, the frequency of coincidences of "adjacent" recombination events in a given "linkage group" was less than expected if recombination was due to random events at specific sites. He concluded that a crossing over at one site of the genetic map interfered with a second event nearby. Thus, "factors behave as though they are joined in a chain; when interchange takes place, the factors stick together in sections according to their place in line and are not interchanged singly" (Muller, 1916, 366). Genetic linkage reflected a physical dependence of markers subject to mechanical exchange between *blocks* or segments of entities along which genes were arranged at fixed loci. Such entities were the chromosomes that pair at meiotic prophase and twist, consequently leading to a tearing and a rejoining of the wrong chromatids. The chiasmatype theory was strongly corroborated (see Weinstein, 1918), but direct experimental evidence was needed.

Direct evidence that crossing over was involved in physical exchanges was obtained only when it became possible to induce chromosome aberrations at will and to select proper chromosome combinations according to experimental design (see Chapter 9). Harriet S. Creighton and Barbara McClintock in maize (Creighton and McClintock, 1931) and Curt Stern in Drosophila (Stern, 1931) combined specific chromosome aberrations that served as *cytological* markers of the two chromosome ends, with a pair of *genetic* markers, presumably linked to the cytologically manipulated chromosome. Individuals heterozygous for the two linked genetic markers and the two cytological markers on the same chromosome (chromosome 9 in maize, and chromosome X in Drosophila) were test-crossed and the genetically recombinant progeny were examined cytologically. Genetic recombination between linked markers was indeed accompanied by physical exchange of the chromosomally marked segments.

6

Mapping the chromosomes

In their classical textbook of 1939 Sturtevant and Beadle noted that:

> [l]inkage was first discovered, in the sweet pea, by Bateson and Punnett in 1906. The interpretation they gave is now discredited; but in the same year Lock suggested that, if homologous chromosomes undergo exchanges of materials (as has been suggested by Correns in 1902 on dubious theoretical grounds), then failure of such interchange might account for linkage – *i.e.*, he postulated that linkage is due to genes lying in a single chromosome pair, and that crossing over is due to exchange of materials between homologs.
>
> Sturtevant and Beadle (1962 [1939], 360)

Morgan's concern for the need to distinguish between the production of apparently unrelated and often variable manifold phenotypes of a gene, and of genotypic coupling, the constantly frequent co-inheritance of distinct factors, which had been confounded in the concept of unit character, is exposed in Sturtevant's description of the "moment of insight" many years later, in his *A History of Genetics*:

> In 1909 Castle published diagrams to show the interrelations of genes affecting the color of rabbits. It seems possible now that these diagrams were intended to represent developmental interactions, but they were taken (at Columbia) as an attempt to show the spatial relations in the nucleus. In the latter part of 1911, in conversation with Morgan about this attempt – which we agreed had nothing in its favor – I suddenly realized that the variations in strength of linkage, already attributed by Morgan to differences in the spatial separation of the genes, offered the possibility of determining sequences in the linear dimension of a chromosome.
>
> Sturtevant (1965, 47)

The insight that *partial linkage* relationships could be translated into topological relationships, i.e., into linear maps of the chromosomes as bearers of the genetic factors, was explicated by Morgan when he introduced the chiasmatype hypothesis of recombination:

> [W]e find coupling in certain characters, and little or no evidence at all of coupling in other characters; the difference depending on the *linear distance* apart of the chromosomal materials that represent the factors. . . . The results are a simple mechanical result of the location of the materials in the chromosomes, and the proportions that result are not so much the expression of a numerical system as of the *relative location* of the factors in the chromosomes.
>
> Morgan (1911, italics added)

Thus, Morgan conceived of linkage to resolve the problems of two or more characters that segregated in proportions which did not accord with Mendel's law of random segregation, as *extending Mendel's particulate theory* rather than refuting it. The relation between particles that were coupled or linked was analyzed in genetic terms, and not merely in anatomical or physiological ones. Mendel's law of independent segregation was a generalized, "ideal scheme." In the heredity of two sex-linked characters,

> [the] ideal scheme is not realized because of a complication that comes in. The complication is due to linkage or a tendency to hang together of the characters that go in together. . . . But I believe we can now offer a reasonable explanation, which shows that we have to do here with an extension of Mendelism that in no sense invalidates Mendel's principle of segregation. It not only extends that principle, as I have said, but gives us an opportunity to analyze the constitution of the germ-plasm in a way scarcely dreamed of two or three years ago. . . . The apparent discrepancy between the expected and the realized ratios is due to the linkage of the factors that went into the cross.
>
> Morgan (1913b, 91–92)

Linkage as a description became linkage as a theory:

> [W]hen factors lie in the same chromosome, they may be said to be linked and they give departures from the Mendelian ratios. The extent to which they depart from expectation will vary with different factors. I have suggested that the departures may be interpreted as the distance between the factors in question.
>
> Morgan (1913b, 92–93)

Morgan "saved the phenomena" (Duhem, 1914) by providing an auxiliary hypothesis to the Mendelian principle of independent segregation, and Sturtevant transformed the empirical linkage data into one-dimensional maps of the chromosomes (Sturtevant, 1913b). Sturtevant turned the phenomenon of linkage into an empirical tool for the analysis of the organization of the factors of heredity, independently of their developmental and physiological effects. This was not a *conceptual segregation* of problems of heredity and development, as had been claimed by many in the early years of genetics, but rather a *methodological distinction.*

LINKAGE GENERATES VIRTUAL GENE MAPS

Linear linkage maps of distinct gene entities among which recombination may take place became the routine of genetic analysis. Preferentially multi-marked hybrids were produced and backcrossed to a homozygote for the recessive alleles of all markers. The rarest recombinants were assumed to be those due to multiple exchanges. From this assumption a consistent linear sequence could usually be constructed (the assumption of interference of adjacent exchange events further helped to resolve numerical inconsistencies). Distances on the virtual maps were given in centi-Morgans, one centi-Morgan (1 cM) indicating an exchange frequency of 1 percent between given markers. However, the resolving power of recombination analysis was limited by the sheer size of the experiments that could be carried out: as a rule, a few scores of progeny in mouse recombination experiments, and not more than some hundreds in Drosophila or maize. This changed in the 1940s when fungi – primarily the bread mold *Neurospora crassa*, but soon also other species, like *Aspergillus nidulans* and yeast (*Saccharomyces cervisiae*) – were utilized. Screening methods were designed in which only (or almost only) an expected class of progeny, such as the prototrophic recombinants between two parental strains, each auxotrophic for a different nutrient, would grow on a minimal medium.[1] Another advantage of recombination experiments with fungi was that one could follow events at specific meiotic divisions by performing tetrad analysis (see below).

---

[1] Prototroph: nutritionally independent cell or individual. Auxotroph: nutritionally dependent cell, individual, or strain whose growth depends on a supplement to the basic, minimal medium on which prototrophs are able to grow.

Maps, however, are abstractions of some physical reality: of the streets of a city, of the ocean's shoreline, or of the countryside. Physical reality precedes its representation in maps. Genetic mapping applied the concept in reverse. But as pointed out by Richard B. Goldschmidt, this is nothing but the standard procedure of analytic geometry. As Goldschmidt emphasized one must keep in mind that linkage maps were merely *virtual maps*; "obviously any proportion may be presented as a distance on a straight line" (Goldschmidt, 1917, 83). Indeed, Morgan was aware of the virtual nature of the linkage maps, in spite of the inspiration received from material, cytological observations:

> The proof of the linear order of the genes is derived directly from the linkage data. It is not dependent on the chromosome theory of heredity. Fortunately . . . there are many facts about the maturation stages of the eggs and sperm that fit in extraordinarily well with the theory of the linear order of the genes, but let me repeat, the proof of the order is not dependent on the chromosomal situation. The evidence for the linear order is furnished by linkage and its correlative phenomenon, crossing over. By linkage is meant that certain factors that enter the cross from each parent remain together in subsequent generations, more often than they separate.
>
> Morgan (1919, 118)

This view of linkage maps as virtual maps was not everybody's view. Although William Castle accepted that Morgan "beyond doubt established the fact that the genes within a linkage system have a very definite and constant relation to each other," he commented "[t]hat the arrangement of the genes within a linkage system is strictly linear seems for a variety of reasons doubtful. It is doubtful, for example, whether the elaborate organic molecule ever has a simple string-like form." Castle did not hide his predilection for a chemical reaction, "the replacement of one chemical radicle [sic!] with another within a complex organic molecule" (Castle, 1919a, 32), rather than a mechanical, brute force mechanism of recombination. Muller's reaction to Castle reveals more than anything the power of "deducting backward" of Mendelian genetic analysis as well as its weakness:

> it has never been claimed, in the theory of linear linkage, that the per cents of crossing over are actually proportional to the map distances: what has been stated is that the per cents of crossing over are calculable from the map distances . . . Whether or not we regard the factors as lying in an actual material thread, it must on the basis of these findings be admitted that the forces holding them linked together – be they physical, 'dynamic' or

transcendental – are of such nature that each factor is directly bound, in segregation, with only two others – so that the whole group, dynamically considered, is a chain. . . . no implication as to the physical arrangement of the genes is intended when the terms 'linear series,' 'distance,' etc., are used.

Muller (1920, 98–101)

Despite the caveat in the final sentence, Muller continued to hold on to the chiasmatype theory as the physico-mechanical model for the cytological correlate of genetic linkage and recombination. And, in line with his confidence, Muller set out on a project to establish the physical nature of the "atoms of heredity" and thus contributed crucially to the experimental program that eventually made the cytological examination of the model possible (see Part IV).

TETRAD ANALYSIS

Constructing linkage maps from recombination data was based on statistical analysis of progeny of mass hybridization experiments. No recombinant could be allocated to a specific meiotic event. This was not the case with Bridges' non-disjunctional daughters, who got two X-chromosomes from the *same meiotic event*; this was a "half-tetrad analysis" for the X-chromosomes. Significant information on the mechanism of crossing over was derived by Bridges when he observed that "[v]ery rarely a [XXY] female which is heterozygous for a recessive sex-linked gene produces an exceptional daughter which is pure recessive." Such a progeny presented an exceptional daughter, "one of whose X chromosomes has undergone crossing over while the other has not! ... It is impossible to obtain exceptions of this type from an XX or XXY oöcyte unless the crossing over has taken place at a four strand stage" (Bridges, 1916, 120–122). This indicated that the exchange took place *after* the homologous chromosomes were duplicated, each into two chromatids, and that only two of the four chromatids, a non-crossover and a crossover chromatid, happened to disjoin (see Figure 6.1).

Although Bridges' analysis allowed him this glimpse into the nature of the recombination process he was aware of the pitfalls of inference from genetic analysis to material cytological conclusions. His conclusion was confirmed by Lillian V. Morgan in her study of the progeny of (cytologically discernible) attached-X Drosophila females. All daughters of such females are matroclinous, receiving from their mothers

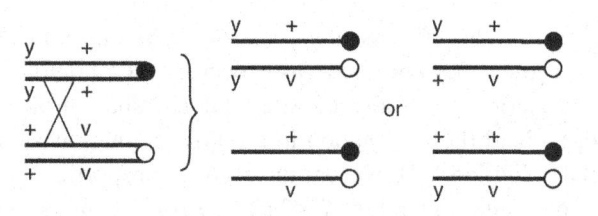

Figure 6.1. Mitotic crossing over between two of four chromatids that may produce homozygosis for distal markers.

both X-chromosomes due to the obligatory non-disjunction. Among the progeny of attached-X females, heterozygous for genetic markers, obtaining some daughters homozygous for the maternal markers confirmed that crossing over occurs between only two of the four chromatids (Morgan, 1922).

This half-tetrad analysis allowed an independent method for constructing virtual chromosome maps, based on the frequency of homozygotization of the markers for which the females were heterozygous: according to the chiasmatype model, the further they were from the site of segregation-determinant the more frequently homozygosis occurred. This allowed putting a *cytological* marker on the map, namely the centromere, or the spindle-fiber-attachment. The topological agreement of the two mapping methods, one based on the frequency of recombinants, the other based on the frequency of homozygotizations, strongly supported the claim for the material basis of the mapping of genes to chromosomes.

Such analysis of individuated meiotic events was extended once recombination in fungi, primarily in the bread mold *Neurospora crassa*, was introduced. In ascomycete fungi each meiotic division is confined to a discrete sac, or *ascus*. In Neurospora the two meiotic divisions are followed by another mitotic division; these eight products are arranged in a linear pattern of *ascospores*. The position of spores in the line indicates the sequence of the division events that produced them. Micro-dissection of asci and testing of their genetic properties, taking care to maintain their linear order, reveal details of the meiotic process.[2] In an ascus

---

[2] The properties studied are usually nutritional needs. Mycelia of the "wild type" fungus grow on a minimal, almost synthetic inorganic medium – they are prototrophs. Mutations in a gene that make the growth of the cells depend on an organic component added to the minimal growth medium – an amino acid, a vitamin, etc. – turn those cells into auxotrophs.

heterozygous for, say, *A/a* usually four spores at one end are *A* and four spores at the other end are *a*, as expected when homologous chromosomes – or, more specifically, the sites of determination-of-segregation of homologs – lead the segregation in the first meiotic division (being a "reductional" division). However, in a given proportion of asci the pattern is not 4*A*:4*a*, but rather 2*A*:2*a*:2*A*:2*a* (or 2*A*:4*a*:2*A*), confirming that recombination occurred after replication, between the site determining segregation and the locus of *A/a*. It involves two of the four chromatids, the second meiotic division distributing "equationally" the products of the first reductional division (see Fincham, 1998; Fincham, Day, and Radford, 1979).

THE CYTOLOGICAL BASIS OF GENETIC LINKAGE

Further support that the linear map represents the topology of loci of particulate genes on the chromosomes came from Bridges' finding of the dominant sex-linked mutation (*Notch*) in Drosophila. In females heterozygous for *Notch* the margins of the wings were notched but no notched males survived (Bridges, 1917). Notched females mated to white-eye males or to facet-eye males (*w* and *fa*, two mutations in complementary but closely linked genes) produced notched-white and notched-facet females, respectively. In line with the chromosomal theory, Bridges interpreted the *Notch* mutation to be a "deficiency" of a short segment of the chromosome, including the loci of *w* and *fa*. Such a deficiency also explained the lethality of the hemizygous males, suggesting that other genes, essential to the fly's life, were also deleted in the implied deficiency. Support for the notion of a material deficiency of *Notch* in females was obtained by Otto Mohr's finding that the frequency of crossing over between the genes lying to the left and right of the "deficiency" was significantly decreased in its presence (Mohr, 1923, 1924). No cytological confirmation was available for this virtual deficiency. However, within a decade it turned out that the choice of Drosophila as a model organism for genetic analysis was most fortunate, when the power of cytogenetic analysis changed dramatically with the elucidation of the patterns of dipteran polytenic chromosomes that allowed a minute and detailed cytological analysis.

In 1930, the Bulgarian geneticist Dontcho Kostoff called attention to the "Discoid structure of the spireme," the thread into which the nuclear framework is gradually transformed in early prophase in *Drosophila melanogaster*. He pointed out that "[t]he linear arrangement of the genes necessitates the assumption that the chromosomes are made up of chemically different components. The discoid structure of the chromosomes ... indicates the existence of such chemical differences in the varying capacity to absorb haematoxylin" (Kostoff, 1930).[3] The spireme in the Malpighian tubules of the fly-like dipteran *Bibio hortulanus* was indeed shown to be a discontinuous structure, the number of components of which corresponded to the haploid number of chromosomes (Heitz and Bauer, 1933; Painter, 1934a, 178). The chromosomes in the salivary glands of *D. melanogaster* too were found to show a definite and a constant morphology which enables one to recognize the same element in any cell (Painter, 1934b, 465).

> Needless to say, this places in our hands a method of studying transloca-tions and plotting cytological maps which is a vast improvement over the study of metaphase plates heretofore used. But what is of far greater importance is the discovery, wholly unexpected, that in the salivary glands of old larvae the chromosomes may undergo somatic synapsis. As a result, homologues pair line for line, band for band, and unite into one apparent element. If, however, one of the homologues has an inverted section, we get typical inversion figures, such as may be seen in meiosis of corn, and by the bands and other landmarks we can tell, cytologically, just where the inversion occurred.
>
> Painter (1934a, 175)

In the following years increasingly detailed cytological maps of the salivary gland chromosomes of *Drosophila melanogaster* were presented, mainly by Bridges (Bridges, 1935, 1938; Bridges and Bridges, 1939). Polytene chromosome maps of other Drosophila species and dipteran species, like that of the fly *Sciara* (Metz, 1935), were also published. The maps of *Drosophila melanogaster* were, however, unique in the detailed genetic data superimposed on them (see Figure 6.2).

In the detailed banding maps the chromosomes of *Drosophila mela-nogaster* were divided into 102 sections, each subdivided into 6 subdivisions, which could be divided further into up to 15 distinct bands (see Lindsley and Grell, 1968). An unprecedented resolution of the mapping power of the chromosomes was obtained, and since the pairing of homologous "bands" was maintained in hybrids for chromosomal

---

[3] As it turned out, the discoid structure of the chromosomes and differences in the varying capacity to absorb hematoxylin are mainly due to differences in the compactness of the packing of the nucleoproteins of the chromosome.

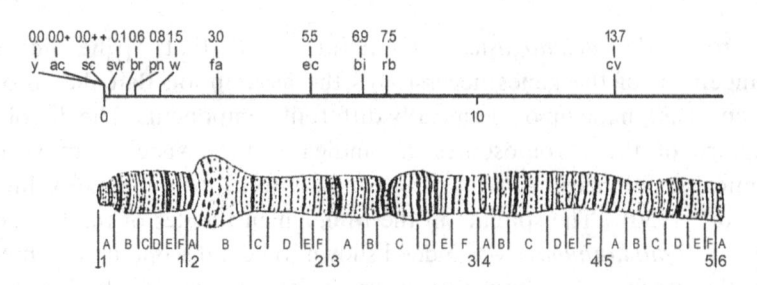

Figure 6.2. "Left end" of the polytenic chromosome-X of *Drosophila melanogaster* with the aligned linkage map above it (after Bridges, 1938).

rearrangements, typical chromosomal configurations, including loops (for inversions), forced Ω-like structures (for deletions like that of *Notch*), or crosses (for translocations), were obtained. This made chromosomal aberrations effective tools for superposing genetic and cytological maps. Thus, striking support for the chromosomal theory of heredity was obtained by the assignment of genes to specific "bands" of the polytenic chromosome map. For example, following a series of white-eyed stocks of Drosophila, which turned out to show small deficiencies of up to several dozen "bands" of the polytenic chromosomes, allowed the assignment of one "band," 3C2, common to all such deletions, as the site of the gene *white* (Demerec and Hoover, 1936).

Analysis of the polytenic chromosomes allowed intensive analysis of the role of the "bands" far beyond the functional organization of the chromosomes themselves, including gene action and the dynamics of populations. However, the results and conclusions derived for chromosome structure, gene function, and population dynamics needed re-evaluation when the resolving power of cytogenetic analysis was further increased by many orders of magnitude above that of the polytenic chromosomes with the introduction of mapping at the level of up to single nucleotide polymorphisms (SNPs) of DNA sequence.

RECOMBINATION BECOMES THE TOOL OF GENETIC ANALYSIS

Although by and large consistent genetic (and cytological) maps provided strong support for the notions of the chiasmatype theory, linkage maps provided solid stepping stones for expanding the power of genetic analysis to the molecular level, and some presumably unrelated observations eventually challenged the mechanistic notion of recombination.

When recombination between very close markers was examined, first in fungi but eventually also in Drosophila, it turned out that far too many double recombination events occurred: instead of the expected increase in *positive* interference with decreasing distance, as observed by Muller and others for sequential exchanges in very short intervals, *negative* interference was observed. When Pritchard (1955) studied crosses between allelic adenine-requiring mutants of *A. nidulans* in the presence of linked marker genes on either side of the adenine locus, such as: $\frac{y\ +\ ad8\ +}{+\ ad11\ +\ bi}$, a significant number of adenine-less recombinants showed also recombination of the adjacent outside markers (e.g., yielding $+ + +$ *bi* or even $+ + + +$ spores rather than the expected $y + + bi$ spores). As it turned out also with other organisms and other markers, recombination of outside markers apparently *increased* the tighter they were linked to the target recombinants. Pritchard suggested that interference reflected the efficiency of the pairing of homologs during meiotic prophase: positive interference was attributed to the spacing of the regions where pairing of homologs was sufficiently effective to allow crossovers to occur. According to this interpretation negative interference was just the artifact of the spacing of multiple events within the effectively paired short regions.

When products of individual meiotic events could be followed, as became possible with tetrad analysis in fungi, a second unexpected finding turned up: infrequent but consistent deviations from the expected 1:1 recovery of the two alleles for which the parent was heterozygous. This phenomenon of "gene conversion," refuting the most basic Mendelian law of segregation, was for a long time ignored as if it were an experimental "artifact" contamination. But eventually genetic analysis had to account for it and for the observation that it was often associated with recombination of "outside markers," to the left and/or to the right of the converted marker. Thus, instead of finding that a heterozygote *A/a* produced 4 *A* spores and 4 *a* spores in the ascus of *Neurospora*, some asci contained 6*A*:2*a* ascospores, or even 5*A*:3*a* ascospores, whereas the nearby outside markers to the left and right of *A/a* segregated normally, giving 4 to 4 ascospores, and were often recombinant (Lindegren, 1953; Mitchell, 1955).

Gene conversion forced a revaluation of the chiasmatype notion of recombination. The physical tearing of the twisted chromatids implied that rupture of the two chromatids that were exchanged occurred *precisely at the same site* in the two chromatids since, as a rule, both products of recombination were recovered with equal frequency: it was argued that if exchange occurred at different sites, the small duplication and deletion obtained would often result in grossly unequal recovery of the two products of recombination, since deletions generally are deleterious or lethal. This, however, need not be strictly the case where breaking the strands at the molecular level is concerned.

It was Harold L. K. Whitehouse who first suggested a model of crossing over in which the breakage points in the two strands of each of the exchanging DNA molecules were staggered, thus no loss of sequence (on at least one of the strands of the two double helix DNA molecules involved) ensued.[4] Rejoining of the DNA molecules took place not by an end-to-end association of broken strands but by lateral association of complementary segments from homologous regions, to give – when exchange of chromatids took place – short *hybrid DNA* sequences (Whitehouse, 1965, 318). Once the idea of lateral association of staggered single strands of DNA was put forward, it followed that enzymatic intervention was required to repair the rejoined homologs. Recombination was now envisioned as an enzymatic process rather than a process of chiasmatypic mechanical breakage and rejoining (Holliday, 1964). According to such models, heteroduplexes – discordances of one or more nucleotide bases in paired strands of the DNA – could occur in any hybrid-DNA segment that calls for the involvement of DNA-repair mechanisms. Such repair, when in the "right" direction, would reconstitute the 2:2 relationships between the products of this meiosis. However, repair in the "wrong" direction would end up in a 3:1 segregation, i.e., in gene-conversion.[5] Thus gene conversion was interpreted as a misrepair in an interaction between chromatids at meiosis. Once these notions were introduced, the mechanical breakage-rejoining model of recombination was soon replaced by a model of recombination as an

---

[4] Without going into the details of the physico-chemical Watson and Crick model of the structure of DNA, suffice it to say that the DNA molecule is composed of two anti-polar strands that are paired thanks to the specific interaction of its purine and pyrimidine nucleotides: adenine (denoted A) pairs with thymine (T) and guanine (G) with cytosine (C) of the opposing strands. This pairing (weak H-bonds) ensures the complementary, or mirror image of the two strands of a DNA molecule.

[5] Suppose that an A-T nucleotide base pair in one chromatid is confronted by a G-C pair in the other, in otherwise identical sequences in both chromatids. If the hybrid DNA segment includes the above mentioned variation, a pair of A-C and G-T heteroduplexes would occur. If one heteroduplex is repaired to become A-T and the other to become G-C, the 2:2 segregation at that site will be reconstituted. If, however, both heteroduplexes are repaired to become A-T, a 3:1 segregation ensues.

enzyme-driven process, amenable to biochemical analysis (see Stahl, 1979, and Chapter 13).

When Lederberg and Tatum embarked on finding out whether bacteria were involved in processes analogous to sexual recombination of Mendelian genes, they did it by testing for linkage between genes, looking for chromosome-like structures (Lederberg, 1947; Lederberg and Tatum, 1946). The discovery of recombination in bacteria "illustrated the experimental concordance of bacterial segregations to a generalized definition of Mendelism," although "[i]n a purely formalistic way, these data could be represented in terms of a 4-armed linkage group." In spite of the formally branched structures that emerged from early recombination data in bacteria, Lederberg *et al.* were aware of the analytic meaning of the virtual maps "without supposing for a moment that this must represent the physical situation" (Lederberg, Lederberg, Zinder, and Lively, 1951, 416–417).

Linkage and recombination was the analytic tool at hand for the genetics of prokaryotes: the non-reciprocal nature of the genetic exchange process, which indicated that a "filterable agent" was being carried from donor to recipient, suggested that a mechanism of "sexual" relations and exchange occurs in bacteria other than the mechanism known in eukaryotes. And a new rationale for interrupted mating was introduced by William Hayes and by François Jacob and Elie Wollman for mapping in bacteria: the kinetics of the process was asymmetric, mediated by an extrachromosoal F fertility-factor, a plasmid, carried by some strains (F^+, or "males") and not by others (F^-, or "females").[6] Bacteria involved in such a conjugation process ("sexduction") which turns F^- into F^+ bacteria do not exhibit genetic recombination, unless an F-factor integrates (by exchange) into the bacterial chromosome. This would turn the F^+ bacterium into an Hfr (high frequency recombination) bacterium that asymmetrically transfers the Hfr chromosome into the F^- bacterium. Genes on the Hfr chromosome may recombine with those of the F^- bacteria as soon as they are introduced to the recipient bacterium. The mapping notion was now changed from recombinant-frequency mapping into temporal mapping by controlled interruption of

---

[6] Plasmid is a generic term for any extra-chromosomal hereditary determinant. The F-plasmid behaves as an infectious particle, which replicates independently of the bacterial chromosome. The scientific-biological as well as the political significance of self-replicating cytoplasmic genetic entities that exhibited non-Mendelian inheritance did not escape the attention of the geneticists of the 1950s (see Creager, 2004, 24–25).

the linear transfer of the "chromosome" (such as shaking the cultures in a blender). This new notion allowed the construction of genetic linkage maps based on the *time of transfer* of marker genes. Conjugation maps with different Hfr strains in bacteria gave each strain a unique consistent non-branched linear linkage map. Such virtual maps derived from different Hfr stocks proved to be formally circular permutations of a single linkage relationship. It was concluded that the bacterial chromosome was a closed circle, which could be "opened" (by the insertion of an F-plasmid that drove the conjugative transfer) at different sites in different Hfr stocks (Hayes, 1964; Jacob and Wollman, 1961). Eventually the virtual circular genetic maps were confirmed "cytologically" by John Cairns by labeling the DNA of *E. coli* – or as it was until then called by skeptical geneticists, its genophore – with tritiated thymidine (Cairns, 1963).

LINKAGE: A REAPPRAISAL

The role of genetic linkage and linkage maps did far more than just describe the genic landscape. Not only was the notion of linkage greatly extended to the cytological and to the molecular levels, but also it served as a tool in the physiological and evolutionary scenes. It became a central notion at the molecular level of DNA strands as well as RNA strands. Linkage gave genetic analysis a tool to investigate spatial as well as functional relationships of chromosomes, and eventually the whole genome, far beyond discrete independent Mendelian entities. It is not an exaggeration to claim that linkage became a central hallmark of genetic analysis.

The detailed maps of Drosophila mutations enabled accurate mapping of specific genes and, as will be discussed in the next parts, also new analysis of problems such as the nature of the gene: were bands (or interbands) the sites of distinct genes? Could chromosome breaks occur within bands or only between bands? How could the phenotypic effects allocated to some bands be affected by changes in chromosome integrity further away? What is the meaning of chromosomal aberrations in natural populations, and how are they related to each other?

Already in 1913, the same year that the first linkage map was published (Sturtevant, 1913b), Morgan's three students gave the abstract notion of linkage a concrete and comprehensive context of relevance. Sturtevant used the notion to discern between linked factors and multiple allelomorphs of the same factor, as in the case of the fur coat pigmentation of rabbits (Sturtevant, 1913a, 1914). In due course, after devising linkage as a tool for constructing virtual maps of the chromosomes, Sturtevant contributed also to extending the notion to other disciplines, such as the fate mapping of embryonic primordia in *Drosophila* (Sturtevant, 1932), the spatial arrangement of the meiotic nuclei in its oogenesis (Sturtevant and Beadle, 1936), and the evolution of chromosomal inversions in *Drosophila* species (Sturtevant and Dobzhansky, 1936). Bridges and Muller, on the other hand, used linkage primarily to establish the physical basis of heredity beyond the virtual mapping, namely to prove the *physical* basis of the chromosomal theory of heredity and the physical mapping of the genes as the particulate entities of the chromosomes. For them genetics became truly cytogenetics (Darden, 1991).

Once DNA-sequencing became feasible, manipulation of linkage relations became a major methodological device. For example, the logic of cytogenetic deletions was utilized to overcome the long distances (in DNA terms) between the molecular marker and the target sequence when sequencing by DNA-walking was introduced (see, e.g., Bender *et al.*, 1983).

The centrality of linkage in the genomic era, whether in the location of genes on DNA restriction maps, or using SNPs (single nucleotide polymorphisms) as anchor markers for sequencing relevant genes, is a direct extension of the chromosome theory of inheritance and the notion of genetic linkage.

The elaboration of the chromosome theory of heredity and the notion of linkage as expounded by Morgan and his students in the 1910s marked the finest days of Mendelian genetic analysis. Contrary to Kohler's description of Drosophila as a "breeder reactor," and the origin of genetic mapping as a means for classifying the large number of mutants this research kept producing (Kohler, 1994), or to Carlson's presentation of the research effort to establish the material basis of genetics and its particulate genes (Carlson, 1966 [1989]), I wish to emphasize the role that linkage studies played in providing genetic analysis with the means to transcend and go beyond the particulate conception of heredity.

7

Cytogenetic analysis of the chromosomes

The chromosome theory of heredity made the chromosomes, their structure, and their functions major targets of genetic analysis. In 1916 Bridges showed how genetic analysis of exceptional aneuploids (organisms or cells that carry an unbalanced number of chromosomes) also provided analytic tools to gain insight into the mechanism of chromosome pairing and recombination, and into the limited role of the Y-chromosome in development. The part chromosomes play in development was also demonstrated by Blakeslee's trisomics in a study on the Jimson Weed (Chapter 5), and by the studies of Lewis J. Stadler and his students Barbara McClintock and Marcus M. Rhoades, who elaborated on the cytology of maize chromosome aberrations and their phenotypic correlates. Once the efficiency of X-rays in inducing aberrations was demonstrated and genetic experiments were designed to isolate proper chromosome constellations, cytogenetic analysis became a central tool for research, especially in those organisms whose chromosomes were large enough or otherwise easy to observe.

EUPLOIDY AND SEX DETERMINATION

Boveri's studies of the development of aneuploid sea urchin embryos indicated the importance of a *complete and balanced set of chromosomes* for development (Chapter 5). The importance of a complete and balanced set of chromosomes was also demonstrated by the normal development of triploid zygotes of organisms that are otherwise diploid, such as triploid Drosophila females.

Many plant species can be shown to be ploidly-related to each other. This is especially significant in cultivated plants, where over the millennia breeders selected (or produced) polyploids. Cereals, like wheat or rye, and many ornamental flowers are tetraploids, hexaploids, even octaploids of corresponding wild-type diploids. The prevalence of polyploidy in plants depends largely on their capacity for "selfing." A new sexually reproducing polyploid when mated to diploids will be semi-sterile (or its progeny will be practically sterile) and selected against, unless positively favored by some special circumstances (such as those provided by breeders).

Analytically, one has to distinguish autopolyploids that have multiple sets of the same genome, which may be due to failure of meiotic division from allopolyploids that originate from a hybrid zygote containing alien chromosome sets. In autopolyploids, multivalent (e.g., trivalents and tetravalents in tetraploids), instead of bivalent pairing at meiotic prophase often results in unbalanced gametes and consequently in low fertility. Allopolyploids may be more fertile because homologs of different parental origin (homeologs) preferentially do not pair, thus resulting in bivalents that segregate normally and produce chromosomally balanced gametes. Such an exclusive bivalent pairing of homologs, diploid-like meiotic behavior of the hexaploid wheat, is further controlled by the gene *Ph1* that suppresses pairing between homeologous chromosomes of the three closely related genomes that comprise the domesticated wheat genome (Feldman, 2001, 15–16; Vega and Feldman, 1998).

Apparent exceptions to the demand for chromosome balance are the sex-chromosomes: two matching chromosomes (XX) in one sex, and only one chromosome with a Y-chromosome (XY) or without it (XO) in the other sex. More elaborate arrangements of sex-chromosomes are also known (see White, 1954). Yet, proper sex-chromosome balance is crucial for development. Unbalanced Drosophila embryos, those with three X-chromosomes (but only two of each of the other chromosomes) or with no X-chromosomes, are lethal. The presence or absence of the Y-chromosome (or an additional chromosome Y), on the other hand, appeared not to affect normal development of either sex, though males with no Y-chromosome were sterile (Bridges, 1916).

Triploid Drosophila females mated to diploid males produced triploid females and intersex progeny (besides normal males and females). No triploid males were recovered. Bridges concluded that in Drosophila sex was determined by the balance between the number of X-chromosomes and that of the rest of the chromosomes, or the autosomes (denoted A). Zygotes with an X:A ratio of 1:1 developed into females, and those with an

X:A ratio of 2:1 produced males. Zygotes with a 2X:3A ratio developed into intersexes. Rarely 3X:2A distorted females were recovered and, according to the balanced-hypotheses, these were denoted "superfemales" (Stern [1959] suggested that "metafemales" was a more appropriate name). The Y-chromosome contributed only to the fertility of males not to their development (Bridges, 1925). Autosomal aneuploidy was not tolerated, except for trisomics or sometimes monosomics of the minute autosome (triplo-IV and haplo-IV) that were subvital. The balanced genomic X:A theory of sex-determination in Drosophila was extended and applied to all other organisms with XX–XY (or XX–XO) chromosomal sex differentiation. Of course, other mechanisms of sex determination were well known: those of *Hymenoptera* in which the fertilized diploid eggs developed into females whereas males were haploids that developed from unfertilized eggs, for example, or those of various orders and species that had more or less complex sets of sex-chromosomes and complicated mechanisms of meiosis which ensured orderly sex determination (White, 1954), or even those of species in which non-genetic conditions, like the kind of nutrition, determined sex differentiation (see, e.g., the notorious location-dependent sex determination in *Bonellia viridis*, Gilbert [1991a, 784]). Also a crucial role for a Y-chromosome in sex determination was demonstrated in at least some cases, such as in dioecious plants like *Melandrium*, in which a conspicuously large Y-chromosome was needed for male plant development (Baur, 1912; Westergaard, 1958). However, the dominant role of genetic analysis in Drosophila imposed the "balance theory" of sex determination for many years, excluding any alternative. This was exacerbated by the difficulties of obtaining clear, analyzable images of mitotic and meiotic divisions in mammalian species, to the extent that the accurate number of human chromosomes was determined only in 1956, with the development of new cytological techniques (see Tjio and Levan, 1956). It was, therefore, a great surprise when in the 1950s it was found that in humans and other mammals XXY individuals developed as (malformed) males and XO individuals as (deficient) females (see, e.g., Stern, 1960b, 410–423).

CHROMOSOME ABERRATIONS AS ANALYTIC TOOLS

By introducing multiple markers on all three X-chromosomes of triploid Drosophila females, Bridges and Anderson followed chromosome pairing, crossing over, and segregation in five segments of a chromosome.

Only exceptional daughters, containing two X-chromosomes derived from single oocytes of their mother (and none from their father) were analyzed. Twenty-eight per cent of their chromosomes were crossovers involving two of the original chromosomes, and 3 percent were crossovers involving all three maternal X chromosomes. Ten percent of the double crossovers were "progressive," in which the second crossover takes place between chromosomes different from the first. "This shows that synapsis generally involves all three X chromosomes and that a crossover in one region does not markedly prejudice which strands may cross over in another region." Many of the exceptional females had identical chromosomes in at least part of their length. "This shows that at the time crossing over occurred each of the original chromosomes had already become split into two strands." The percentage of equationals was lowest for the right end of the linkage map and increased progressively toward the left. If indeed the separation of the strands is controlled by the spindle fiber attachment site (the centromere), it is located to the extreme right of the map, "at which point the first division would always be reductional, and that equationals occur as a result of crossing over" (Bridges and Anderson, 1925, 440).

The power of chromosome rearrangements as tools of genetic analysis was amply demonstrated by the sporadic gains of the experiments with attached-X females recovered by L.V. Morgan (1922; for an amusing description, see Srb and Owen, 1953, 92; see also Chapter 5). However, toward the end of the 1920s, when Muller demonstrated the mutagenic effects of X-rays (Muller, 1927a), experimental production of aberrations became possible, allowing properly designed screening procedures selected for aberrations that would allow specific genetic analysis. Compound chromosomes of Drosophila and maize and also of other species were produced at will. Changed linkage relationships became major tools for the detection of chromosomal rearrangements, more and more of which would eventually be confirmed by cytological observations not only in organisms with relatively large chromosomes, like maize, but even in the small chromosomes of Drosophila.

> In all cases so far found in which the cytological picture has disclosed that a section of a chromosome has become broken off from its original connection, and either lost or attached elsewhere, and in which at the same time a genetic analysis has been made, it has been found . . . that the genes involved form a "block," that is, *they constitute a coherent section of the linkage map.* . . . There is no escape, then, from the conclusion that the previously constructed genetic map represented the genes as correctly

distributed with respect to the two sides of the breakage point . . . There can, therefore, no longer be room for doubt that the arrangement (that is, the order, but not necessarily the relative distance) of the genes as given in the linkage maps is the same as their actual, physical arrangement.

In view of this, it follows too that the interchange which occurs between homologous groups of genes prior to segregation is an interchange of entire morphological sections of the homologous chromosomes . . . But to say this is to say that the so-called "mechanical theory of crossing over" is correct – the theory which holds that the chromosomes, in their normal process of interchange, become broken at one or a few points, and reattached in their sequels of the previously homologous members.

Muller and Painter (1929, 197–199)

Major chromosomal aberrations, especially paracentric inversions (inversions of chromosome arms that do not include the centromere) were found to be common components of variability of natural populations of Drosophila species, such as *D. pseudoobscura* (Sturtevant and Dobzhansky, 1936). Genetic analysis as well as direct cytological observations had shown that most recombination events within such inversions lead to bridges between segregating chromosomes during meiosis and to acentric segments. Thus, products of exchange were expected to be non-viable. This turned such inversions into very useful tools in the design of experiments and in the maintenance of stocks with genetic marker combinations of interest. Muller's *C*-factor for "crossing over inhibitor," such as that of the *ClB* chromosome of Drosophila, pivotal in the heuristics of screening methods, proved to be an inversion (see Chapter 9). But how could such laboratory tools proliferate in natural populations? This prompted a cytogenetic analysis by Sturtevant and Beadle (1936), who showed that because the polarity of the meiotic spindle is perpendicular to the oocyte membrane, the meiotic products end up in a linear arrangement, the outer being those that segregated from the inner at first meiotic division, and the second meiotic division further aligning the nuclei into the linear arrangement. Since crossing over occurs at the first meiotic division between two of the four chromatids, a bridge produced at intra-inversion exchange events keeps the recombined nuclei in the central positions of the alignment. The furthest two nuclei in the alignment are thus the non-crossover product (unless two recombination events, involving three or all four chromatids, occur within the inversion). Since the outermost nucleus of the four is usually the one to become fertilized, paracentric inversions, they concluded, efficiently suppress recombination without causing deleterious effects (Sturtevant and Beadle, 1936) and such inversions may be established in natural populations with almost no adaptive deleterious consequences.

CYTOLOGY IN THE SERVICE OF GENETIC ANALYSIS

Parallel to the genetic analytical effort to establish the cytological physical basis of the laws of inheritance, once the role of chromosomes as causal agents in the processes of inheritance was recognized, a major research effort was directed at establishing the mechanics of the chromosomes as the foundations of the laws of inheritance. Cyril D. Darlington in the two editions of his *Recent Advances in Cytology* turned cytology into a systematic and powerful instrument of genetic analysis in its own right (Darlington, 1937). He was drawn by Weismann's speculation from 1887 that there had to be a process of reduction and by the Boveri–Sutton hypothesis of the individuality of chromosomes, to Morgan's "Fly Room" chromosomal theory of inheritance. The chiasmatype hypothesis that provided the conceptual framework for the genetic analysis of the chromosomal theory of inheritance was taken up by Darlington for observational and experimental analysis of chromosome organization and meiotic mechanics (see Harman, 2004).

In 1896 in *The Cell* Wilson wrote that heredity was accomplished through the union of chromosomes of cells produced by reduction division and defined the salient conclusions as the necessary foundations for a cytological theory of heredity that would be developed in the first decades of the twentieth century. What remained unknown were the precise mechanics, and ultimate theory, behind the chromosome movements during cell division. How did the homologous chromosomes actually find each other in order to pair up during metaphase cell division? How, wrapped around each other in chiasmatic formations, did they exchange genetic material? And how had the cell developed two forms of division for two separate functions – somatic replication and sex-cell formation – both necessary for life? (Harman, 2004, 40–41). Darlington began working on chromosomes in 1926, and was faced with two fundamental problems: the problem of chromosome pairing, and the problem of crossing over and chiasma formation. He espoused a religiously materialistic view of the chromosomes' role in heredity, and his method of genetic deductions was based on cytological assumptions – where possible observed, where not, inferred. In a sense,

Darlington, like Mendel, had a hypothesis to prove. Genetic premise and hypothesis would come first; the observed corroboration was almost invariably second (Harman, 2004, 44 and 53).

Two rival theories of chromosome pairing prevailed: the *parasynaptic* theory, claiming that during meiotic metaphase the chromosomes were pairing side by side in preparation for segregation and reduction division, and the *telosynaptic* theory, claiming that chromosomes were pairing end to end in preparation for being cut crosswise, which would result in segregation and reduction. The Morgan school argued for the side-by-side pairing. Janssens suggested that the points of contact, or chiasmata, which he observed between homologous chromosomes at meiosis might represent the sites of crossing over – of exchange of genetic material between those chromosomes. But Janssens' chiasmatype theory was based on speculation. What he had seen were merely homologous chromosomes (each apparently duplicated into two sister chromatids) twisted around one another at pachytene and chiasmata, non-sister chromatids that remained in contact between homologs that seemed otherwise separate at diplotene and later meiotic stages (diakinesis and metaphase I). Many cytologists, in fact, took the chiasmatype speculation of twist-induced breakage and mis-joining of chromatids to be false, adopting instead the "classical theory" that did not require breakage and rejoining of chromatids, but rather simply the exchange of partners between the four chromatids (see Chapter 5).

When Frank Newton, the cytologist in Bateson's lab and Darlington's mentor, discovered a triploid variety of the tulip, he realized that if a third chromosome paired alongside the other two, telosynapsis would be refuted. The tulip experiment proved parasynapsis to both Newton's and Darlington's satisfaction. Yet, cytologists continued to deny any causal relationship between chiasmata and crossing over.

If likeness were the sole condition of pairing at meiosis, in tetraploids only sets of four were expected. Yet, in *Primula sinensis* some chromosomes nearly always appeared paired in twos or in threes, and in *Hyacinthus* tetraploids the four chromosomes were always paired two by two. This behavior was not explicable by an affinity theory, whereby chromosomes with similar structural properties were those that paired. Why, Darlington asked, should one get certain chromosomes that failed to pair while others that were equally similar to their partners, paired? A causal relationship between pairing and chiasmata was suggested.

If chiasmata were the structural evidence for genetic crossing over, it follows that there should be a correspondence between the *frequencies* of chiasma formation and of crossing over, as well as between the *position* of the chiasmata on the chromosome and the corresponding distances of genic factors known from linkage mapping. Rigorous evidence was eventually obtained by comparing the frequency of chiasmata within a major inversion with the frequency of crossing over of that chromosome of *Lilium formosanum*. Because of the heterozygosity for paracentric inversions practically all crossing-over exchanges in the experiment led to cytologically observable events, namely dicentric chromatid-bridges between the two divided nuclei and loose acentric fragments at meiotic anaphase (see also above). There was an excellent agreement between chiasma frequency and the frequency of fragments, bridges (or loops) expected in heterozygotes for paracentric inversions on the chiasmatype hypothesis (Brown and Zohary, 1955). However, as a rule, the number of observable chiasmata was reduced toward diakinesis, compared to their number at diplotene.

Two different kinds of associations between homologs were observed at diakinesis: there were the Janssens-type chiasmata, and there were certain terminal associations. Darlington concluded that the observed chiasmata were only the leftovers of exchange events. Once homologs were pulled apart like opening a zipper starting at their centromeres' attachment to the spindle fibers, chiasmata arising from earlier crossing-over exchanges of chromatid events were pushed toward the chromosome ends, in a process that Darlington called *terminalization* (Harman, 2004, 59). The observed terminal association was accordingly the accumulated chiasmata that had not yet fallen off the paired homologs. Chiasmata were merely the observed consequences that were left over after the Janssens' kind of physical exchange had taken place when homologs were tightly paired at pachytene. The further terminalization proceeded, the fewer discernible chiasmata would remain. Such an interpretation would explain observations of correspondence between chiasmata frequencies at diplotene and exchange frequencies between genetic factors. It would also explain the reduction in the number of chiasmata between prophase and metaphase.

Thus, at pachytene parasynaptically paired chromosomes begin to repel each other and are pulled in opposite directions. Towards metaphase all that holds homologs together are the terminalized chiasmata that have not yet fallen off, hence the end-to-end image of telosynapsis. Chiasmata had to be the *condition* whereby chromosomes that had been paired should *remain* paired at metaphase. Longer chromosomes, Darlington argued, have more chiasmata and consequently, in spite of terminalization, have a better chance of remaining paired than shorter ones with fewer chiasmata between them. Shorter chromosomes are more often observed to be telosynaptic merely because terminalization has already been obtained.

On the assumption of the hypothesis that observed chiasmata were the result of crossing over rather than the cause, Darlington further analyzed his hyacinth triploids. In some of the trivalents with chiasmata involving all three homologs, two chromosome elements were associated with two chiasmata (A and B) and the third-chromosome element was connected to one of the first two elements by a chiasma C, *intercalary* between A and B. To account for such a configuration, it was necessary to assume that at the earlier pachytene stage there had been changes of pairing partners. Since the "classical theory" of exchange pairing chromatid partners had been refuted (see Chapter 5), breakage and rejoining were necessary at least in the formation of the intercalary chiasma C. Since at least chiasma C was due to breakage–rejoining of non-sister chromatids, it made sense to infer that also the non-diagnostic chiasmata A and B, and as a rule all chiasmata, are crossing-over events that are the *result* of exchange rather than its *cause*, in accordance with "chiasmatype theory."

Darlington defined what seemed to him to be two universal laws: (1) attraction between chromosomes is always in pairs, and (2) chiasmata are the condition for orderly segregation in cell division. Once the role of chiasmata in the orderly segregation of chromosomes was established, the attraction of homologs and pairing became the focus of interest.

Throughout mitotic prophase, chromosomes are composed of pairs of chromatids that are held together. This attraction appears to be turned into repulsion at metaphase. On the principle of Occam's Razor, Darlington suggested that meiosis and mitosis were on a similar mechanical plane of attraction–repulsion, and the difference was in the *timing* at which chromosomes split into two chromatids. In mitotic prophase the doubleness of the chromosomes, each divided into two chromatids, satisfied attraction until, at metaphase, it became repulsion. According to Darlington's "precocity theory," at meiotic prophase chromosomes seemed to be single as late as the pachytene. Due to the singleness of the chromosomes entering prophase, attraction causes pairing between homologs. This is reversed at pachytene into repulsion after duplication of the paired homologs, each into two chromatids. By now, paired tetrads are kept together only by chiasmata, the aftermath of

116

homologous exchange. Because of repulsion of centromeres chiasmata slip to the chromosome ends, or undergo terminalization, but by this time homolog disjunction is secured (Darlington, 1931).

As it turned out DNA content was doubled in the interphase preceding each mitotic nuclear division (Swift, 1950; Taylor, 1953). At meiosis only one DNA replicate was followed by two meiotic divisions, and the replication of DNA occurred in the interphase before the nuclear division began (see Chapter 4). Darlington's speculations on the timing of chromosome duplication and the processes of meiotic homolog pairing, though not sustained, stimulated much discussion and experimental activity with the introduction of molecular methods of analysis.

FUNCTIONAL ORGANIZATION OF CHROMOSOMES

Cytogenetic analysis established that the chromosome segments which stained differently with laboratory-dyes like hematoxylin or Giemsa in anaphase and interphase, were more densely compacted. These chromosome sections, designated by Heitz (1929) as heterochromatin, were most extensive at the centromeric and also at the end-sections of the chromosomes. Alignment of linkage maps with cytological maps suggested that few if any genes were located in these heterochromatic regions: microscopically observable major deletions of heterochromatic segments showed no or only little phenotypic effects, and contrary to the highly specific band-to-band pairing in euchromatic sections of polytenic chromosomes of Drosophila, heterochromatin tended to fuse nonspecifically, forming a "chromocenter" (Zacharias, 1995). Once hybridization methods were extended to *in vitro* hybridizations between polynucleotide sequences, it was demonstrated that *in situ* the radioactively labeled DNA probes were concentrated in the heterochromatic segments adjacent to the centromeres of mouse chromosomes (Pardue and Gall, 1970). This indicated that these chromosome segments were composed of highly repetitive DNA sequences. Other heterochromatic segments also turned out to be rich in repetitive sequences and their typical staining properties reflected the compacted coiling of their DNA strands around the nucleosomes.[1] These studies suggested that these

---

[1] Nucleosome: repeated flat units of a histone core around which the DNA is folded in eukaryotic chromosomes.

specific chromosomal regions were allocated to chromosome household functions rather than to storage of genetic information and its activation in the development and function of organisms.

From detailed analysis in polytenic chromosomes of aberrations and of their effect on the phenotype of Drosophila flies Muller concluded that chromosome ends were distinct in their properties from all other sites of the chromosomes (see Muller and Herskowitz, 1954): all along their length the chromosome structures are "bipolar," so that when broken ends are produced they comprise "sticky sites" that may combine with any other "sticky site." At the ends of chromosomes, however, he concluded that there are special "monopolar" structures that do not attach to other broken sites. He called these "telomeres" (Muller, 1940). Similar evidence of the monopolar nature of chromosome ends was obtained by Barbara McClintock in her work on deficiencies in maize tissues. She followed mitoses in cells with aberrant ring chromosomes (having no free ends). She found that these become double-sized dicentric interlocking rings upon replication. At anaphase a bridge is formed and torn apart; the broken ends being bipolar and "sticky." These attach into new rings (often slightly shorter than the previous one) that may be torn apart again at the next mitotic division. The process may go on for several cycles of Breakage–Fusion–Bridge (BFB) formation. It may also serve as a kind of somatic genetic mapping procedure: BFB cycles sequentially delete cells of the topologically distal to proximal (in relation to the centromere) segments, sequentially exposing markers for which the cells were heterozygous (McClintock, 1938; McClintock, 1951).

Biased as some cytological analyses may have been by the logic of the genetic analysis, molecular analysis only further confirmed the telomeres' unique structures. Details of enzymatic DNA replication revealed that an RNA primer is needed to initiate DNA synthesis. The removal of this RNA primer at the nascent DNA strand (that ends with a 5′ end) at the telomere would leave a gap corresponding to the length of the primer. Thus, it became obvious that with each nuclear division chromosomes would shorten at their ends, unless a special mechanism was available to compensate for this loss by providing telomere-specific DNA synthesis. Indeed a special enzyme that synthesizes short, tandem repeats onto telomere ends was discovered by Greider and Blackburn in 1985 and termed telomerase (Greider and Blackburn, 1985). The repeated DNA sequence synthesized by telomerase is consistent with the telomeres' "heterochromatic" properties. The activity of telomerase is absent or highly restricted in human somatic cells to allow at least some progressive shortening of telomeres with age. When telomeres become critically short, they arrest the proliferation of cells. Thus telomeres set a limit for the number of cell divisions possible and provide a tumor-suppressing mechanism. Many cancerous cells overcome this mechanism by upregulating telomerase expression (for a review, see Verdun and Karlseder, 2007). Telomeres were also implicated in other functions, such as the tethering of the chromosomes to the nuclear membrane and the pairing of homologous chromosomes during meiosis.

Eventually molecular analysis indicated the functional organization of the DNA sequences of the centromere and the adjacent regions. An important cytogenetic finding, using molecular techniques on chromosomal aberrations of Drosophila, was the evidence that chromosomes, from one end to the other, comprised one continuous double-stranded DNA molecule. Kavenoff and Zimm (1973) measured the viscoelastic retardation times of detergent-Pronase systems of Drosophila cells. Four Drosophila species with different normal karyotypes, *Drosophila melanogaster, D. Hydei, D. virilis* and *D. americana*, were examined, as well as stocks with pericentric inversions, X:A translocations and deletion in the first two species. Estimates of the molecular weights of the largest (and accordingly presumably unbroken) molecules ranged between 20×10^9 and 80×10^9 depending on the species, independent of the metaphase shapes (metacentric or submetacentric) but proportional to the DNA content of the chromosomes in the case of translocations and deletions. The results indicated that the DNA molecules run uninterrupted for the full length of the chromosomes. Kavenoff and Zimm concluded that "*In its simplest form one chromosome contains one long molecule of DNA*" (Kavenoff and Zimm, 1973, 24). Contrary to the strict reductive notion of genetics and the role of DNA, the results indicated that any qualifying properties of the chromosome were due to its interaction with other molecules, whether with respect to its functions in cell inheritance or in cellular performance.

Of special significance were the centromeric segments involved in spindle-fiber-attachment and the mechanisms involved in the proper segregation of the chromosomes at meiosis, such that each of the four meiotic end products would comprise a complete genomic set. Geneticists and cytogeneticists conceptualized this specific chromosome pairing, or *synapsis*, as a precondition for segregation.

Analysis of chromosome segregation in heterozygotes for major autosomal translocations indicated that of the possible patterns of segregation those of homologs predominated (Dobzhansky, 1929, 1931). According to the "competitive pairing hypothesis" both crossing over and disjunction were predetermined by the intimacy of synapsis between homologous loci, the intimacy of pairing being the critical requirement for disjunction, exchange being secondary to it.

Frequently at early meiosis, the chromosomes are very elongated and slender, hence the term "leptotene," and they form a highly polarized "bouquet" orientation with all their ends directed towards a small area on the nuclear membrane. The bouquet, first described by Gustav Eisen in 1900 in salamander meiosis, is highly conserved among eukaryotes. This peculiar arrangement of a definite orientation and polarization appears to be a property of the ends or the telomeres (White, 1973, 81). Carl Rabl had observed already in 1885 arrangements of chromosome terminals and of centromeres on opposite sites of the nucleus, and Boveri considered it "conceivable that this remarkable agglomeration [of meiotic chromosomes] represents the mutual searching of the homologous chromosomes" (Scherthan, 2001).

In the 1950s electron microscopy techniques revealed the arrangement of the paired chromosomes at meiotic pachytene in the form of the synaptonemal complexes, a tripartite ribbon consisting of two proteinaceous parallel "core" strands running along homologous chromosomes. Both cores are connected through a third "central element" of transverse filaments that align the core strands. The complex is firmly attached by the telomeres to the nuclear envelope. Although high-resolution fluorescent imaging and ultrastructural and genetic analysis have shown that bouquet formation is independent of both synapsis and recombination, as expected on the assumption that telomere clustering is the instigation of chromosome interaction towards stable homolog pairing, bouquet topology dissolves during zygotene in most organisms that have synaptic meiosis (Scherthan, 2001). Molecular studies of nuclear arrangement indicate that chromosome terminals (telomeres) actively attach to the inner nuclear envelope to form a meiotic bouquet. It seems, however, that the bouquet only makes homologous pairing more efficient (Harper, Golubovskaya, and Cande, 2004).

Molecular analysis identified proteins involved in the tethering of the telomeres to the nuclear membrane and the catalytic activity of bouquet formation that promotes the physical proximity of components. Notably, physically small chromosomes suffer more from the absence of bouquet formation, presumably because they will have less opportunity for chance encounters with their homologs. A meiosis-specific protein,

NDJ1, which is required for bouquet formation, was found in yeast (*S. cerevisiae*). An *NDJ1* mutation disrupts the bouquet organization of the telomeres and shows a considerable delay in homolog pairing (Scherthan, 2001). It appears therefore that bouquet formation makes homologous pairing more efficient (Harper *et al.*, 2004) but that this is only one step in a dynamic process of an intensive pairing competition between strands that pair, dissociate and pair again, until stable partners are established (whether on the basis of homology, or of length) and the partners zip to form synaptonemal complexes on which "recombination nodules" may or may not be formed (Liebe, Alsheimer, Höög, Benavente, and Scherthan, 2004; Scherthan, 2007; Schmitt *et al.*, 2007; see also Davis, 2000).

The mechanical attachment of the chromosomes to the spindle determines their delivery to the daughter cells. Chance encounters of microtubules with back-to-back arrangement of partner kinetochores (the cytologists' term for geneticists' centromeres) secure the proper attachment of most chromosomes to opposite poles and proper detachment (Nicklas, 1997). By micromanipulation of living grasshopper spermatocytes Nicklas showed that a faulty attachment of fibers is unstable and repeatedly changes until, by chance, the proper attachment is hit upon. It appears that the bipolar tension which slightly stretches the bivalent secures the stability of the proper detachment orientation (Nicklas and Koch, 1969). This upholds Darlington's claim that the repulsion force leads to univalent formation at anaphase unless a chiasma is present to maintain a physical link between homologs. In retrospect, perhaps the most significant contribution of Darlington's precocity theory was its emphasis on the functional role of chiasmata in segregation.

Chromosome pairing may be a precondition for proper segregation. However, specific homolog pairing followed by crossing-over exchange is not universal. Even in Drosophila there is no exchange in male meiosis, and also in females, chromosomes without a single chiasma do occur, e.g., in the minute chromosome IV, or in the large chromosomes, when major aberrations make homolog pairing difficult. Rhoda F. Grell suggested that at least in Drosophila, another pairing process occurs after the early prophase *exchange pairing*, namely the *distributive pairing*, which is apparently dependent on size similarity rather than on homology.

In its broadest sense, distributive pairing is defined as any achiasmate association and coordination of chromosomes that determines a subsequent segregation pattern. This usage includes the primary spermatocyte of *D. melanogaster*. . . . *Distributive pairing* in the oocyte is a postexchange process and is confined to chromosomes that have not undergone exchange with an independent homologue. It could occur at any time after exchange and preceding anaphase I . . .

Grell (1976, 436)

Normally it works inconspicuously in Drosophila females since the two minute chromosomes are the only candidates for the non-chiasmatic distributive pairing. The two major autosomes are long enough to always be engaged in at least one chiasma (though in heterozygotes for major aberrations in which no euploid products of crossing over occur homologous arms might depend on distributive pairing). Also, although some of the intermediate-length chromosome-X remain with no exchange they segregate preferentially from each other (approximately 5 percent of tetrads are non-crossovers, but only 0.05 percent of the Xs show non-disjunction), rather than from the small chromosome IV: "Chromosomes of equal length pair and disjoin more frequently than those of unequal length; the more disparate the chromosomes are in size, the greater is their non-disjunction frequency. Consequently, distributive pairing occurs preferentially between homologues" (Grell, 1976, 436). Genetic analysis of experimental introduction of short X-chromosome fragments that would compete with the normal chromosome IV for distributive pairing induced high frequencies of non-disjunction of the latter. Not only the presence of a Y-chromosome but also hetero-zygosity for an X-autosomal translocation caused a high frequency of X-chromosome non-disjunction in XXY oocytes. This refuted the *ad hoc* speculation that the observed pairing of the X and Y chromosomes reflected an obligatory compensatory double exchange that is not detectable genetically because of the absence of genetic markers in the relevant segment of the chromosomes.

Evidence that at least in Drosophila females exchange-dependent homolog segregation was insufficient was also obtained from muta-tions affecting segregation. In the *c*(3)G mutation, which practically eliminated meiotic exchange, chromosome assortment was far from random. Another Drosophila mutation, *nod*, considerably increased non-disjunction of chromosome IV and also somewhat that of chromo-some X, without affecting the frequency of recombination. Such obser-vations support the function of a distributive pairing process, without affecting the exchange pairing that these mutants effect (Grell, 1976,

437–438). Speculations on the significance of regular Mendelian segregation in evolution and that of exchange were unavoidable. Obviously homolog pairing as a precondition for regular disjunction of genetic factors was evolutionarily more basic than exchange. Exchange turned crucial once multi-gene chromosomes became the entities of cell division, so as not to lose evolutionarily essential genetic variability. The organization of the genome into chromosomes undoubtedly greatly increased the efficiency of disjunction of regular factors. However, only when exchange was established was genome evolution greatly enhanced, as more chromosomes could be manipulated and these could become longer. An increasing size of the genome could be involved in an increasing complexity of living organisms.

In spite of abundant experimental evidence, Rhoda Grell's distributive pairing model remains an *ad hoc* phenomenological explanation.

IV

Genes as the atoms of heredity

> I think what interested me most in the history of science is the relationship between ideas held at different times, couched in similar terms, yet obviously having different contents and meaning. The view of matter as composed of elementary corpuscles, atoms, preceded by two millennia the development of atomic history. What, if anything, does the second concept owe to the first? How, if not derived from the first, did the second arise?
>
> Dunn (1965, xvii)

A century ago, "genes" endowed with the instrumental reductionist status of the Mendelian *Faktoren* were introduced to overcome the notion of the unit character. The gene was an empiric, functionally defined entity. For Johannsen, gene was a word derived from genotype that should express "only the simple imaging" that "through 'something' in the gametes a property of the developing organism is conditioned or is, or could be co-determined" (Johannsen, 1909, 124).

Yet, the meaning of the gene, whether as an intervening variable, a heuristic device for the study of inheritance, or a hypothetical construct, an entity in the physico-chemical world, and what kind of such an entity, became a major issue of genetic analysis. From Richard Goldschmidt's "physiological" perspective genes were merely centers of gravity along the integral units of the chromosome. For Hermann Muller genes were discrete structural entities along the chromosomes. Others, like East, Morgan, and Stadler adopted an operational approach of "instrumental reductionism" (Falk, 1986, 141). In Morgan's phenomenological conception genetic analysis was adequate for elucidating the mechanics of hereditary transmission and the outlines of the involvement of such operationally defined entities in development. Goldschmidt in the context of *Naturphilosophie* conceived of genes from the perspective of organisms as functional entities. Muller, ever directing his attention also to the social, hence population-evolutionary aspects of heredity, initiated a research program to uncover the physico-chemical properties of genes as the atoms of evolution. Admittedly, "to grind genes in a mortar and cook then in a beaker," was at the time only a hope, but he set out to turn phenomenological genetic analysis into an instrument for the elucidation of the material properties of genes (Muller, 1922). Heuristically he utilized that property of genes that made them unique, namely changing without losing their autocatalytic capacity, in other words, mutating. This property made them the cornerstones of Darwinian evolution (Muller, 1927a).

Asking similar questions but in physico-chemical terms, Watson and Crick's structure of the genetic material apparently adumbrated the failure of Muller's research program to elucidate the nature of genes by phenomenological genetic analysis. Still, their model was established on precisely the same conceptual demands of the genetic material as Muller's and his challengers. The detailed phenomenological analyses of mutagenesis and linkage that culminated in establishing the complexity of fungal genes and eventually of bacteriophage genes, and the colinearity between genetic sequences and polypeptides' amino-acid sequences providing biological meaning to the physico-chemical model, must be considered the culmination of the genetic analyses of their forerunners.

By elucidating the details of the distinct mechanisms of transcription and translation (Rheinberger, 1997), and by clarifying the involvement of "genes" in regulating transcription of other genes (Jacob and Monod, 1961), determinist genetic analysis reached its climax with the "Central Dogma of molecular biology" (Crick, 1958). Genetic analysis turned from phenomenological to biochemical and molecular properties. But these achievements only stressed the problem of the identity of genes as the entities of analysis. Were they structurally identifiable discrete entities of DNA in the chromosomes? or at least functional units that uniquely directed polypeptide sequences, all possible interactions being secondary consequences of such determinations?

The findings that large proportions of the DNA sequences were highly repetitive; that transcription could involve both strands, not necessarily from the same starting point, and not necessarily ending on the same chromosome; that the coding sequences were not transcribed into continuous open reading frames (ORF) translatable to

polypeptides, but rather needed splicing and editing processes to provide for translatable messenger-RNA sequences; and all this from
continuous DNA sequences from origin to end in the bacteriophage as
well as in the chromosomes of Drosophila, indicated that cells are
determinants of the properties of DNA as much as DNA is a determinant of the properties of cells. DNA is as much a phenotype as are
RNA, proteins, enzymes, and yellow pea seeds. Towards the start of the
twenty-first century, when it became clear that the mechanisms of
translation and transcription are just two very complex and diverse
processing stages and that there is no way to uniquely read from a DNA
sequence its product, not at the level of the polypeptide and not even at
the level of RNA, genes became generic terms of hereditary transmission (Falk, 2000b, 2004). In the post-genomic era the gene became a
metaphoric entity (Griffiths and Stotz, 2006).

8

Characterizing the gene

Johanssen wished to discern the pheno*type* from the geno*type* (Johannsen, 1909). "Genes," however, became what geneticists concluded Mendelian *Faktoren* had been. Were genes hypothetical constructs, autonomous structural entities or loci of differential functional emphases along an integral chromosome? Or, did all that not matter and genes were simply intervening variables, helpful entities for experimental work (see Chapter 4)?

Strictly speaking, there are no hypothesis-free concepts, and the distinctions between hypothetical constructs and intervening variables that proved very helpful historiographically, are more problematic when one attempts to apply them to personal conceptions. Whether we wish it or not, our concepts "are constructions in thought representing historically an immense amount of intellectual work ... the scientific hypothesis does not come after the numerical data but *before* them" (Woodger, 1967, 366). Morgan, like Johannsen, insisted on presenting genes as intervening variables, with no hypothesis about the nature of the "something." As late as in his Nobel talk he asserted:

> Now that we locate [the genes] in the chromosomes are we justified in regarding them as material units; as chemical bodies of a higher order than molecules? Frankly, these are questions with which the working geneticist has not much concerned himself, except now and then to speculate as to the nature of postulated elements. There is no consensus of opinion amongst geneticists as to what genes are – whether they are real or purely fictitious – because at the level at which the genetic experiments lie, it does not make the slightest difference whether the gene is a hypothetical unit, or whether the gene is a material particle.
>
> Morgan (1934b, 315)

Was this position, which I call "instrumental reductionism" (Falk, 1986, 141 and footnote 30) really above speculation and preconceived notions? Muller explicitly conceived of genes as hypothetical constructs, material atoms of heredity. Richard Goldschmidt went further, claiming that the skepticism of Johannsen "was completely sterile ... even if it turned out now that [his] sterile skepticism in the face of overwhelming facts had happened to put them after all on the right side" (Goldschmidt, 1938a, 268). Was not Goldschmidt's acceptance of genes as functional entities, in spite of his denial of genes as structural entities, also a notion of genes as intervening variables? Or at least one of operationalism, like that of Lewis Stadler, who accused geneticists of adopting the dictum of Humpty Dumpty in *Through the Looking Glass*, "When I use a word, it means just what I choose it to mean – neither more nor less" (Stadler, 1954, 813)?

IF MENDEL'S LAW IS TRUE FOR DISCRETE TRAITS IT MUST ALSO HOLD FOR CONTINUOUS TRAITS

When William Castle adhered to the notion of the unit character as the only one which was instrumentally meaningful, rejecting Muller's claim that it was not "in harmony with the results of Johannsen" (Castle, 1915) or that "[i]n no known case do the variations of a gene ... form a probability curve" (Muller, 1914a), he remarked that Muller's "use of the word 'gene' in this sweeping statement safeguards the author, since no one, so far as I know, claims ever to have seen a 'gene' or to have measured it" (Castle, 1915, 38). For the breeder Edward M. East, the nature of the gene and "*a gene terminology*" were adequate as long as they expressed "all known varieties of inheritance phenomena" (Castle, 1919c, 130). East consistently argued that genes were only intervening variables. Asking "[h]ow far may we carry this conceptual notation?" he answered, "just as far as the notation interprets the facts of breeding and is helpful" (East, 1912, 635; and Chapter 4). Accordingly the "Mendelian notion" was "a description of physiological facts":

> As I understand Mendelism it is a concept pure and simple. One crosses various animals or plants and records the results. With the duplication of experiments under comparatively constant environments these results recur with sufficient definiteness to justify the use of a notation in which

129

theoretical genes located in the germ cells replace actual somatic characters found by experiment. ... Mendelism is therefore just such a conceptual notation as is used in algebra or in chemistry.

East (1912, 633)

Mendelism, however, traditionally was concerned with qualitative, readily distinct traits. However, when it was claimed that such traits were inherently different from most other traits that varied continuously, i.e., that Mendelian variability was not methodologically but immanently different from that of most relevant natural variability, East protested. Once the concept of the unit character according to the morphogenist tradition was replaced by the gene of the hybridist tradition, East pointed out, Mendel's choice of specific characters must be considered only as one of practical, experimental considerations:

Since quantitative characters were the ones that could be divided into definite categories. ... One by one they were analyzed. ... Qualitative characters however form a very small proportion of the characters in animals and plants. The numerous characters are the quantitative, the size characters. If Mendel's law is to be worth anything as a generality, therefore, it must describe the inheritance of these characters.

East (1912, 637)

East himself observed, as did Nilsson-Ehle, that properties which in some crosses behaved like simple units that segregated as monohybrids, in others "there were two definite independent Mendelian unit characters, each of which was allelomorphic to its absence." In one of Nilsson-Ehle's crosses, involving the color of wheat seeds, "a very old red variety from north of Sweden – the ratio in the F_2 generation was 63 red to 1 white." As East noted, the results were so close to the theoretical calculation that they quite convinced him "that he was really dealing with three indistinguishable but independent red characters, each allelomorphic to its absence." Furthermore, the distribution curve of differences in the intensity of the red color of the F_2 seeds provided a good approximation to a normal distribution curve. In East's own experiments with endosperm colors in maize, there was some evidence "that there are three and possibly four independent red colors in the pericarp, and two colors in the aleurone cells" (East, 1910, 66). All this suggested an instrumental "Mendelian interpretation of variation that is apparently continuous." Interestingly, although East consistently adhered to formal presentation of the results of his experiments as corresponding to the expectation of a multi-hybrid Mendelian segregation – no material factors intended – he did bother to speculate on the physiological meaning of different genotypes, say *AA Bb Cc* and *aa BB CC*, producing the same phenotype. "It may be that there is a kind of biological isomerism, in which, instead of molecules of the same formula having different physical properties, there are isomers capable of forming the same character, although, through difference in construction, they are not allelomorphic to each other" (East, 1910, 71). This is typical of the "instrumental reductionism" strategy of formally analyzing hybridization results in purely phenomenological terms, yet speculating on a physiological, even molecular bottom-up mechanism causing them.

Mendel's experimental procedure of selecting among the variable characteristics at his disposal those that would best allow him to examine his hypothesis, namely characters with unequivocal binary variation that is maintained over numerous generations, relieved him from indulging in the problem of the relation between genotype and phenotype, and hence also in the nature of his *Faktoren*. The nature of the genes was explored, however, once Mendelian traits turned out to be only the markers of entities whose properties were defined by a hypothesis.

THE MATERIAL BASIS OF THE GENE

In a programmatic paper Hermann J. Muller introduced his conception of genes as the fundamental entities for understanding the physiology of growth and function of living creatures which evolved along the principles of Darwinian evolution:

> ... besides the ordinary proteins, carbohydrates, lipoids, and extractives, of their several types, there are present within the cell *thousands* of distinct substances – the "genes"; these genes exist as ultramicroscopic particles; their influences nevertheless permeate the entire cell, and they play a fundamental rôle in determining the nature of all cell substances, cell structures, and cell activities. Through these cell effects, in turn, the genes affect the entire organism.
>
> Muller (1922, 32)

Muller entitled this paper "Variation due to change in the individual gene" and was "concerned rather with problems, and the possible means of attacking them," that were made possible by "the fundamental contribution which genetics has made to cell physiology within the last decade." Although it was not "mere guesswork to say that the genes are ultra-microscopic bodies" and there "must be hundreds of such genes within each of the larger chromosomes," admittedly the "chemical composition of the genes, and the formulæ of their reactions, remain as yet unknown" (Muller, 1922, 32). Still, the genes' main properties may be deduced from the experimental analysis:

> The most distinctive characteristic of each of these ultra-microscopic particles – that characteristic whereby we identify it as a gene – is its property of self-propagation. . . . But the most remarkable feature of the situation is not this oft-noted autocatalytic action in itself – it is the fact that, when the structure of the gene becomes changed, through some "chance variation," the catalytic property of the gene may become correspondingly changed, in such a way as to leave it still *auto*catalytic. . . . Since, through change after change in the gene, this same phenomenon persists, it is evident that it must depend upon some general feature of gene construction – common to all genes – which gives each one a *general* autocatalytic power – a "carte blanche" – to build material of whatever specific sort it itself happens to be composed of.
>
> Muller (1922, 33–34)

These properties of autocatalysis and change, without the loss of auto-catalytic capacity, reside in the structure of the genes themselves, and the outer protoplasm does little more than provide the building material. This peculiar property of the genes "lies at the root of organic evolution."

> Inheritance by itself leads to no change, and variation leads to no permanent change, unless the variations themselves are heritable. Thus it is not inheritance *and* variation which bring about evolution, but the inheritance *of* variation, and this in turn is due to the general principle of gene construction which causes the persistence of autocatalysis despite the alteration in structure of the gene itself. Given now, any material or collection of materials having this one unusual characteristic, and evolution would automatically follow.
>
> Muller (1922, 35)

This powerful analysis reduced the problem of inheritance to that of the nature of "the gene," the atom of inheritance. Muller set out to characterize these atoms although, unfortunately, "a gene can not effectively be ground in a mortar, or distilled in a retort."

According to Muller, three properties *defined* genes: independent primary function, location at specific sites along chromosomes, and autonomy in mutations. The first property was verified experimentally by "complementation tests." Heterozygotes for two (recessive) characteristics that segregated according to Mendelian rules identified two alleles of the same genes if they did not complement each other, and were denoted: $a_i \backslash a_j$. Heterozygotes for both that were wild type identified different genes, each carrying also the corresponding wild-type allele in the double heterozygote (denoted: $a\,b^+ \backslash a^+ b$, or in short: $a + \backslash + b$). As a rule, a gene could be located experimentally to a specific site or *locus* on the linkage map, and there was no recombination between alleles of a given gene. Finally, as a rule, many mutations affected exclusively a given gene.

Although these criteria were quite effective, there were exceptions that occupied the attention of Muller and his colleagues. An early one was that of "truncate wings" (Altenburg and Muller, 1920, later called "dumpy") that showed complex allelism relationships. Other complex loci were found. But at least some of the single-gene mutations were assumed to be "point mutations," or mutations that changed the (chemical?) qualities of the gene entity without inducing (physical) aberrations such as deletions in the continuity of the chromosomal structure. Mutations that affected multiple, usually linkage-neighbor genes were considered to be chromosome rearrangements (even before it was possible to verify this at the cytological level).

To a great extent, Muller molded the ethos of genetic analysis by reducing the problem of inheritance to the structure and properties of the physical entities predicted by such an analysis, and by following the properties of these entities by experimental hybridization analysis of the phenomenological changes of the phenotypes.

Goldschmidt, like Muller, was hypothesis-guided. He too saw in the elucidation of the nature of the gene a central problem of genetics: "As long as genetics has existed, the ultimate problem has been the nature of the gene, its reduplication and mutation" (Goldschmidt, 1950, 336), but he denounced Muller's reductionist philosophy and what he called the "classic theory of the gene." Experimental data pointed only to loci along the chromosomes. Adding to these observational correlates any assumptions about the properties of these loci was "hyperatomism," which was destructive:

> The statistical basic philosophy tries to interpret every generalized set of facts by the introduction of more and more units for statistical treatment ... [It] tries to explain all basic features of genetic phenomena by introducing more genes ... In this way, a system is finally established, ... which I must call hyperatomism and hyperselectionism.

> Although [the physiological] approach accepts, naturally, the basically
> statistical tenets of genetics, it tries, actually within the rule of parsimony,
> to avoid looking for explanations in terms of unproved, additional systems
> of units for more and more genic permutations.
>
> Goldschmidt (1954, 704)

Goldschmidt's outlook of a top-down biologist, like the bottom-up
biologist's outlook of Muller, conceived of all problems from the
perspective of evolution, but he maintained that the insistence on
the autonomy of the genes had, in the best case, only heuristic advantages, and that it was the organism and its chromosomes that should
be conceived as the fundamental entities of inheritance. Goldschmidt
argued that mutants were not so much of interest beyond being
tools that revealed something of the wild-type alleles. On general
principles, Goldschmidt claimed that there are *no autonomous* structures along the chromosomes, and that "ordinary and typical so-called
gene mutations are actually the results of rearrangement" (Goldschmidt, 1938a).

> Apart from purely genetical considerations, there is also a very general
> reason for the attitude I have taken. This is the fact that the hierarchical
> order is clearly essential in living nature, although it also exists in inanimate
> nature ... each higher member of the hierarchy being composed of the
> lower ones but different in its qualities from a mere sum of these. It is clearly
> not the sum but the orderly relationships of the components that are
> responsible for the actions at the different levels of the hierarchy. Therefore,
> at these different levels also, new types of interrelationships appear ... In
> view of such facts a biologist, studying a clearly hierarchical system of
> activities like that of chromomere, chromosomal segment, chromosome,
> genome ... would rather expect to find a still more complicated relationship
> in which the parts, in a hierarchical order work together via spatial rela-
> tionships, orders, patterns.
>
> Goldschmidt (1954, 709)

Goldschmidt's argument that there were no particulate discrete entities
along the chromosome meant that all mutations were actually rearrangements, changes in the sequence of the chromosomes. In other words,
observed mutations were position effects, changes in functional properties as a consequence of relative positions of loci on the chromosomes,
rather than in the autonomous properties of presumed autonomous
genes. Position effect was first described by Sturtevant for the Bar-eyes
mutation of Drosophila (B) as caused by the spatial relation of sites on
the chromosome as a result of unequal crossing over: duplication of a
polytene band in the same chromosome has a different effect than

the two bands located opposite each other in homologous chromosomes ($B\backslash B \neq BB\backslash+$) (Sturtevant, 1925). Goldschmidt analyzed common spontaneous mutations, for which it could be shown that chromatin rearrangements were involved. "If this can happen there is no reason to doubt that in due time all other mutants will appear the same way." He concludes "that gene mutation and position effect are one and the same thing. This means that no genes are existing but only points, loci, in a chromosome which have to be arranged in a proper order or pattern to control normal development. Any change in order may change some detail of development, and this is what we call a mutation" (Goldschmidt, 1938a, 271).

It appeared that the argument on the nature of apparent point-mutations would become redundant once the whole chromosome was conceived as a continuous DNA molecule on which the "genetic information" is contained in the nucleotide *sequence* in a specific stretch along this sequence. Goldschmidt declared: "There is no longer a gene molecule but a definite molecular pattern in a definite section of the chromosome, and any change of pattern (position effect in the widest sense) changes the action of the chromosomal part and thus appears as a mutant" (Goldschmidt, 1955, 186). But genes are assigned to the molecular chromosomes today as they were in pre-DNA times. In the pre-DNA era heavy theoretical as well as empiric arguments were brought against Goldschmidt's position. Central was the argument that according to Goldschmidt's theory mutations are practically all losses of genetic material and no Darwinian theory of evolution could be sustained on the mere reshuffling of preexisting materials; some qualitative differences must occur. On the empirical level, multiple alleles existed for many genes and, more importantly, "back-mutations," primarily from mutant to wild type (but also from one allele to another, such as $w \rightarrow w^+$, but also $w \rightarrow w^a$), indicated that mutations were more than quantitative changes such as losses of genetic matter. Goldschmidt insisted that although his notion of a hierarchical system of functional organization along the chromosome was more cumbersome than the reductionist conception of genes, he could counter any theoretical or empirical argument with the system's insight:

> It is the way we are looking at the facts that is different. But this difference does not necessarily mean that one is a better thinker than another. Rather it means, in many cases, that the general way of thinking, of analyzing facts and of putting them into categories, is different in different minds.
>
> Goldschmidt (1954, 703)

A reductionist experimental approach is no doubt simpler to follow, but it oversimplifies the essence of life as a complex interactive system that depends on its complex system's properties to survive and evolve.

THE *ClB* SCREENING METHOD

To examine the assumptions on the nature of the gene required a powerful analytic instrument that would allow quantitative examination of the qualitative contrasting conclusions. Muller aimed at the property that he considered to be unique to evolving living matter, mutagenicity. To make mutagenesis an efficient analytic tool it must be amenable to simple and unequivocal experimental designs. True to the Mendelian tradition, Muller chose a *class* of mutations, irrespective of their specific effects, for the quantitative analysis of mutations. Screened were all recessive *lethal* genes of the target chromosome (the X-chromosome of *Drosophila melanogaster*), genes which, when mutated, were fatal to their carrier (Rieger *et al.*, 1976). A lethal mutation in any of the genes (or loci) along the tested chromosome would eliminate all males that are hemizygous for that chromosome, whereas females heterozygous for such mutations would remain alive to be further tested. A crucial idea was to mark the *tested chromosome* as the one that did *not* carry the marker genes of the *tester chromosome*. Then, in order to maintain the tested chromosome as an entity in which a lethal mutation in any of its sites would be detected, it was necessary to prevent recombination between the tested chromosome and the homologous tester chromosome. A tester chromosome identified by the dominant mutation (*B* for Bar eyes) also carried a factor that would inhibit crossing over of the X-chromosomes (the *C*-factor. Later it turned out to be a major inversion of segments of the chromosome. Crossing over in a heterozygote for such an inversion will nearly always end in non-viable progeny, see also Chapter 7). Finally a lethal mutation (denoted by *l*) was inserted in the tester chromosome, in order to increase the efficiency of screening for new mutations in the tested X-chromosome. The tester chromosome was called the *ClB* (Muller, 1928), and the model of the screening procedure was called the *ClB*-method.

Drosophila males were the experimental objects. Since the frequency of (induced) lethal mutations in the X-chromosome was tested, no pre-existing mutations could bias the results. Tested males were mated to females heterozygous for the *ClB*-chromosome. Each Bar-eyed daughter

136

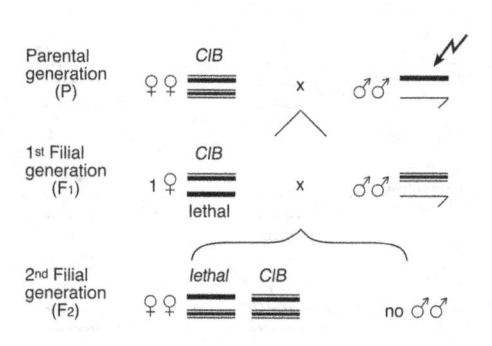

Figure 8.1. The *ClB* mating scheme, screening for mutations on the X-chromosome of *Drosophila melanogaster.*

of these males was heterozygous for a specific individual non-*B* X-chromosome of the sample of sperm cells produced by the experimental males. The tested X-chromosome was recognized by being the non-*B* chromosome, and the *C*-factor provided for the chromosome's integrity, such that no mutation on the tested chromosome would cross over to the tester chromosome (or vice versa). The daughters were allowed to mate with their brothers, and then each daughter was bred individually in a new culture vial, each vial representing one tested X-chromosome of the experimental sample. Since the *ClB*-chromosome contained a recessive lethal (l), all that was needed was to inspect each vial for male progeny. Vials with no male progeny were those in which a new recessive lethal had occurred or was induced in the sampled X-chromosome. The frequency of new lethal mutations was equal to the frequency of male-less vials out of the total number of vials (new visible, say eye-color, mutations were detected in all males in a vial showing the new phenotype). Females from the male-less vials could be further mated to verify the genetic or developmental properties of the new lethal mutation (Muller, 1927b) (see Figure 8.1).

The notion of the *ClB* method was extended to the screening of other chromosomes and other types of mutation (although the procedure was often more cumbersome). The autosomes of Drosophila were screened by mating males to females both of whose homologous chromosomes carried different dominant markers, different sets of inversions, and different recessive lethals. Typical is the *Cy/Pm* stock for screening Drosophila's 2nd-chromosomes (see Figure 8.2). Male flies were mated to *Cy/Pm* females. Individual curly winged (*Cy*) male progeny were backcrossed to virgin *Cy/Pm* females and the curly progeny of each of the individually tested experimental males were interbred. The absence

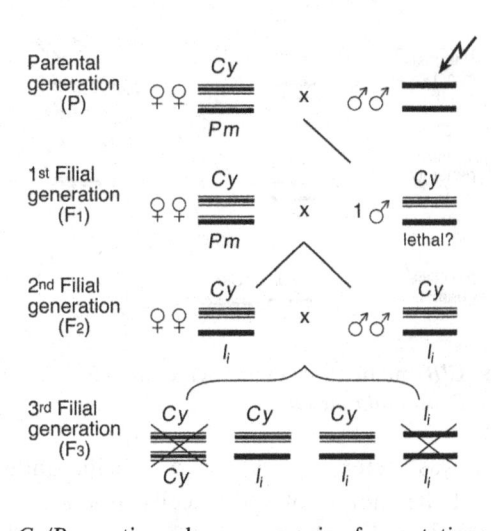

Figure 8.2. The *Cy/Pm* mating scheme, screening for mutations on chromosome II of *Drosophila melanogaster*.

of non-curly progeny indicated that the tested chromosome carried a lethal mutation. The appearance of non-curly flies with a changed phenotype indicated that the chromosome carried a mutation for the specific phenotype. The numerical proportion of non-Cy to Cy flies indicated the presence of viability reducing, or "detrimental" non-lethal mutations on the examined chromosome (in a normal, non-mutant culture the expected proportion of curly to non-curly flies is 2:1. Notice that homozygotes for the Cy-chromosome are lethal. So also are homozygotes for the Pm-chromosome and consequently the Curly/Plum flies may be maintained as a "balanced stock" producing only Cy/Pm progeny). Similar tests were developed for other autosomes and for other species. Of special significance was the extension of this method to studies of cytologically observed chromosomal variability (mainly inversions) in natural populations, first of *Drosophila pseudoobscura* (and *D. persimilis*) and then in many other Drosophila species, by Dobzhansky and his coworkers (see Lewontin, Moore, Provine, and Wallace, 1981).

The availability of such methods for quantitative genetic analysis of whole specified chromosomes made Drosophila a favorable and most useful organism for a wide range of genetic problems. The screening heuristic was extended to other phenomena. Thus, an efficient method for screening translocations was to mate experimental Drosophila males to females homozygous for recessive markers on two different chromosomes, such as *cn bw* (cinnabar and brown producing white eyes) on the 2nd-chromosomes and *e* (ebony body color) on the 3rd-chromosome. Individual male progeny, each representing an individual haploid genome of the experimental flies, were backcrossed to *cn bw; e* females. The absence of flies with "white" eyes (and wild-type body color) and of flies with "dark" body (and wild-type eye color) indicated that a 2nd–3rd chromosome translocation had been induced. Also, the absence of recombination between sex and one or another chromosome marker indicated the presence of a Y-chromosome–autosome translocation.

Although no such sophisticated auxiliary methods were available in organisms other than Drosophila, the *ClB*-design served as a model for detecting chromosome mutations and aberrations in other organisms too. As we shall see, a completely different screening principle was applied once rare genetic events in fungi, and later bacteria and viruses, were tested. Yet the essence of genetic analysis was the principle of designing screening procedures where only (or nearly only) the desired events survive.

THE INDUCTION OF MUTATIONS

While for East genes were theoretical units that "replace actual somatic characters found by experiment" (East, 1912), for Muller the gene was "the basis of life" (Muller, 1929), and he set out to study its properties through the quantitative analysis of mutations (Muller, 1922). Decades later he introduced his Pilgrim Trust Lecture stating:

> The gene has sometimes been described as a purely idealistic concept, divorced from real things, and again it has been denounced as a wishful thinking on the part of those too mechanically minded. And some critics go so far as to assert that there is not even such a thing as genetic material at all, as distinct from other constituents of living matter.
>
> However, a defensible case for the existence of separable genetic material might have been made out on *very general considerations alone.*
>
> Muller (1947, 1, italics added)

Starting in 1916, Muller published successive estimates both of the number of genes in the genome and of their dimensions. However, although repeated attempts were made to induce mutations by various agents, such as temperature, nutrition, ultra-violet radiation, even X-rays, there was no experimental method to systematically study mutagenicity. Mutations are by definition rare events; any intensive study of the nature of mutations needed a system that could yield not only qualitative differential effects but also quantitative differences. For such an analysis it is necessary either to use an agent that increases mutation frequency significantly above that of the "spontaneous" frequency and/or to devise a screening procedure that would effectively pick up rare events. It was the introduction of the *ClB* method that allowed the genetic analysis of rare events. High-energy penetrating X-rays provided an efficient mutagenic agent.

The efficiency of the *ClB* method for retrieving X-ray-induced lethal mutations in Drosophila was demonstrated in 1927 at the International Congress of Genetics in Berlin (Muller, 1927a). In the same year Lewis J. Stadler demonstrated the mutagenic effects of X-rays (and radium) by irradiating germinating barley seeds and maize ears at or shortly before fertilization and following the appearance of somatic mutations in only part of the progeny plant and of mosaic patches in the endosperm, respectively. Cytological examination showed that many of these were apparently chromosome deficiencies, non-disjunction, and chromosome eliminations, but some mutations, including chlorophyll defects, were apparently single gene mutations (Stadler, 1928a, 1928b). The *ClB* mass-screening methods, whose products could be further analyzed in the stocks established from such mutations, had obvious advantages over Stadler's methods of induced somatic mutations that were subject only to cytological analysis (see Chapter 9).

9

Analysis of the gene by mutations

Upon irradiation, the X-ray's energy absorbed in a tissue causes electrons to be expelled from atoms. These high-energy primary electrons eventually expel secondary electrons from atoms, leaving ionized molecules in their track through the tissue. It is the secondary ionization(s) at the end of the primarily induced electrons' tracks that are "biologically effective." Quantitative analysis of X-ray-induced recessive lethal mutations in the X-chromosome of *Drosophila melanogaster* revealed a reasonable linear increase with the dose of the radiation given to spermatozoa. In roughly 3 percent of the X-chromosomes lethal mutations were induced per 1000r (r = roentgen unit of absorbed radiation: 100r = 1 Gray unit). The rate of induced mutations was independent of the intensity of radiation (dose/time) and – when extrapolated – apparently with no threshold dose (above that calculated to be due to natural cosmic gamma-radiation). This indicated that X-ray-induced mutations were basically single-hit events. If a single hit was sufficient to induce a localizable lethal mutation, the target must be discrete and the mutation must have been a "point mutation" rather than a (minute double-hit) aberration. On the other hand, the kinetics of the induction of rearrangements, like translocations – obviously a multiple-hit event – was, as expected, of a higher order than the linear relationship. In reality, the aberration induction curves were only to the 3/2 power of the dose (up to a dose of circa 2000r). Several *ad hoc* interpretations were put forward to explain such discrepancies from the single ionization per break/mutation hypothesis (such as a small cluster of ionization rather than a single effective ionization at the target site). Further support for the hypothesis of single-hit events was obtained when neutron or alpha-particle radiation was applied to induce rearrangements. With such dense-track ionizations the dose–effect curve for aberration induction became fairly linear, as expected when numerous effective ionizations were produced per track.

Whether a single induced ionization event was necessary and also sufficient to break a chromosome or to induce a mutation became a central question at several levels. If true, this would refute Goldschmidt's claim that all point mutations are "chromosomal rearrangements with a break near a definite locus" (Goldschmidt, 1950, 337), since it takes two hits to obtain a rearrangement, whereas only one would suffice for a Muller-style "intramolecular" point mutation. Another crucial question was the level of the deleterious effects of radioactive radiation, especially in the late 1940s and the 1950s with the increase in the use of ionizing radiation in commerce and medicine, and especially in warfare (see Chapter 10). The extrapolation of the dose–effect curve for single-hit mutations all the way down to a single ionization, and the deduction from this that there is no threshold for the mutagenic effects of ionizing radiation made ionizing effects unique: many chemicals (used in industry or medicine) had well-defined threshold concentrations under which they were harmless. Muller, always the politically conscious fighter, also took a central position in the campaign against nuclear testing and, as a scholar of evolutionary theory, struggled to warn the scientific community as well as the public at large of the dysgenic effects of radiation. His paper "Our load of mutations" (Muller, 1950b) was influential among wide circles of scientists as well as policy makers.

THE TARGET THEORY

The "target theory" was developed to enable scientists to estimate the size of microscopic corpuscles from the dynamics of their inactivation by X-rays (Lea, 1962). The amount of induced radiation energy necessary to inactivate targets was calculated, and quite good estimates of the size of enzyme molecules and even of bacteria were obtained when these were compared to those obtained by conventional methods. Once genes were envisaged as molecule-like corpuscles, it was straightforward to extend the inactivation-energy target-size theory to genes. On the assumption that mutations were caused by single pairs of the small clusters of ions, Timofeéff-Ressovsky, Zimmer, and Delbrück estimated the dimensions of the genes, assuming that they comprised discrete, quantifiable molecule-like, approximately globular entities, like beads along the chromosomes (Timofeéff-Ressovsky, Zimmer, and Delbrück, 1935).

The integration of genetics with cytological research has shown that the gene, originally a simple symbolic representative for a Mendelian unit, could be localized in space and its movement followed. The refined analysis in Drosophila has led to [estimates] of gene sizes which are comparative to those of the largest known molecules . . . when we speak of [genes as] molecules we are . . . thinking . . . more generally of a well-defined union of atoms [than of chemically defined molecules].

Timofeéff-Ressovsky *et al.* (1935, 225)

This appeared to be the triumph of Muller's research project of determining the *physics* of the genes through the unique *phenomenology* of the genes' mutagenesis. This conception of the genes as the basic units of life, constructed into a tentative "atomic physical model" of the gene, induced Erwin Schrödinger to suggest in his booklet *What Is Life?* (Schrödinger, 1962 [1944]) that genes are "aperiodic crystals" of chromosomal structure, in which mutations were "quantum jumps" of states of matter from one stable state to another, as a result of thermal agitation or the absorption of radiation energy. The Timofeéff-Ressovsky, Zimmer, and Delbrück analysis had an important effect on the thinking of the non-biological community with regard to biological issues. It has sometimes been called the TZD experiment in analogy to the physicists' EPR crucial experiment of the same year, named after its authors, Einstein, Podolsky, and Rosen. However, although the physical approach of Timofeéff-Ressovsky and his colleagues, and that of Schrödinger that followed, had an important impact on genetic research in persuading physicists to direct their research programs to biological systems, the role of the "target theory" in elucidating the nature of the gene was short lived: in the words of Elof A. Carlson: "The Target Theory: A Successful Failure" (Carlson, 1966 [1989], 158).

One of the implicit assumptions of the theory was that the effective mutagenic ionization hit the genes proper. However, it turned out that it was in fact the conditions under which the radiation was applied that significantly affected the efficacy of the mutagenesis. It was implausible that the "gene-molecule" swelled and dwindled with changing circumstances. A particularly favorable material for the study of the dynamics of induced chromosome breaks were the large mitotic chromosomes of various plant species, like those of *Vicia faba*, or species of *Liliacae*, but also those of insects (especially grasshoppers), and the mouse. In particular the rate of oxygenation of the irradiated tissue was relevant: reduced oxygen pressure during irradiation protected the tissue against radiation effects, or in terms of the target theory, deoxygenation reduced the radius of the target (Thoday and Read, 1947). It was concluded that the target comprised a considerably larger diameter than the gene itself, and that not only is the ionization itself mutagenic but also highly reactive (short-lived) free radicals in the vicinity of the genes may induce mutations and breaks in nearby chromosomes (see, e.g., Bacq and Alexander, 1955, Chapters 3 and 8; Muller, 1954, 555–564).

Drosophila and mouse males were generally mated shortly after irradiation, thus effects induced in spermatozoa or late spermatids were the ones examined. The frequency of induced mutations recovered from premeiotic cells, when irradiated males were mated considerably later, was significantly lower. Many mutations, notably aberrations that caused chromosome entanglements at meiosis, were cell-lethals and eliminated before screening. Also, presumably because the premeiotic cells were metabolically active, failures in cell functions further eliminated many mutations. Still, experiments indicated that the mutagenic efficiency of given doses of X-rays varied according to the developmental and metabolic stages of the cells, independently of eliminative processes at or prior to meiosis. Thus, K. G. Lüning (1952a, 1952b) showed, by mating irradiated Drosophila males successively to different batches of females, that the efficiency of a given dose of radiation depended on the stage even in post-meiotic spermiogenesis in which, apparently, gene action was completely or nearly completely silenced: spermatozoa were more sensitive to radiation than early and late spermatids, and spermatocytes were most sensitive.

CHEMICAL MUTAGENESIS

The accomplishments of the quantitative analysis of the mutation-induction process by physical means encouraged Muller and other research workers to go one step further in the effort to "grind the gene in a mortar," and to extend the genetic analysis of mutagenesis to chemical mutagens. The properties of the chemicals that could induce mutations and the specificity of these mutations, it was hoped, would provide information on the chemistry of the genetic material in general and of the properties of specific genes in particular. However, contrary to the effective penetration of the high-voltage X-rays, any chemical applied to living organisms, or even to single cells, such as spermatozoa or pollen cells, must be potent enough to reach the cell nucleus and then also to interact specifically with the genetic material in the chromosomes, and not with other cell components (note that at that time proteins were believed to be the building blocks of the genetic materials).

This problem of the mutagen reaching the target cells had already been a major one with attempts to use other agents such as ultra-violet radiation (UV). Stadler and his colleagues succeeded in applying UV radiation to thin layers of seeds (Stadler and Sprague, 1937) or pollen grains, and Altenburg succeeded in inducing UV mutations in Drosophila sperm by squeezing the abdomen of the irradiated flies between two thin quartz plates (Altenburg, 1933). But although it was generally believed that UV and chemically induced mutations more faithfully represented natural, "spontaneous" mutagenesis than those induced by X-rays, the empirical problems were for many years a constant obstacle to their intensive application.

During the 1940s, in conjunction with the war effort, chemical mutagenesis was tried out in Germany by Friedrich Oehlkers, in the USSR by J. A. Rapoport, and in the United Kingdom by Charlotte Auerbach: the last two were disciples of Muller. All early chemical mutagens were quite powerful cell poisons. Auerbach and Robson's study of the mutagenic effect of mustard gas was the first one that became public (Auerbach and Robson, 1947) hoping that "the knowledge of the reagents capable of initiating this process [of mutagenesis] should throw light not only on the reaction itself, but also on the nature of the gene" (Auerbach, Robson, and Carr, 1947).

An early observation of the effect of chemical mutagens was that many mutations were recovered as "mosaics," i.e., when induced in Drosophila spermatozoa the mutations frequently detected in the progeny flies affected roughly only half of the body, the other half being non-mutant. This indicated that mutations were induced when the chromosomes were already duplicated (or at least double structures) and the chemical mutagen could affect one of these structures.

Eventually, chemicals believed to interact more specifically with nucleic acid, such as the alkylating agent ethyl-methanesulfonate (EMS), were examined for their mutagenic effects in plant pollen cells and in Drosophila spermatozoa. However, attempts to differentially correlate the mutation patterns with the chemical properties of mutagens – a strategy that yielded evidence on the nature of mutations when applied to bacteriophage mutagenesis (Freese, 1959) – failed. The hope of differentiating between chemically induced "point mutations" and breakage-related mutations also gave equivocal results: all chemical mutagens induced at least some major chromosome rearrangements (see, e.g., Lifschytz and Falk, 1969a, 1969b). Experiments with spores of Neurospora, where the application of the mutagen appeared to be less of a problem, yielded similar findings (Malling and de Serres, 1968).

In retrospect, Auerbach asserted that:

> [T]he hopes which my colleagues and I set on the new field of chemical mutation research were expressed as follows. 'If, as we assume, a mutation is a chemical process, then knowledge of the reagents capable of initiating this process should throw light ... on the nature of the gene. ...' The chemical nature of the gene has not been elucidated by research on mutation but in entirely different ways.
>
> Auerbach (1967; see also Falk, 1986, 152, footnote 72)

Notwithstanding, although the program to reveal physico-chemical properties of the gene did not fulfill Muller's expectation, his research program was crucial in the effort "to furnish a biological setting that should be of use in the coming *chemical* attack on the nature of the gene, on the mechanism of its self-duplication, its mutation, its behavior in meiosis and its action on the organism" (Muller, 1947). Direct genetic analysis, it was hoped, would be more informative in unraveling the functional organization of the genes.

VARIEGATED POSITION EFFECT

Goldschmidt's claim that what seemed to Muller to be "point mutations" in specific discrete genes was actually local rearrangement in the chromosome, or position effects, needed experimental analysis. This was offered by a special type of chromosome rearrangement that involved changes in the juxtaposition of euchromatic and heterochromatic chromosome segments. As it turned out, these rearrangements did not settle the issues and were interpreted by all involved according to their preconceived convictions.

Cytologically, heterochromatic chromosome segments were more densely packed than euchromatic segments, and in polytenic cells of Drosophila they were packed into one undifferentiated "chromocenter," instead of the highly specific pairing of "bands" in the euchromatic segments. Genetically, heterochromatin segments were extremely scarce in specific genes. Indeed, heterochromatin provided a completely different cytogenetic environment in the hierarchical organization of the cells. A "mottled" phenotype was characteristic for mutations at several genes that involved euchromatin/heterochromatin rearrangements. These were interpreted as variegated (V-type) position effects caused by fluctuations in the border heterochromatic condensation at the hetero-chromatin/euchromatin junctions that irregularly suppressed adjacent normal gene expression. This assumption was corroborated by the "spreading effect": the further away the (euchromatic) gene from the junction, the less extreme the variegated phenotype.

Evidence that the change was only in the position of the wild-type locus of the gene and not in its structure was shown by experiments such as those of Panshin with a white-mottled rearrangement (w^{M4}). White-mottled flies were irradiated and their progeny screened for both "reversions" to wild type and "extrimations" to (non-variegated) white phenotypes. Cytological examination revealed that all these were *re*-rearrangements: the intensification of the loss of eye color was pro-portional to the increase of heterochromatic segments next to the locus of w^+ (irrespective of the chromosomal source of the heterochromatin) and the intensity of reversion to wild type was proportional to the loss of heterochromatic segments next to the w^+ locus (Panshin, 1936, 1938).

That rearrangements of the organization of the chromosome *might* produce phenotypic effects similar to gene mutations did not refute Muller's notion of autonomous functional as well as structural discrete entities. The apparent specific phenotypic response of each gene to the spreading effect of the heterochromatin only strengthened his claim for discrete structures involved in specific functions. But the study of posi-tion effects led him to reconsider his hypothesis-driven interpretations in more instrumental terms.

In 1929 his Russian colleagues Serebrovsky and Dubinin had raised the possibility that the gene might be divisible into "subgenes" (Raffel and Muller, 1940, 541). Muller translated this suggestion into the question of whether in recovered rearrangement-mutants of a given gene, chromo-some breaks are restricted to "sensibly the same" position (Raffel and Muller, 1940, 569). His intensive cytogenetic studies of the distribution of breaks in recovered rearrangements affecting the *scute* gene (involved in the bristle production on and about the scutellum region of the fly's thorax) eventually convinced him that there are only a restricted number of breakage sites. This indicated the structural – and not merely the functional – discontinuity along the chromosome (Muller, 1956).

In view of the above considerations, it is not far fetched to imagine that the "gene for scute," as recognized by the test of allelism of its mutations, may nevertheless consist of an undetermined number of parts ... These parts, then, might themselves be denoted "genes," and the whole a "gene-complex," or the parts might be called "sub-genes" or something equivalent, and the whole a "gene," depending upon the taste of the writer and upon the criterion which he prefers ... for defining the limits of a "gene."

Raffel and Muller (1940, 570)

Although Muller somewhat relaxed his position, he insisted that evidence from proven cases of position effects was not evidence for events at the level of the gene: position effects are significant at one level but of no significance at a different level. To save his gene concept Muller admits:

Although we differ from Goldschmidt ... we must nevertheless agree in the more general proposition, central to [Goldschmidt's] theme and ours, that the "gene," in a rather loose sense in which it has so long been taken for granted by most geneticists, may *perhaps* be genetically further divisible, even into many genetically linearly arranged points of semi-autonomous character (autonomous in the sense of their being able to reproduce when in other linear arrangements), and that, contrariwise, neighboring "genes" often or usually co-act, in a manner made possible by their juxtaposition, so as to produce character-effects that depend upon overlapping regions, somewhat larger than a single "whole gene," as formally conceived.

Raffel and Muller (1940, 576)

The logic of Muller that exceptions do not prove the rule may not be very convincing, yet his arguments for genuine gene mutations, hence autonomous genes, and again, his more elegant reductionist experimental system versus that of Goldschmidt, appeared to convince geneticists. Furthermore, it was probably the suggestion of Beadle and Tatum of the one-to-one relationship between genes and enzymes, the "one gene – one enzyme" slogan, that strengthened for many the belief in the structural autonomy of genes.

THE GENE: AN OPERATIONAL VIEW

Like Muller, Stadler too believed that "genic substance" appeared to have properties unique to living matter, and thus, that "knowledge of the nature and properties of the genic substance might give clues to the distinctive physical mechanisms of life." He agreed that "the properties of the genes may be inferred only from the results of their action" and considering this method of reverse genetic induction, that it is "gene mutations [that] are essential for such comparisons" (Stadler, 1954, 811). Stadler, however, considered the Mullerian project of forward deductions that culminated in the "target theory" model of the physical properties of the gene to be "an impressive picture [that] has no valid relationship to the experimental data from which it was derived." Many of the X-ray-induced mutations may "result not from a structural change in a gene but from some alteration external to the gene, such as physical loss or rearrangement of a segment of the gene-string" (Stadler, 1954, 812). However, contrary to Goldschmidt's notion, Stadler's criticism of the gene concept was "from below." Discrete functional genes do exist; it is the inferences made about their structure that should be modified.

Stadler directed his attention to the mutagenic effects of UV radiation on maize pollen. Ultraviolet radiations are of much lower energy than X-rays and special experimental procedures were needed to allow their absorbance at or near the genetic material. Combining genetic and cytological efforts, Stadler and his colleagues found that the frequency of chromosome changes after UV radiation was not as high as that induced by an X-ray dose that was comparable in the production of viable "point mutations." Also cytological investigations by others, such as those utilizing pollen tubes of *Tradescantia*, have shown that UV radiation differs from X-rays in that the radiation produced only chromatid breaks and no isochromatid breaks or chromatid exchanges. The kinetics of dose activation curves of ultraviolet radiation in spores of various fungi (which are smaller than plant cells and therefore the calculations of the radiation energy absorbed in or at the chromosomes are more reliable) was that of one hit for the induction of mutations and of chromatid breaks (Muller, 1954, 531–542). UV-induced mutations simulate spontaneous events; put otherwise, induced mutations that might provide "criterions for identifying gene mutations of evolutionary significance" are not the ones produced in X-ray induction experiments. Whereas the former are mainly "point mutations" the latter are mostly involved with aberrations. "The analysis of induced mutations . . . indicated that the accepted definitions and criterions related to genes and gene mutations needed reconsideration" (Stadler, 1954, 813).

Once it became clear that "various extragenic alterations might produce the effects considered characteristic of typical gene mutation . . . [t]he working definition of mutation necessarily differs from the ideal definition" (Stadler, 1954, 813). Since there was no test to identify mutations due to a change within the gene, Stadler emphasized that "it was simply *inferred* that the mutants that could not be identified as the result of specific mechanical causes were, in fact, due to gene mutation in the ideal sense" (Stadler, 1954, 813, emphasis added). Stadler's concern was that *gene mutation* was used with two distinctly different meanings: "a change in the constitution of a unit of the genetic material, producing a new gene with altered gene action" and "the occurrence of a mutant character inherited as if it were due to a change in a gene" (Stadler, 1954, 813). Therefore, Stadler suggested adherence instead to Percy W. Bridgman's operationalism: "an object or phenomenon under experimental investigation cannot usefully be defined in terms of assumed properties beyond experimental determination but rather must be defined in terms of the actual operations that may be applied in dealing with it" (Stadler, 1954, 813).

> The significant ambiguity is not in our definition of gene mutation, but in our definition of the gene itself . . . Operationally, the gene can be defined only as the smallest segment of the gene-string that can be shown to be consistently associated with the occurrence of a specific genetic effect . . . The term *gene* as used in current genetic literature means sometimes the operational gene and sometimes the hypothetical gene, and sometimes, it must be confessed, a curious conglomeration of the two.
>
> Stadler (1954, 813–814)

Operationally we overshadow relevant effects when we average out classes of mutations. Furthermore, Stadler warned that we inevitably study mutations that are relatively easy to detect, and consequently "we must confine ourselves to mutations of relatively large effects." Thus it is quite possible that "the sharply distinct mutations identified in our experiments may be exclusively the result of extragenic phenomena" (Stadler, 1954, 816), as suggested by Goldschmidt. A deficiency cannot automatically be identified with lethality. Observations showed the existence of "cytologically demonstrable deficiencies viable in haploid tissue or in hemizygous individuals, or viable in homozygotes in diploid individuals." Also, the assumption that genes, as a rule, are individually essential to life lost its plausibility.

Another assumption that had been made consciously, and had to be questioned was that "multiple alleles are variant forms of a single unit" and that accordingly mutations in these genes could not be due to loss. The empiric fact that we cannot (yet) separate the functional gene by linkage experiments is no proof against this possibility: "On the hypothesis that they represent different mutations in a complex of closely linked genes, we

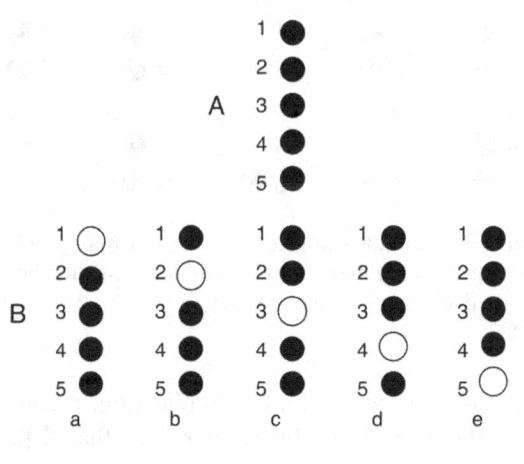

Figure 9.1. A. Scheme of a multi-site gene or a nest of genes so closely linked that no crossing over can be detected between them. B. A mutation in each site of the gene/nest of genes (Stadler, 1954).

could account for mutation to different levels by loss of different segments of the chain" (Stadler, 1954, 816). Stadler stressed that the test for allelism, *per definition*, "rules out the existence of the nest of closely linked genes only on the assumption that each mutation must be an alteration of a single number of the group" (Stadler, 1954, 816). He suggested that operationally there is no reason to exclude a bottom-up gene concept as a structurally multi-unit entity (see Figure 9.1).

In diametric contradiction to Muller, who – in order to avoid peculiarities that might be due to specific properties of particular genes (among other reasons) – directed his efforts to the study of a *class* of genes that mutate, Stadler arrived at "the unpleasant fact that significant progress in our understanding of gene mutation requires the investigation of the mutation of specific genes . . . because, as we have seen, it is hopeless to identify and exclude the spurious or extragenic mutations in experiments on mutation rates at miscellaneous unspecified loci." Notwithstanding, there is also some advantage in this. "The chief advantage in focusing the study on the single gene is that this makes it possible to substitute the direct experimental analysis of specific mutants for the application of generalizations assumed to apply to mutations at all loci" (Stadler, 1954, 814–815). He chose to study the *R* and *A* genes in maize that affect endosperm characters. Both have many known spontaneous mutations to alleles that are apparently of such trivial effect physiologically that their mutants survive with no detectable loss of viability.

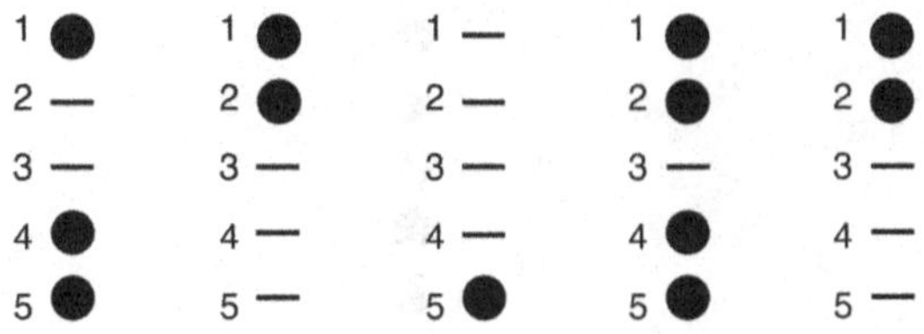

1 ●	1 ●	1 —	1 ●	1 ●
2 —	2 ●	2 —	2 ●	2 ●
3 —	3 —	3 —	3 —	3 —
4 ●	4 —	4 —	4 ●	4 —
5 ●	5 —	5 ●	5 ●	5 —

Figure 9.2. Scheme of a multi-site gene, or a nest of tightly linked genes. Sites may be deleted, which would make the operational distinction between one gene or a nest of genes irrelevant (Stadler, 1954).

For Muller the gene was a unit of replication, recombination, and mutation as well as of function. But Stadler notes that "[i]f, instead, each mutation were a loss of one or more contiguous numbers of the group, the fact that crosses between them might commonly show them to be allelic would not rule out the 'compound gene' as the basis of the multiple allelic series" (Stadler, 1954, 816) (see Figure 9.2).

He agreed that "the 'compound gene' is in a sense a contradiction in terms, for the hypothetical gene is unitary by definition. But the gene identified in our experiments cannot be made unitary by definition" (Stadler, 1954, 816–817). In other words, contrary to the hypothesis-driven conceptions of the gene, operationally driven considerations of experimental results "clearly show that expression effects may be the actual cause of apparent gene mutations, even when the mutation observed shows no indication of a change of position or of any associated chromosomal aberration" (Stadler, 1954, 818). The gene may be a discrete unit, occupying a stretch of the chromosome that functions as one entity. Operationally, the structure may define the function, but the function identifies the structure.

THE COMPOUND GENE: PSEUDOALLELISM CHALLENGES
THE ATOMIC GENE CONCEPT

The dialectic of Stadler (1954) had a profound impact on the logic of genetic thinking and experimentation. The notion of a one-to-one relationship between genes and enzymes, the "one gene – one enzyme" conception (Beadle and Tatum, 1941a, 1941b), was, strictly speaking, an operational reference to genes as functional entities along the chromosomes. So were notions of population genetics that operated with

152

mathematical models of genes and their alleles, irrespective of what the genes might constitute (see Chapter 10).

Toward the 1950s, with the improvement of the resolving power of genetic recombination experiments, the unitary image of the genes started to falter. Even before then, Altenburg and Muller (1920) had already called attention to *Truncate*, a complex trait of Drosophila. Of the three phenotypes affecting wing pattern (later denoted *o* for *oblique*), thoracic bristle pattern (*v* for *vortex*), and the recessive lethal (*l* for *lethal*), different combinations of *o*, *v*, and *l* identified various allele-combinations of this "complex locus" (e.g., $dp^{ov}\backslash dp^{vl}$ were viable, no-oblique, vortex flies). Its complex nature, exhibiting combinations of three presumably independent phenotypes, was not resolved at that time (Carlson, 1959; and Chapter 8). Edward B. Lewis studied various "nests" of closely linked but recombinable loci of similar functions in Drosophila, such as *Stubble-stubbloid*, *Star-asteroid* or *bithorax-Ultabithorax-bithoraxoid*. Since the loci recombined they comprised different genes (or different centers of gravity along a continuous chromosome). Lewis conceived of the possibility that they were due to ancient duplication of one gene (probably due to unequal crossing over) followed by the functional diversification of the couple with evolutionary time. Lewis tried to extend this notion when recombinants were detected in apparently non-complex loci such as the Drosophila *white* and *lozenge* eye-color genes. By definition, mutations were referred to the same gene if they did not complement each other in heterozygotes for both mutations. But, when recombinants were detected between two mutants that did not complement each other in compound *trans* a_1/a_2 heterozygotes, formally they should be considered to involve two genes, presumably an $a_1 +/ + a_2$ genotype. The phenotype of the recombinant in which the two mutations were now on the same homolog chromosome (in *cis*: $a_1\ a_2\ /\ ++$) was, as expected, wild type. Lewis called such mutants "pseudoalleles," and to maintain the unitary gene hypothesis, he suggested that pseudoalleles belonged to distinct adjacent genes, one producing the substrate for the function of the other. He suggested that the products of these genes that were induced in close proximity to their respective genes could diffuse only as far as the adjacent gene on the *same* chromosome, but not as far as the gene on the *homologous* chromosome. This explained how the *cis* constellation, but not the *trans* constellation, would result in complementarity, while preserving the notion of discrete and distinct genes (Lewis, 1951). True to his notion of genes, Muller pointed out, such an interpretation demanded a change in gene nomenclature: pseudoalleles should be denoted as being of *different* genes. Thus, when the apricot allele of white eyes, w^a, turned out to recombine with other alleles, he suggested that it be denoted *apr*.

Similar studies in other "classical" Drosophila genes (Chovnick, 1961; Green, 1963) suggested that they too were divisible into two (or three) distinct subunits, rather than being single genes with mutation-sites distributed all over. This notion was finally overtaken by Arthur Chovnick *et al.*'s study of the *rosy* gene of Drosophila, in the footsteps of Benzer's genetic analysis of the *r*II locus of the T4 bacteriophage of *Escherichia coli* (see Chapter 11). In these studies of the gene *rosy*, the mutants of which turned out to be xanthine-dehydrogenase-less, selective media were developed on which only the rare recombinants (and some other discernible rare events) survived. Eventually a gene of Drosophila could be mapped as a continuous linear entity with recombinable mutation sites all along the gene (Chovnick, 1989; Chovnick, Schalet, Kernaghan, and Kraus, 1964).

Although in the 1950s, the scale of recombination experiments with Drosophila was upgraded considerably – achieving resolution between mutants of 0.01 cM apart and less – the introduction of experimental organisms, in which more efficient screening for relevant recombinant progeny were attainable, was imperative. Guido Pontecorvo impressively increased the resolving power of recombination experiments to about 10^{-5}, several orders above that obtained for the best Drosophila experiments by following mutations of the mold *Aspergillus nidulans*. He hybridized independently obtained non-complementing auxotrophic mutations that were by definition of the same gene, and screened for rare recombination by plating the progeny spores on a minimal medium on which the auxotrophic spores could not grow. Inter-allelic (or, inter-pseudoallelic) recombinants could be discerned from (about equally rare) back-mutations by checking for recombination of outside markers – those to the left and right of the experimental target mutants. It soon became clear that auxotrophic mutations in given genes could be mapped into consistent linear sequences (Pontecorvo, 1958). The genes turned out to be compound, linear structures, more like "molecules" than "atoms" of inheritance.

Yet Pontecorvo's assertion that resolving chromosomes into their "linear array of genes" by crossing-over experiments could be pushed "down to interatomic distances" (Pontecorvo, 1952, 129) took on a significance quite different from what its author had intended. Pontecorvo had published his review just before the appearance of the Watson–Crick double helix. By the time Seymour Benzer conceived of the screening system for selecting rare events of recombinants in the "hybridization" of rII mutants of T4 bacteriophages – viruses that grow on ("eat") bacteria – increasing the resolving power of recombination experiments up to 10^{-8} and more – the new model of DNA had already altered the discourse around the question of whether genes were megamolecules (Holmes, 2000 and see also Chapter 11).

This was the triumph of phenomenological heuristics which systematically brought the analysis of the Mendelian factor, or the gene, down to the physico-chemical molecular level. Muller accepted that "modern evidence supports the conception of a linear arrangement of the genetic material all the way down to its ultimate components, the nucleotides," and "we arrive at a conclusion in harmony with the model that has been presented to us by Watson and Crick" (Muller, 1956).

Pontecorvo, Muller's previous student, believed that the traditional picture of a gene as a "sharply delimited portion of the chromosome – the 'corpuscular gene'" had lost its heuristic value (Holmes, 2000, 119). Muller, however, continued to argue "the case for a segmented chromosome" of structurally discrete and distinct genes, on both experimental – most chromosome breaks are between genes – and theoretical grounds:

> Our best evidence for intragenic mutation . . . consists of the evidence in all its great sweep and diversity for the theory of evolution itself, taken in conjunction with the evidence for the existence of individual genes. So long as we retain both concepts we must admit that at the bottom evolution has been built up out of intragenic mutations, together with some super-structure of intergenic structural changes.
>
> Muller (1956, 134)

FROM PHENOMENOLOGICAL TO MOLECULAR ANALYSIS

Petter Portin in his retrospective review defined the extension of Muller's "classic concept of the gene," the mapping of the fine structure of the genes first in Drosophila then in *Aspergillus* and finally in the bacteriophage T4, as the "neoclassical concept of the gene" (Portin, 1993). This was adopted as a statement of conviction by many researchers who entered genetic research at the molecular level and conceived of the notion of the phenomenological research tradition as an outdated burden of past times. For them it was the beginning of a new era, that of molecular genetics. As Holmes pointed out, some modern textbooks do not even mention Benzer's analysis: "the recapitulation of Benzer's analysis of the *r*II region is no longer mandatory to the presentation of molecular genetics" (Holmes, 2000, 116). My claim is that the new era of molecular genetics adopted the conceptions as well as the heuristics of phenomenological genetics, and successfully extended them.

Detecting recombination frequencies as rare as 10^{-8} wild-type recombinants, convinced Benzer and other geneticists that it was "technically possible to study, *by genetic means*, the structure of the ... 'gene' down to the individual nucleotide" (Holmes, 2000, 126, my emphasis). The analysis of the gene "down to the ground" could now be conceived as an extension of phenomenological classic linkage analysis to what appeared to correspond to a DNA sequence. Unfortunately, the protein coded by the *r*II gene of the T4 bacteriophage was not known for many years. But phenomenological high-resolution analyses were extended to genes of bacteria, such as the histidine-metabolism genes of *Salmonella typhimurium* (Hartman, Loper, and Serman, 1960) or that of the tryptophan-synthetase enzyme of *E. coli* (Yanofsky and Lennox, 1959), which allowed direct inference from the linkage map of the presumed DNA sequence to the sequence of amino acids in the protein product and the establishment of the concept of the colinearity of the DNA's genetic code and the primary polypeptide product (Yanofsky, Carlton, Guest, Helinski, and Henning, 1964; Yanofsky, Drapeau, Guest, and Carlton, 1967). And once *in-vitro* DNA reassociation methods had been worked out (Hoyer, McCarthy, and Bolton, 1964), linkage became a major component of molecular genetic analysis (see Chapter 17). For much of modern genetic analysis, SNPs (single nucleotide polymorphisms) replace the wrinkled peas and the white-eyed flies as markers.

But the dialectic between the instrumental and the material conception of the gene was maintained. The betting competitions of practicing molecular scientists on the number of human genes on the one hand (see, e.g., Oliver and Leblanc, 2004), and the plethora of structure-functions that are (or are not) designated as genes (such as overlapping genes, antisense overlapping genes, antisense trans-splicing genes, and many more) on the other hand, are witness to the entrenchment of the gene concept. Clearly there is no one DNA sequence that *on its own*, independent of the context of the product's function, can be unequivocally defined as a gene (Falk, 2000b, 2004; Griffiths and Stotz, 2006).

Philosophers of science have attempted to establish the continuity of genetic theory by formally reducing Mendelian genetics to molecular genetics, but failed (Rosenberg, 1985; Schaffner, 1976; see also Sarkar, 1998). However, as pointed out by Ken Schaffner, the models discussed represented an ideal standard for accomplished reductions, and do not characterize the research programs of molecular biologists (Schaffner, 1976). Furthermore, are not the DNA sequences the "most basic of all phenotypes" from which "we read off the genotype" (Griffiths *et al.*, 1999, 576)?

To quote Sahotra Sarkar:

> ... an analysis, falling back on developments in the analysis of whole genomes and of "forward genetics," which purports to predict the function of sequences directly from their sequence, underestimates the fact that throughout the billions of years of organismic evolution, no novel *structurally discrete* DNA entities evolved. It has been a dialectical, philosophically loose discourse, which allowed functions to instrumentally parse DNA and refer to sequences as genes.
>
> Falk and Sarkar (2006, 339)

10

From evolution to population genetics

Mendelian nomenclature of symbolic representation of genes and their alleles makes it readily amenable to algebraic presentation of population events. Mendel himself was the first to take notice of this when he explained Gärtner's and Kölreuter's observations "that hybrids have a tendency to revert to parental forms" in algebraic terms:

> If one assumes, on the average, equal fertility for all plants in all genera-tions, and if one considers, furthermore, that half of the seeds that each hybrid produces yield hybrids again while in the other half the two traits become constant in equal proportions, then the numerical relationships for the progeny in each generation follow from the tabulation . . .
>
> Mendel, in Stern and Sherwood (1966, 16)

In the last line of the tabulation Mendel generalized the results for the nth generation of selfing that follows mono-hybridization: the ratio of "$A:Aa:a$" would be $2^n-1: 2: 2^n-1$.

Bateson adopted Mendel's hypothesis upon reading de Vries's paper because he believed that it was consistent with his notion of evolution by discontinuous steps. This placed him in opposition to the claim of the mathematical statistician Karl Pearson in favor of Galton's "Law of Ancestral Heredity" of continuous Darwinian evolution (Provine, 1971; and see Chapter 3). In a discussion following a talk by Bateson's colleague Punnett, it was suggested that "a dominant allele, once introduced into a population, would increase in frequency until reaching stability at 0.5, given the usual phenotypic ratio 3 dominant: 1 recessive thereafter" (Provine, 1971, 133–134). This did not make sense to Punnett, who took the problem to the mathematician Hardy. The legend is that Hardy (who bragged "I have never done anything 'useful'," Hardy, 1940, 150) at once gave him the solution to the problem on the back of a cigarette pack: given that the frequencies of two alleles A and a of a gene are p and q respectively ($p+q=1$), the frequency of the genotypes AA, Aa, and aa will be p^2, $2pq$ and q^2, respectively, within one generation provided that matings in the (infinitely large) population are at random and no selection or migration takes place (Hardy, 1908). Punnett called it the Hardy Law. However, when it turned out that Wilhelm Weinberg came to the same result in the same year, it became known as the Hardy–Weinberg Law (Stern, 1943) although a third geneticist, William Castle, also formulated the law at the same time.

Of course, no real population fulfilled all the conditions for the Hardy–Weinberg formula. Yet, for many rare monogeneic human diseases such as Tay–Sachs syndrome, phenylketonuria, etc. or for genes of slight adaptive differences between the phenotypes such as of the M-N blood groups, and even that of the ABO blood groups, the Hardy–Weinberg Law provided good enough approximations to estimate populations' allele frequencies. More importantly, the Hardy–Weinberg Law was a scaffold generalization that served well for examining the effects of each constraint separately on the idealized relationships in a population, and then in combination provided an efficient analytic instrument for examining the genetic dynamics of populations in the light of Darwinian insights.

R. A. Fisher's paper of 1918, on "the correlation between relatives on the supposition of Mendelian inheritance" (Fisher, 1918), was a successful theory that put forth the relationship between biometry, Darwinian evolution, and Mendelian inheritance (Tabery, 2004, 82). Fisher provided a conceptual reduction of the biometric law of ancestral inheritance to the Mendelian principles of inheritance: the statistical law was just a special case of the physiological law (Sarkar, 1998, 106). This analysis offered the framework for the exploration of the mathematical consequences of the introduction of Fisher's s, the selection coefficient, of Haldane's m, the mutation rate, and Wright's F, the inbreeding coefficient; and, eventually, the involvement of more than one variable gene and the interactions of variables, on Mendelian heredity in populations. Together with the field observations of Chetverikov, Timofeéff-Ressovsky, Dobzhansky and others this culminated in the formulation of the New Synthesis of the 1940s (Huxley, 1943). The theory of Darwinian evolution was reduced to that of changes in gene (allele) frequencies (see Falk, 2006).

The validity of the theoretical models was examined through empirical studies. One of the most significant conclusions of these studies was that there existed a considerable amount of concealed genetic variability in natural populations. This was also revealed by direct observations such as effects of inbreeding in plants as well as in animals and in human populations. However, the choice of materials and the design of the experiments as well as their interpretation were heavily biased by the schools of genetic theory to which the authors belonged. This was largely a consequence of the "epistemological paradox," namely the contradiction between what we wish to measure and what we can measure: "In contradistinction to the discreteness of genotypic classes demanded by Mendelian analysis lies the quasi-continuous nature of the phenotypic differences that are the stuff of evolutionary change." The paradox, "What we can measure is by definition uninteresting and what we are interested in is by definition unmeasurable," has plagued population genetics from its foundations (Lewontin, 1974, 20–23).

Thus, whereas the school of E. B. Ford, following Fisher, examined the effects of particular ecological niches on the adaptive value of specific traits with rather distinct individual variation, assuming infinite populations (Ford, 1964), Sewall Wright emphasized correlation pathways ("path-coefficients") that considered the interactions of many genes and of multiple environmental conditions on the continuous adaptive struggle for the peak of limited, even small breeding groups (Wright, 1958).

As a rule, research was directed at the sources and causes of genetic polymorphisms in natural populations. Clearly much of this was "balanced-polymorphism," maintained by mutations, selection, mating-system, and migration. Of special significance were studies by Ford and his colleagues on the phenomenon of transient polymorphism, such as industrial melanism: the shift in the pigmentation of species, such as moths, over the century of the industrial revolution from multi-colored (often "peppered") patterns to increasingly dark to black patterns, in direct correlation to the extent of the soot in the environment. Field observations and designed experiments (Kettlewell, 1955, 1956) with the moth *Biston betularia*, which is grayish white sprinkled with black dots, and its two melanotic variants, *Carbonaria* and *Insularia*, convinced Ford and his associates that birds act directly as selecting agents on the moths that rest over the daylight hours on the trunks of trees that became increasingly dark with the advance of industrialization. They also followed the changes in the banding patterns of the shells of the snail *Cepea nemoralis* with the seasonal changes in the landscape, by inspecting the site of stones where Song Thrushes break the shells ("thrush anvils"). Of special importance were also the studies of Clarke and Sheppard of the evolution of mimicry of *Papilio dardanus* in Africa:

The results have demonstrated incontestably the truth of the explanation put forward by Fisher in 1927 which involves the selective adjustment of the mimetic forms, operating within the scope of one or more major-genes which may act as switches in controlling them when multiple. The polymorphism so produced exhibits the tendency for the creation of super-genes and for the establishment of heterozygous advantage.

Ford (1964, 246)

Across the Atlantic Theodosius Dobzhansky published in 1937 his influential *Genetics and the Origin of Species* (Dobzhansky, 1937). His attention was directed especially at the concealed genetic variability in populations of Drosophila species detected by cytological studies of their polytenic chromosomes: laboratory hybridizations of flies from different localities allowed the detection of many aberrations, primarily paracentric inversions. The most intensive studies were those of the polymorphism of inversions of chromosome III in populations of *Drosophila pseudoobscura*. By comparing partially overlapping inversions it was possible to follow the phylogenetic chart of these "chromosomal races" and their segregation into two new incipient species, *D. pseudoobscura* and *D. persimilis*. This method was eventually applied to the study of the evolution of many other Drosophila species, such as those of *D. willistoni* in South America and several species on the Hawaiian archipelago. Dobzhansky's series of forty-three papers on "Genetics of Natural Populations" (Lewontin *et al.*, 1981) deals with the empirical evidence of natural and laboratory populations of Drosophilidae and other species. Hampton Carson's studies of the planitibia group of Hawaiian Drosophilids reveal considerable convergence between the phylogeny based on the cytological pattern and that based on the electrophoretic polymorphism of ten enzymes and indicate the paths that the flies followed from the continents and then from one island to the other in correlation to the age of the islands over the last five million years (see Raff and Kaufman, 1983, 93ff.). What maintained that much genetic variability in natural populations?

The phenomenon of heterosis or "hybrid vigor" was developed by breeders into an effective industrialized agricultural technique for many crops, but especially corn. Breeding companies selected highly inbred lines that were inferior, but when cross-bred yielded high uniform crops. Attempts to further breed from these hybrids resulted in a fast decline of the crop, typical of "inbreeding effects" (Gowen, 1952). Once it was established that quantitative traits (like size of the corn plants and their cobs) were explainable by the action of many genes (East, 1910), two

161

theoretical explanations were suggested (see Muller and Falk, 1961; Paul, 1992): heterosis was the consequence of different inbred lines being homozygous for deleterious or subvital recessive alleles in many but mostly different genes. The F_1 heterozygotes were heterotic because in (most) genes there was at least one dominant allele that "covered" the specific deleterious effect of the recessive allele (Bruce, 1910; Keeble and Pellew, 1910). This reductionist genetic interpretation was juxtaposed with that of East and Shull, who claimed that it was the specific inter-action of alleles at various loci that was superior to that of both alleles in homozygotes: according to this model heterozygosity *per se* was adaptive (East, 1911). As was pointed out by Lewontin and Berlan, these hypotheses also had strong socio-economic consequences: the hetero-zygosity *per se* explanation meant that there is no way to cultivate a strain of corn that will keep the properties of vigor from one generation to the next, making cultivators immanently dependent on yearly fresh supplies of seeds from the breeder companies. On the alternative, dominant model of heterosis, there was a point in striving to breed an "ever-heterotic" strain (Lewontin and Berlan, 1986, 1990).

This dispute gained new vitality in the 1950s when the deleterious effects of radiation, at the time primarily from nuclear experimentation fall-out but also from lax control of industrial and medical radiation absorption by humans, were considered. In "Our load of mutations" Muller (1950b) claimed that new mutations are usually deleterious compared with their wild-type (that had been previously selected) allele. Furthermore, on the average, lethals are not completely reces-sive; the viability of heterozygotes may be only 1–2 percent lower than that of the homozygote for the lethal allele, but since the frequency of the heterozygotes is much higher than that of the lethal homozygotes their effect on the viability of the *population* is considerably higher than that of the homozygotes (whose effect on the *individual* is no progeny).[1]

Considering that besides deleterious mutations that are lethal in homozygotes there are even more mutations whose effect in homozygotes is subvital, the net result is that the adaptive value of natural populations

---

[1] Consider a lethal allele of a given gene being 1/1000 as frequent as the normal allele of that gene. According to the Hardy–Weinberg Law the ratio of heterozygotes and lethal homozygotes would be 2000:1. A 1% disadvantage of the individuals heterozygous for the lethal allele (compared to 100% lethality of the homozygote individuals) would impute on the *population* a disadvantage 20 times higher than that of the death of homozygous *individuals*.

is considerably below what it could be if there were no mutations at all. Mutations are, however, the price that populations had to pay for long-range Darwinian survival: only populations that carry a concealed genetic variability may be able to respond adaptively to changing conditions. Muller called this the "mutation load," and calculated that each human being carries on the average four hidden lethal equivalents: "We say 'lethal equivalents' because one cannot distinguish between 4 full lethals and 8 genes with 50% probability of causing death, or any system where the product of the number of genes and the average effect of each is 4" (Crow and Kimura, 1970, 77). Haldane came to similar conclusions on the long-range advantage of keeping a load of mutations that might provide the necessary adaptive combination in future constellations. Considering the time required for changing the gene frequency, he calculated the "replacement load": the number of deaths that a population would suffer for the replacement of a new mutation for a prevailing allele (Haldane, 1957; see also Crow and Kimura, 1970, 244ff.). Note that all these considerations reduce the problem to the alleles of single genes, with no reference to genic interactions.

Dobzhansky, on the other hand, saw in the prevalence of inversions and their evolution evidence for the role of coadaptation of genes along the chromosomes as well as between alleles of given genes, and supported the notion that heterozygosity *per se* was advantageous. The adaptive advantage of a heterozygote over both homozygotes imposes a load on a population, by segregating each generation of homozygotes that are adaptively inferior to the heterozygote; the heterozygosity will, however, be maintained in a population if the advantage of the heterozygote is greater than the "segregation load" incurred by the homozygotes (see Crow and Kimura, 1970, 303–308). This notion was supported by Michael I. Lerner (1954), who followed breeding in chickens that proved a diminishing-returns curve with respect to crop (body weight, egg production) and an increment of sterility with increasing inbreeding coefficients: "a degree of heterozygosity above a certain minimum level may be requisite for high fitness in normally outbreeding species" (Dobzhansky, 1970, 180). This conception of the advantage of genomic heterozygosity *per se* was considered by Muller to be a return to "mysticism." In accord with the New Synthesis the evolutionary forces that shaped populations were reduced to those of autonomous determinant genes, and no notion of the genes' functions being determined or constrained by the cellular or organismic system was envisaged.

To resolve the conflict between these genic notions of the forces that maintain the variability of natural populations, experimental evidence on the effect of newly induced mutations was needed: Dobzhansky's coworker Bruce Wallace induced a low rate of mutations by irradiating *Drosophila melanogaster* males with 500r of X-rays and then checked the viability of heterozygotes for irradiated 2nd chromosomes by a modification of the *Cy/Pm* method (see Figure 8.2). Wallace found a slight improvement in viability of the heterozygous flies for the newly low-dose induced mutations (Wallace, 1958), supporting the hetero-zygosity *per se* hypothesis. This was in contrast to the finding of Falk, who worked with Muller: they irradiated the chromosomes with very high doses but collected only chromosomes from cells that were at the spermatogonia stage when irradiated, in an attempt to eliminate aberrations from their sample. Besides lethals, many subvitals were also induced in many chromosomes. Each of the irradiated chromosomes was combined with an (unirradiated) marker chromosome and the viability of these heterozygotes was compared with that of the het-erozygotes for the corresponding unirradiated chromosome and the marker chromosome. The viability of the heterozygotes for the irra-diated chromosomes was lower than that of the unirradiated controls (Falk, 1961). Further experiments confronted for 7–12 months each of almost 200 heterozygotes for an irradiated chromosome with het-erozygotes for a corresponding unirradiated chromosome in small population cages (the design of the experiment avoided other selection-competitors in the cages). The frequency of heterozygotes for lethal chromosomes was reduced on the average by 7 percent per generation, and that for subvitals by 2 percent. In the control cages the competing chromosomes maintained their relative frequencies throughout the experiment (Falk, 1967). Direct evidence for the partial dominance of recessive lethals in natural populations of Drosophila was also obtained by Crow and Temin (1964). However, conflicting experimental evi-dence for what Dobzhansky called the "balanced theory" and what was denoted the "classic theory" could not settle the conceptual differences (see, e.g., Muller and Falk, 1961; and Dobzhansky, 1970). One problem was that whereas screening was at the chromosomal level, the inter-pretation was at the genic level. While Wallace estimated that 50 per-cent or more of the loci of a given individual in an open, natural population may be occupied by dissimilar alleles, Crow stated that on theoretical considerations it was "highly improbable that more than a minute fraction of the inbreeding effect [in finite populations] is due to alleles maintained by balanced selective forces" (see Hubby and Lewontin, 1966, 577–578).

Richard Lewontin noted that with the exception of a few major single-locus polymorphisms, the attempt to measure the difference in fitness between homozygote and heterozygote at a single locus is frustrating:

> First, most single-locus fitness effects must be very small and, second, it is impossible to know whether the results of the two crosses differ only at the locus in question or in a whole block of genes surrounding the locus ... Genes five units on either side of a marker will retain 50% of their correlation with the marker for 14 generations of backcrossing even without selection. Thus, the attempt to measure single-locus fitness effects for most loci is doomed to failure until methods of analysis several orders of magnitude more sensitive than we now have are created.
>
> Franklin and Lewontin (1970, 731)

The problem was that of detecting "so called isoalleles ... of such small effect that they cannot be detected by simple Mendelism" (Hubby and Lewontin, 1966, 579). The solution lay in extending phenomenological genetics to the molecular level (see Lewontin, 1991, 227). "It appeared certain that the sequence of the nucleotides that make up the structural gene is translated with a high degree of accuracy into a sequence of amino acids making up a polypeptide chain" (Lewontin, 1974, 99):

> Any mutation in a structural gene should result in the substitution, deletion, or addition of at least one amino acid in the polypeptide produced by the gene. Such a substitution, deletion, or addition will, in some fraction of cases, result in a change in the net electrostatic charge on the polypeptide. ...we can expect that electrophoretic differences in enzyme proteins will segregate as single Mendelizing genes.
>
> Hubby and Lewontin (1966, 579)

As it turned out, all the electrophoretic differences were single Mendelizing entities, a fact that freed the method from any *a priori* assumptions about gene action. Still no electrophoretic variation of a protein was not an indication that there was no variability at the level of the gene. From their first study, of eighteen randomly chosen loci of *Drosophila pseudoobscura*, Lewontin and Hubby concluded that the average population is polymorphic for 30 percent of all loci. The proportion of all loci in an individual's genome that are in a heterozygous state was estimated to average at 11.5 percent. An independent study of electrophoretic isoalleles of enzyme polymorphism in humans gave essentially similar results (30 percent and 9.9 percent, respectively)

165

(Harris, 1966). The technique could be applied to any organism, whether or not the organism could be genetically manipulated. In an increasing number of species it was shown that "[a] typical species population for most organisms is polymorphic for about 1/3 of its loci that code for enzymes and other soluble proteins, and an average individual is heterozygous (or, for haploid organisms, has a probability of nonidentity with another individual) at about 10%" (Lewontin, 1991, 658–659).

The apparent ubiquity of polymorphic loci in a variety of organisms implies that many pairs of polymorphic loci must be closely linked (Franklin and Lewontin, 1970, 708). This means that the frequency of an allele at a locus "may be overwhelmingly determined by the average effects of other loci forming a linked complex with the [locus] in question" (see "linkage disequilibrium" in Chapter 20).

> If we take it as given that balancing selection is rare and that natural selection is nearly always directional and "purifying," how can we explain the observed polymorphism for electrophoretic variants at so many loci? We can do so by claiming that the variation is only apparent and not real. That is, we can suppose that the substitution of a single amino acid, although detectable in an electrophoresis apparatus, is in most cases not detectable by the organism. . . . They are "genetic junk," revealed by the superior technology of the laboratory but redundant physiologically. From the standpoint of natural selection they are *neutral mutations*.
>
> Lewontin (1974, 197)

The New Synthesis *sensu stricto* – by now a dogma – could not be maintained.

This did not mean that "nearly all mutations are neutral or that evolution proceeds without natural selection . . . On the contrary, the claim is that many mutations are subject to natural selection, but these are almost exclusively deleterious and are removed from the population." But there exists a second class of redundant or neutral mutations, "and it is these that will be found segregating when refined physiochemical techniques are employed" (Lewontin, 1974, 197–198). Of course, rare favorable mutations do occur, and they will be fixed by natural selection (and also occasional heterotic mutants might arise). "Thus the so-called neutral mutation theory is, in reality, the classical Darwin–Muller hypothesis about population structure and evolution, brought up-to-date" (Lewontin, 1974, 198). In 1968 Motoo Kimura explicitly directed the discussion to "the molecular level," by noting that "calculating the rate of evolution in terms of nucleotide substitutions seems to give a value so high that many of the mutations involved must be neutral ones" (Kimura, 1968).

The role of random process in the fate of natural populations had been emphasized by Wright as early as the 1920s: in finite small populations random fluctuations in gene frequencies may overrule their adaptive value, and when occasionally the fluctuation reaches an extreme value (0 or 1) such "random drift" may fix an allele in a population, independently of its adaptive value (Wright, 1931). Kimura (1968) and King and Jukes (1969) suggested, however, that much of the concealed genetic variability in natural populations was selectively neutral and accordingly a great deal of evolution in terms of changing gene-frequencies was "non Darwinian."

King and Jukes remind us that "patterns of evolutionary change that have been observed at the phenotypic level do not necessarily apply at the genotypic level," but add also at "molecular levels," thus identifying the genotype with DNA sequences. Contrary to the rule at the phenotypic level, they claim that autonomy of DNA exists at the genotypic level; it is as such that DNA does not necessarily follow the action of Darwinian natural selection. "Evolutionary change is not imposed upon DNA from without; it arises from within." It is precisely the Darwinian notion that natural selection being the editor, rather than the composer of the genetic message, does *not* remove changes which it is unable to perceive (King and Jukes, 1969, 788). And this is the case when, because of the degeneracy of the genetic code, 134 of the 549 possible single base-pair substitution changes (one-fourth) are without effect on protein structure. "As far as is known, synonymous mutations are truly neutral with respect to natural selection" (King and Jukes, 1969, 789).[2] Data indicate that indeed, changes in amino-acid sequences occur much more slowly than changes in total DNA. Furthermore, they note that "it is probable that not much more than 1 percent of mammalian DNA codes for proteins" (King and Jukes, 1969, 790). By introducing the cleft between the phenotype and DNA King and Jukes accommodate the New Synthesis view that no cellular or organismic function affects the autonomy of DNA sequences:

> ... two views are expressed regarding the number and distribution of amino acid replacements in the evolution of homologous proteins. The first is that of the protein chemist ... It is the necessary properties of the

---

[2] This was found later not to always hold. See "codon bias" observations in Kreitman and Antezana (2000).

protein that dictate its primary structure. This view tends to push DNA, as the driving force in evolution, into the background.

The second view, to which we subscribe, is that the protein molecule is continually challenged by mutational changes . . . Natural selection screens these changes. The fact that some variable amino acid sites are more subject to change than others in a set of homologous proteins is an expression primarily of the random nature of point mutations and only secondarily of protein function.

King and Jukes (1969, 791)

Kimura calculated the average rate of amino-acid change per 10^7 years for three human proteins (hemoglobin, cytochrome c, and triphosphate dehydrogenase) to be approximately one substitution in 28×10^6 years for a polypeptide chain consisting of 100 amino acids. He showed that "this evolutionary rate, although appearing to be very low for each polypeptide chain of a size of cytochrome c, actually amounts to a very high rate for the entire genome" (Kimura, 1968, 625). Assuming that the haploid human chromosome complement comprises about 4×10^9 nucleotide pairs he calculated "that in the evolutionary history of mammals, nucleotide substitution has been so fast that, on average, one nucleotide pair has been substituted in the population roughly every 2 years" (Kimura, 1968, 625). Considering Haldane's substitutional load, it "becomes so large that no mammalian species could tolerate it." Kimura concluded that "if the neutral or nearly neutral mutation is being produced in each generation at a much higher rate than has been considered before, then we must recognize the great importance of random genetic drift due to finite population number in forming the genetic structure of biological populations" (Kimura, 1968, 626).

It appeared that the identification of the genotype with DNA sequences had saved the Central Dogma: events at the level of DNA were controlled by different rules than those at the level of proteins and upwards. But this could be maintained neither at the genotypic nor at the phenotypic level: molecular biology increasingly brought to light the dependence of DNA, its transmission and translation as well as its replication and packaging, on the cellular system as a whole, and phenotypic analysis revealed that the organism as a developmental system played a role in these processes no less than the proverbial genotype. The studies of DNA sequences are important "because we can read off the genotype directly from this most basic of all phenotypes" (Griffiths *et al.*, 1999, 576). The acceptance of Gould and Lewontin's critique of the adaptationist program and the constraints of "The spandrels of San Marco"

(Gould and Lewontin, 1979) established the theory of developmental systems evolution over that of evolution as a change of gene frequencies. The genes again were the intervening variables that merely express the simple idea that a property of the developing organism is or may be conditioned or co-determined, through "something" in the gametes (see, e.g., Neumann-Held and Rehmann-Sutter, 2006, and Chapter 20).

V

Increasing resolving power

Continuity between the early reductionist ethos and the late
anti-vitalist sentiment of Francis Crick, Jacques Monod and
Linus Pauling . . . is suggested in the areas of fine structural
genetic analysis, as in Benzer's wish to "translate linkage
distances, as derived from genetic recombination experiments into molecular units"; the development of the operon
theory . . . and the genetic code, as in Crick's legislative cod-
ification of molecular biological reductionism in the Sequence
Hypothesis and the Central Dogma.

Fuerst (1982, 268)

Toward the 1940s genetics became a self-confident autonomous discipline. The Nobel Prize awarded to Morgan in 1933 was a symbol of this
autonomy and it certainly also added to its self-confidence.

During the first decades of the twentieth century, the methodology of
hybridization analysis of discrete traits was established, and a conceptual
framework for the mechanics of inheritance was formulated: genes were
inherited entities that (within given environmental circumstances)
determined the properties of morphological, physiological, as well as
behavioral traits of living organisms; the chromosomal theory of inheritance situated the genes along the chromosomes and mapped them; the
analysis of mutations combined with cytogenetic observations indicated
that there was a material basis for the Mendelian *Faktoren* and demonstrated the mechanics of their inheritance. Geneticists in Morgan's
group and elsewhere[1] now invaded other disciplines and explored them

[1] Morgan and his group moved in 1929 from Columbia University in New York to CalTech
in California, and in the 1930s Morgan turned his attention back to embryology.
Sturtevant inherited Morgan's professorship.

more aggressively: Dobzhansky's *Genetics and the Origin of Species* (Dobzhansky, 1937) declared his intent to expand into the sphere of evolutionary research. And George Beadle, another member of the Morgan–Sturtevant group, joined forces with embryologist Boris Ephrussi to combine the methods of *Entwicklungsmechanik* with those of gene-mechanics for the study of the genetic control of embryogenesis (Beadle and Ephrussi, 1936).

The triumph of the reductionist conception of genetics seemed to be complete, notwithstanding some heretics and more than a few others who advocated a more careful instrumental or operational interpretation of the phenomena that were observed by the empirical reductionist methodology. Yet, a better resolution of the phenomena was now needed, preferably in physico-chemical terms. It was time to try to grind the genes in a mortar as anticipated two decades earlier by Muller (1922). For a while it was hoped that the "Target Theory" (Timofeéff-Ressovsky, Zimmer, and Delbrück, 1935) would provide a foundation for such an analysis. Delbrück, who wrote the theoretical section of the paper, rephrased his conclusions in the language of quantum mechanics, and saw his gene model in terms of "atomic associations" (*Atomverbände*) and electron states defined within contemporary physical theory. Indeed, the physicist Erwin Schrödinger was intrigued enough by these ideas to suggest a physical model of the organization of the genetic material as aperiodic crystals, the metastable energetic levels of which were responsible for the different mutation rates of various alleles of genes (Schrödinger, 1962 [1944]). Consistent with his research program, Muller suggested extending the research of induced mutations beyond that of physical agents like X-rays and UV to chemical mutagenesis (see Chapter 9).

The genetic analysis with Drosophila, corn, and mice was pushed to its limits, and although the analysis of Neurospora indicated a way out, this too was not enough. The introduction of bacteria into the cycles of genetic research allowed a further increase of many orders of magnitude in the resolution power of genetic analysis and also upgraded the phenomenological analysis to the biochemical and eventually molecular levels. For a while bacteria and viruses became dominant in genetic analysis, overshadowing all other model organisms. However, for this to happen profound conceptual developments in the field of microbiology were necessary.

Up to the 1870s it was believed that the different shapes that bacteria assumed were varieties of the same organism; by then, the doctrine of

Louis Pasteur and Ferdinand Cohn promoted the concept of bacteria being true to type. The methods Robert Koch developed to isolate and grow microbes in pure culture were crucial support of such a germ theory. However, the doctrine suppressed the study of inheritance and variation in bacteria. Cohn was convinced that bacteria were primitive plants which could only reproduce by asexual means (Zuckerman and Lederberg, 1986). But by 1944, René Dubos criticized germ theory bacteriologists for their "blind acceptance" of the doctrine of monomorphism because it "discouraged for many years the study of the problems of morphology, inheritance, and variation in bacteria" (Moberg, 2005, 69–70). Dubos accepted Claude Bernard's notion of "man of research" who concentrates on fine facts of science as a research heuristic but rejected Bernard's "man of science" who creates a pattern or imposes an order on the confusing accretion of facts (Moberg, 2005, 72). This combination of conceptually accepting bacteria as just one form of life amenable to the principles of all living cells, with methodologically accepting reductive research as the most productive means for analysis, was an important stimulant to extending genetic and molecular analysis to the bacteria and their viruses.

There can be little doubt that "Molecular biology has indeed transformed our understanding of heredity. The recognition of the structure of DNA, the understanding of gene replication, transcription and translation, the cracking of the genetic code, the study of regulation, these and other breakthroughs have combined to answer many of the questions that baffled classical genetics" (Kitcher, 1984, 335–336). But according to many historians and philosophers of science, not to mention many of the research workers involved, there were now two theories which addressed the phenomenon of heredity: "One, *classical genetics*, stemming from the studies of T. H. Morgan, his colleagues and students, is the successful outgrowth of the Mendelian theory of heredity rediscovered at the beginning of this century. The other, *molecular genetics*, derives from the work of Watson and Crick" (Kitcher, 1984, 336).

The split, however, was on socio-political grounds more than on conceptual-philosophical arguments. Molecular biology was the brain child of Warren Weaver, director of the Natural Sciences Section at the Rockefeller Foundation, who promoted a new branch of science "in which delicate modern techniques are being used to investigate ever more minute details of certain life processes" (Weaver, 1970). This "new biology" was reductionist in conception, as succinctly expressed in 1931 in a letter of Otto Warburg to Weaver: "The most important problem in biology is to obtain an understanding in physico-chemical terms of the processes – and the substances which take part in these processes – that occur in the normal living cell" (Fuerst, 1982, 255). Nonetheless, the roots of molecular biological research were not in biology but rather among physicists as well as several "clusters" of chemists who maintained the attitude that all biological phenomena could ultimately be accounted for in terms of conventional physical laws (Fuerst, 1982, 250–253). With the increasing involvement of molecular biological researchers in the problems of genetics, they obviously considered these to be *their* problems, and since the conceptions were highly reductionist, the issue became whether the "classic" aspects could be reduced to those of the "molecular" approach (see also Olby, 1990).

Heroic efforts by Ken Schaffner to formally reduce classical genetics to molecular biology failed (Schaffner, 1976), and Philip Kitcher showed that "even after considerable tinkering with the concept of reduction, one cannot claim that classical genetics has been (or is being) reduced to molecular genetics" (Kitcher, 1984, 336). Schaffner concluded that reductionism is peripheral to molecular biology and Kitcher believed that the separation of the two theories is not "a temporary feature of our science" but reflects an immanent organizational difference between the two theories (Kitcher, 1984, 371). In variance to this anti-reductionist claim, Alexander Rosenberg held that the arguments against the adequacy of reductionist explanations merely reflect our cognitive limitations. Furthermore, he wanted to maintain a reductionist position because "antireductionism supports an instrumentalist approach to biological theory" (Rosenberg, 1979). Nevertheless, the derivation of the laws of classical genetics from principles of molecular biology would not do because, in effect, explanation is a *nontransitive* activity:

> We can explain any laws of classical genetics by deriving them from various cytological principles. These cytological principles may in turn be explicable by deriving them from chemical laws . . . But . . . "[w]hat is relevant for the purposes of giving one explanation may be quite different from what is relevant for the purposes of explaining a law used in giving that original explanation." In the case of cytological explanations, the molecular details of the various mechanisms and processes involved are typically irrelevant and "would not deepen our understanding of the transmission law . . ."
>
> Gasper (1992, 657–658, quoting from Kitcher, 1984)

Gasper suggested that multiple supervenience, rather than reduction would best describe the case: "We have a case of multiple supervenience when the same physical substrate realizes a variety of different higher-level properties. . . . [T]he same molecular process underlies a multiplicity of properties at the cytological level" (Gasper, 1992, 668).

Extending the phenomenological reductionist heuristics to the biochemical and molecular level had given genetics an enormous thrust that reflected also on conceptual genetic determinism. Considering the methodological achievements, Fuerst suggested that for

> early molecular biology, it seems justified to define reductionism as a *belief system* which incorporates a belief in the possibility of the explanation of biological entities in terms of their physical and chemical components, and which *may* incorporate a denial of any limitations to such a physico-chemical approach to biological knowledge, or a denial, for such an approach to succeed, of the need for special laws outside 'state-of-the-art' physics and chemistry.
>
> Fuerst (1982, 247–248)

The triumphant declarations of Crick's Central Dogma of Heredity (Crick, 1958) and Monod's Chance and Necessity (Monod, 1972) mark the peak of this belief system.

In the introduction to his volume *A History of Molecular Biology* Morange explains:

> Molecular biology is not merely the description of biology in terms of molecules . . . Rather, molecular biology consists of all those techniques and discoveries that make it possible to carry out molecular analyses of the most fundamental biological processes – those involved in the stability, survival, and reproduction of organisms. . . .
>
> Molecular biology was born when geneticists, no longer satisfied with a quasi-abstract view of the role of genes, focused on the problem of the nature of genes and their mechanism of action.
>
> Morange (1998, 1–2)

Morange believes that the justification for a dichotomy, for a distinct status, is the *"new way of looking at organisms* as reservoirs and transmitters of information" (Morange, 1998, 2, italics added). This is not a new way, but rather a molecular extension of the phenomenological determinist notions of genetic scientists. Genetic transmission formulated in terms of information theory and cybernetics is a misleading metaphor. The mathematical-cybernetic notion of information theory deals with the probabilistic *reliability* of transmission of signs, whereas the information-metaphor of genetic analysis refers to the comparison and transmission of *semantic* information, as obtained by the methodology of hybridization, whether that of organisms or of nucleic acids (Falk, 2007, 296). Sterelny and Griffiths (1999, 101–102) refer to the latter as "intentional" information, which, being immanently context-independent, promotes genetic determinism (Griffiths, 2006, 183–188).

From its beginning, phenomenological genetic analysis explored the possibility of relating genetic analysis to molecular organization. It explored the nature of the entities of inheritance, as well as their function in the development of individuals and the evolution of populations. It resolved the "atomic" gene concept into a linear sequence all the way down to its molecular dimensions. The increased resolution was that which indicated "abnormalities" in the process of recombination and it was the study of these "gene conversions" that finally refuted the mechanistic model of crossing over and made way for the analysis of recombination at the molecular level. The identification of the mutation in sickle-cell anemia with a specific single amino-acid substitution in the β-chain of the hemoglobin molecule was one of the highlights of the application of biochemical methods to phenomenological genetic analysis. The notion of "one gene – one enzyme" provided the foundation for the specification of molecular transcription and translation, and in due course for the regulation of their production. The extension of genetic analysis to mitotic crossing over in somatic tissues opened the way for new attacks on cellular determination (through phenomena such as "transdetermination") and the expansion of methods of molecular genetic analysis to developmental biology. A most fruitful extension of genetic analysis was that of hybridization at the molecular level: double-stranded DNA could be dissociated and when mixed with other single-stranded DNA (or RNA) strands reannealed. This allowed genetic analysis to overcome its inherent limitations of intra-specific hybrids and opened the way for the extension of the analytic power of hybridization both to different somatic tissues of the same organism, and to other organisms, irrespective of their systematic status. As phrased by Kitcher: "Even though reductionism fails, it may appear that we can capture part of the spirit of reductionism by deploying the notion of explanatory extension. The thesis that molecular genetics provides an explanatory extension of classical genetics embodies the idea of a global relationship between the two theories" (Kitcher, 1984, 365).

But the triumphs of molecular genetic analysis only further emphasized the inherent dialectic of genetics since its establishment as a discipline. Encouraged by the successes of the reductionist methods at the molecular level, researchers extended their ambitions to go further to the most basic elements, namely the sequencing of whole genomes. Genetic determinism is epitomized by declarations such as Watson's statement that "we used to think out fate was in our stars. Now we know, in large part, our fate is in our genes" (*Time Magazine*, 10 March 1989, 67). On the other hand, bewilderment at the complexities of the processes of transcription and translation of the DNA sequences and their dependence on cellular functions, the realization of the dependence of evolutionary processes on cellular and organismal constraints and the facts of the evolutionary coordination of developmental processes, encouraged developmental system theories that rejected biological determinism as a conception, to the extent of exploring possibilities of avoiding the application of reduction as a methodology (see, e.g., Oyama, Griffiths, and Gray, 2001).

As noted by Garland Allen, molecular genetics grew out of the attempt to determine the nature of the gene and how it works. It grew out of the attempt to apply functional and informational questions to the highly elaborate but formalistic structure of the classical Mendelian chromosome theory. As such its area of concern involved those traditionally limited to genetics or biochemistry. "It would appear ... that molecular genetics represented a sharp break from classical genetics ... In fact, molecular biology represents a strong element of continuity with its past." Classical genetics "had raised questions about the nature of the gene and gene action that could not be answered in the three decades following the discovery of *Drosophila* in 1910." Thus, "what Morgan and his co-workers had done was to erect an elaborate, logically consistent, but formalistic concept of genetics that was wholly detached from any biochemical foundations" (Allen, 1975 [1978], 226).

11

Recruiting bacteria and their viruses

The application of biochemical processes as phenotypic *markers* of Mendelian factors is not new. As early as 1901 the physician Archibald Garrod cooperated with Bateson in collecting "extraordinarily interesting evidence . . . regarding the condition known as 'Alkaptonuria'" (Bateson and Saunders, 1902, 133–134). Alkaptonuria was one of several diseases that appeared to be markers of Mendelian factors and were later called by Garrod "inborn errors of metabolism" (Garrod, 1908). Systematic studies of markers of inherent biochemical variability were carried out by J. B. S. Haldane and his associates on the synthesis of anthocyanins that are involved in flower color in various plant species, such as *Pelargonium* (Haldane, 1954, 52–58). Also in Drosophila notions of the biochemical basis of genetic differences were investigated in studies such as that by Sturtevant (1929) on the eye-color mutant vermilion.[1]

Although many investigators conceived of the role of genes in developmental and biochemical terms, Boris Ephrussi and George Beadle made a breakthrough when they applied the classical methods of developmental mechanics to problems of genetic analysis in Drosophila (Beadle and Ephrussi, 1936). They transplanted the imaginal discs[2] of the eyes from larvae of one genotype into larvae of another type, which enabled them to inspect the transplant's eye color in the host-imago's abdomen. The autonomy or non-autonomy of the transplant's eye color provided an indication of the function of specific genes in the *developmental* pathways of eye-color pigmentation. Demonstrating the

---

[1] Morgan and Goldschmidt were more interested in the physiological-embryological aspects of the genes, whereas Muller, following Troland's suggestion (1917), believed for some time that genes were enzymes, or at least acted like enzymes (Muller, 1922).

[2] Distinct clusters of cells in specific sites of the larvae that become in imagoes specific organs upon metamorphosis.

feasibility of that analytic approach, the problem was obviously more biochemical than developmental. Two independent synthetic pathways of eye pigments were detected: The *bw* mutants had brown eyes because of a deficiency in the synthetic pathway of the red pigment, while the *v*, *cn*, *cd*, and *st* mutants were all red, being deficient in the synthetic pathway of the brown pigment. Reciprocal transplantations of the latter mutants established the sequence of the involvement of these genes in the pathway.

Very similar experimental work was carried out with the moth *Ephestia kühniella* during the same period in Germany by Ernst Caspari (before he emigrated to the US) and Alfred Kühn. Chemical analysis of hormone-A extracts from Ephestia that overcame the mutant effect of *v* in Drosophila identified the product of that gene as kynurinine. Further physiological insights were gained from the cooperation of zoologist Kühn and biochemist Adolf Butenandt (Kühn, 1941; Rheinberger, 2000). Beadle, on the other hand, associated with Edward Tatum and carried out genetic analyses of biochemical reactions with a simpler organism, the mold *Neurospora crassa*, which could be studied *in vitro*, on chemically defined media. This became an exceedingly successful research project because Beadle conceived of using genetic analysis as a tool to discover biochemical details of metabolic pathways rather than just using biochemical properties as efficient genetic markers (Beadle, 1945; Beadle and Tatum, 1941a, 1941b). Years earlier B. O. Dodge had pointed out to Morgan the advantages of Neurospora over those of Drosophila as an experimental organism for genetic research, but only one student, Carl Lindegren, took up the challenge: in Neurospora it was possible to systematically follow products of specific meiotic events (tetrad analysis, see Chapter 6). By adding controlled nutrients to the growth medium it was possible to save many lethal mutations and employ biochemistry and genetics as different aspects of the same discipline (Davis and Perkins, 2002).

Spores that were treated by X-rays or other mutagenic agents were grown on a nutrient-rich medium and the mycelia of their vegetative spores (conidia) were tested on "minimal medium" containing only inorganic components (with sugar as a carbon source and the vitamin biotin). Conidia that did not grow on this medium potentially carried auxotrophic mutations. Adding systematically varied amino acids or vitamins to the minimal medium of the plates revealed the metabolic failure of the presumptive mutations. The monogenic nature of the auxotrophy could be verified by showing a 4:4 segregation of ascospores in the hybrid of the mutant and the wild type. Combining two auxotrophic mutations for the same metabolic pathway may reveal whether they complement each other. This may be done by traditional mating of the two and checking the ascospore products of meiosis, or more straightforwardly in a spurious hybrid, by fusing their mycelia into a heterokaryon. Neurospora mycelia are composed of multi-nuclear cells. If two adjacent mycelia fuse, cells with numerous nuclei of different genetic content – heterokaryons – are retained. The growth of a heterokaryon of two auxotrophic mycelia on minimal medium indicates that the relevant mutations affect different genes, the products of which, even if involved in the same metabolic pathway, complement each other intracellularly. Checking the tentative precursors of the missing end-product to each mutation culture may identify the sequence of the affected metabolic pathway. Some mutations excrete an accumulated blocked precursor that may be tested for its capacity to feed other mutations in the same metabolic pathway, indicating that the feeding mutant affects a step later in the pathway than the fed mutant.

Beadle and Tatum conceived that *each* gene is a determinant of the structure of a *specific* enzyme functional in the metabolic pathways of cells, and accordingly formulated the catchphrase "one gene – one enzyme." This one-to-one notion of genes specifying enzymes introduced at one fell swoop problems of biochemical interest as central issues to genetic analysis and turned phenomena at the biochemical level into major "marker" variables of genetic analysis (Beadle, 1945). It opened the way to a more direct (or less indirect) study of the phenomena at the biochemical level, which allowed an increased resolution of a gene's structure by many orders of magnitude. Not less significant, it established an efficient experimental methodology that specified how genes are involved in cellular function. The fundamentals of screening for the recovery of rare genetic events, introduced by Muller's *ClB*-method were fully developed now in Neurospora by applying chemically defined media that allowed specific selective growth of some types, while excluding others. This method was further elaborated once genetic analysis of bacteria and their viruses was introduced.

Fifty years later, a pioneer in biochemical genetic research called the introduction of Neurospora as an experimental organism of genetics "The Neurospora revolution" (Horowitz, 1991). However, the extension of genetic research to Neurospora, and even more so to bacteria as model organisms of genetic analysis, was embedded in the "taxonomic status" of these systems. As has already been emphasized, although all geneticists were keen to apply reductionist methodologies, they were in conflict about how reductionist genetic conceptions should be applied to the study of "real," complex living beings that developed and differentiated. Horowitz himself adopted an uncompromising reductionist biochemical position with regard to the meaning of one gene – one enzyme. He noted that according to Sturtevant "geneticists were disinclined to accept simple ideas of gene action because they were convinced that development was too complex a process to be explained by any simple theory" (Horowitz, 1991, 634). Horowitz blamed this on the impact of Edmund B. Wilson who, Horowitz said, took an exceedingly murky view regarding the role of the genes. Horowitz quoted Wilson, complaining that many writers had treated the chromosomes as the actual and even the exclusive "bearers of heredity" to which all else was subsidiary. Wilson opposed those who referred to the chromosomes or their components as "determiners" of characters. From Horowitz's perspective, "in a more important" sense, Wilson's statement "is altogether wrong. . . . The individual gene in some way determined the specific enzyme, although it was not yet seen how" (Horowitz, 1991).

Beadle and Tatum no doubt overcame any rudiments of the Wilsonian conceptual hurdle by extending their notions of heredity to a mold, an organism with apparently no complex development and differentiation (in all fairness, other "simple" organisms, like Paramecium, had been the objects of genetic research for many years). This step conceptually and methodologically opened a new era that led reductionist genetic analysis from phenomenological variability to biochemical processes and eventually to a molecular level of resolution. Extending genetic research to bacteria posed, however, the empirical question of whether bacteria obey the same rules of the life sciences as worked out for "higher" organisms.

BACTERIA OBEY THE RULES OF BIOLOGICAL TRANSMISSION

Neurospora and Aspergillus, though having a simple life cycle when compared to the classic organisms of genetic research, mice, Drosophila, maize, Datura, snapdragon etcetera, were still organisms that had an observable cellular nucleus and their life cycle included sexual reproduction, involving nuclear fusion and meiotic reduction. This was not so for bacteria, which appeared to reproduce asexually by fission, and even less so for viruses (for some time Muller considered viruses to be naked genes, see Muller, 1922). Furthermore, the Pasteur–Cohn doctrine that adopted monomorphism ignored the phenotypic plasticity of micro-organisms induced by culture conditions. No wonder genetic analysis was not extended to these problematic forms of life. The story changed dramatically, however, in 1943 when Salvador Luria and Max Delbrück proved that hereditary changes do occur in pure bacterial cultures, and that such changes may be pre-adaptive. Culture conditions may only expose (and enlarge) preexisting inheritable variants (Luria and Delbrück, 1943):

> When a pure bacterial culture is attacked by a bacterial virus, the culture will clear after a few hours due to the destruction of the sensitive cells by the virus. However, after further incubation for a few hours, or sometimes days, the culture will often become turbid again, due to the growth of a bacterial variant which is resistant to the action of the virus. This variant . . . will in many cases retain its resistance to the action of the virus even if subcultured through many generations in the absence of the virus. . . .
>
> The nature of these variants and the manner in which they originate have been discussed by many authors, and numerous attempts have been made to correlate the phenomenon with other instances of bacterial variation.
>
> Luria and Delbrück (1943, 491)

Considering the alternative hypotheses of the origin of hereditary differences, the experimental strategy to differentiate between them was formulated:

> Do the original variants trace back to mutations which occur independently of the virus, such that these bacteria belong to a few clones, or do they represent a random sample of the entire bacterial population? . . . If we find that a bacterium survives an attack [by viruses], we cannot from this information infer that close relatives of it, other than descendants, are likely to survive the attack. . . . On the mutation hypothesis, the mutation to resistance may occur any time prior to the addition of virus. The culture therefore will contain "clones of resistant bacteria" of various sizes, whereas on the hypothesis of acquired immunity the bacteria which survive an attack by the virus will be a random sample of the culture.
>
> Luria and Delbrück (1943, 493)

Luria and Delbrück's "fluctuation experiment" consisted of splitting a culture of bacteria (sensitive to the virus) into numerous subcultures, each starting with one or few bacteria. After a while all subcultures were exposed to the virus and then bacteria from each subculture were spread on culture plates with the virus, so that clones of individual bacteria could be examined for resistance to the bacteriophage: If resistance was acquired by the presence of the virus, the variation in the number of resistant bacteria between subcultures was expected to fluctuate at random. However, the variation between the numbers of resistant bacteria in the subcultures was orders of magnitude higher than that expected by the Poisson distribution for random events, indicating that resistant bacteria were present in the original culture before and independently of the addition of the viruses to the subcultures.

Luria and Delbrück considered "the above results as proof that in our case the resistance to virus is due to a heritable change of the bacterial cell which occurs independently of the action of the virus." They added, however, that it "remains to be seen whether or not this is the general rule" and whether there is reason to suspect that "the mechanism is more complex" in some other cases (Luria and Delbrück, 1943, 509). But their study was adopted as unequivocal evidence that bacteria undergo pre-adaptive mutations at a fixed small chance per time unit for each bacterium. The pre-adaptive nature of genetic variants of bacteria provided the conceptual basis for extending genetic research from mouse, Drosophila, maize and other experimental organisms to bacteria, and shortly also to their viruses.

Elegant as the fluctuation test was, Delbrück pointed out that it was necessary to show pre-adaptation at the level of individual cells rather than as a statistical effect of populations of bacteria. This was taken care of when two other, more direct analyses were developed to prove the pre-adaptive nature of the hereditary changes.

Joshua and Esther Lederberg spread individual bacteria from a colony, probably originating from a single bacterium, on a non-selective "master" plate and then "replica plated" the pattern of colonies by pressing a wooden block over which a piece of velvet cloth was stretched, first to the surface of the master plate and then to a series of Petri dishes with media designed to allow only specific bacteria to grow, say, those resistant to a given virus. The sites of most of the virus-resistant colonies in the different replica plates overlapped, indicating that these colonies had already been resistant to the virus when they were grown on the master plate that had never been exposed to the selective agent. The colonies grew from pre-adaptive changes to resistance (some resistant colonies were specific to only one plate or another, indicating that the change to resistance could also occur later than that on the master plate) (Lederberg and Lederberg, 1952).

H. B. Newcombe experimented by allowing individual bacteria to become discrete colonies on agar plates with no selective agent. After a day the bacterial colonies on some of the plates were redistributed. When the selecting agent, say the bacteriophage, was then added to the plates, typically few resistant colonies grew on the non-redistributed plates, whereas resistant bacteria grew confluently as a rule on the redistributed plates. This indicated that the few pre-adaptive resistant discrete colonies on the original plates were broken up and distributed all over the plates in the disturbed ones (Newcombe, 1949).

The pre-adaptive nature of bacterial mutations, following the expectations of the "Darwinian" rather than the "Lamarckian" adaptive doctrine, became the foundation for extending genetic analysis to bacteria. It was therefore somewhat of a shock to the geneticists when in 1988, more than four decades later, John Cairns and colleagues challenged the founding experiments of Luria and Delbrück.

> Our purpose is to determine how many of these variants are arising as a direct and specific response to selection pressure (would not have occurred in its absence) and how many are 'spontaneous' (would have arisen even in the absence of selection).
>
> Cairns, Overbaugh, and Miller (1988)

Cairns *et al.* argued that it takes several cell generations for a genotypic mutation to be expressed phenotypically: Cells had to dilute the distribution of virus-receptors on the cell walls before turning resistant, or dilute the concentration of an enzyme before becoming auxotrophic, and bacteria were not given a chance to do so in the founding experiment. Even more damaging, they showed that in cultures that were *lac⁻* (i.e., incapable of replacing lactose for glucose as an energy source), although some of the mutations to *lac⁺* that restored the capacity to utilize lactose were *pre-adaptive*, others appeared later in a direct (*adaptive*) response to the presence of lactose (rather than the standard glucose) in the medium: When *lac⁻* strains of *E. coli* were plated out on media containing limiting amounts of lactose, "papillae" of lactose-fermenting mutants appeared. In strains bearing a specific (amber) mutation at the *lacZ* locus, revertants to *lac⁺* occur during the logarithmic growth of bacteria independently of the presence of lactose; yet more mutations occur at the late, stationary phase in response to the distress of carbohydrate source (no parallel effect is observed in these cultures at another locus, Val^R). It stands to reason that mutation induction (or repair of pre-mutations) is also prone to evolutionary changes. Apparently populations of bacteria may have acquired some *specific, adaptive* response of producing (or selectively retaining) the most appropriate mutations. The net effect is one of "directed mutagenesis." This does not refute the observation that mutations in bacteria as in "higher" organisms, occur as a rule, independently of their adaptive value.

Cairns *et al.* stressed that the main purpose of their paper was "to show how insecure is our belief in the spontaneity (randomness) of most mutations. It seems to be a doctrine that has never been properly put to the test" (Cairns *et al.*, 1988, 145). Although there was no breakdown of the accepted Darwinian theory (Stahl, 1988) and the appearance of adaptive, directed mutations was eventually understood to involve another source of hereditary changes, these genetic analyses emphasized the complex nature of the production of the phenotype, and the vulnerability of the determinist approach to the genotype–phenotype relationship. They emphasized that in biological research there are no laws of nature in the sense that there are purported to be in physical (and chemical) theory. There are no *a priori* laws of life (of course, besides the laws of physics that the contingent forms of living systems obey). In an important sense, life is nothing but the history of increasing constraints and opportunities of increasingly complex systems.

GENETIC ANALYSIS IN BACTERIA

Luria and Delbrück demonstrated that bacteria obey the rules of evolution that had been established in "higher" organisms; but they did not show that bacteria are amenable to genetic analysis by hybridization. In 1946 Tatum noted at the Cold Spring Harbor symposium that "the main attribute lacking in bacteria which would make them ideal material for combined genetic and biochemical investigation is their apparent lack of a sexual phase." Lederberg and Tatum set out to establish whether that crucial step needed to further turn bacteria into objects of genetic analysis on a par with Drosophila and maize existed. Inspired by the prospect of Avery and his colleagues' experimental results on the possibility of transforming the properties of bacteria by deoxyribonucleic acid extracted from another strain of bacteria, Lederberg examined whether an analog to a sexual process takes place in bacteria. He tested whether one could find bacterial hybridization that might lead to reassortment of their hereditary markers. He showed that a process analogous to sexual mating, followed by chromosomal recombination, may take place in bacteria and that linkage maps may be constructed for them (Lederberg, 1996).

Mixing bacteria of two different auxotrophic strains in a nutrient broth and spreading the (washed) samples on minimal-medium agar plates, would allow only prototrophs to multiply and produce visible colonies. Such an experimental system would easily detect one recombinant in millions. To avoid the risk of confusing recombinants with rare back-mutations, multiple auxotrophic strains were needed. Auxotrophs are slow to grow on minimal-medium agar plates enriched with some nutrient broth. Lederberg and Tatum picked up slow and late colonies with sterilized toothpicks and examined their nutrient needs on a series of plates with minimum medium, each plate enriched with all but one of the essential amino acids or vitamins (to examine the capacity to utilize different sugars as a source of energy the cells were grown on a minimal enriched medium in which glucose was replaced by another source of energy). *Escherichia coli* bacteria of the strain K12 with different multiple auxotrophic needs were produced, mixed, and grown on plates with minimal growth medium (sufficient for mating but insufficient for colony formation). The frequencies of prototrophic bacteria – presumably recombinants – were too high to conform to the possibility of multiple-site reverse mutations. When Bernard Davis separated the bacterial strains in a U-shaped tube provided with a filter through which only molecules less than 0.1μ in diameter were forcefully pumped, no recombinants appeared when the bacteria were spread on minimum-medium plates. This excluded artifacts such as cross-feeding or migrating particles as agents that might have led to prototrophy. The insensitivity of the results to the presence of DNase in the medium further excluded the possibility that this was an Avery-like process of transformation (see Chapter 12).

In the spirit of classical Mendelian hybridization experiments it was also necessary to examine whether the recovery of each marker was independent of the experimental selection procedures. Thus crossing, say, Bio^- $Met^- \times Thi^-$ Leu^- and checking for colonies on leucine-supplemented plates did not discriminate (select) auxotrophy or prototrophy for leucine. Rechecking all colonies that grew for leucine-less (Leu^-) proved leucine-auxotrophs to be very rare in relation to the leucine-independent colonies (Leu^+) (see Table 11.1: the ratio Leu^-/Leu^+ is 1/10.4 rather than the expected 1/1 ratio), whereas thimine-less

Table 11.1. *Relative frequency of unselected characters in the mating of two multiply marked* E. coli *K12 strains: Bio⁻ Met⁻ Thr⁺ Leu⁺ Thi⁺ (×) Bio⁺ Met⁺ Thr⁻ Leu⁻ Thi⁻ (from Lederberg, 1947)*

Unselected character	Ratio of unselected character $(-/+)$ among recombinants
Bio	0.17
Thr	0.24
Leu	0.096
Thi	9.88

(*Thi⁻*) colonies were much more frequent than thimine-independent colonies on thimine-supplemented plates (*Thi⁻/Thi⁺* ratio was 10/1). This indicated that the genes are linked (and that the sequence of the linkage map was *not* as that of the Table heading, *Bio Met Thr Leu Thi*).

Although Luria failed in repeated experiments with another stock of bacteria (*E. coli* B), and experiments were for many years limited to the one stock of *E. coli* K12, the art of genetic analysis by hybridization was extended to bacteria, which seemed to indicate that bacteria too possess chromosomes that are involved in meiotic processes like those observed in higher organisms (Lederberg, 1947). More marker genes were introduced and recombination and linkage appeared to involve the whole genome of *E. coli*. But the analysis led to serious difficulties in interpretation since it was impossible to construct a consistent linear linkage map (Stent, 1971). Lederberg proposed constructing a branched map while he emphasized that this virtual *formal* solution of recombination data may not reflect the *physical* organization of the bacterial chromosomes: Mating and meiotic processes may be different in bacteria and consequently end up, contrary to the situation in higher organisms, in observed discordance between the genetic analysis and the physical state. Further hybridization analysis carried out by plating the mated stocks on a medium that eliminated one of the participants after conjugation (by a high dose of streptomycin, to which it was sensitive; the other partners being streptomycin-resistant) implied that genetic recombination in the bacteria was a directional process, in which one strain acted as the donor and the other the recipient of genetic material (Hayes, 1953).

The "branched" linkage map was, indeed, constructed: bacterial chromosomes had a circular structure unlike the linear structure of eukaryotic chromosomes; also no meiotic-like process occurred in these (haploid) cells. Sexual-like processes in bacteria were due to special mutation-like events ("sexduction") at which an episome or plasmid (that turned out itself to have a mini-circular chromosomal structure) was inserted by recombination into the circular chromosome. This transformed the bacterium into a high-frequency-recombination (Hfr) cell. Since the episome was inserted at different sites and directions it provided for different linkage maps that were constructed from the time-sequence of the transfer of markers for each Hfr. All proved to be permutations of a consistent sequence, which was established as the bacterial chromosome's linkage map (Jacob and Wollman, 1961; Wollman, Jacob, and Hayes, 1956; and see Chapter 6).

Genetic analysis in bacteria made a qualitative leap forward in resolving power. However, a still more powerful resolving tool for genetic analysis was offered by viruses that feed on bacteria. Bacteriophages (in short: phages) are scored by seeding them together with their hosts, like the *Escherichia coli* bacteria, on agar plates on which the host grows. When there are far fewer phages than bacteria, the bacteria grow into a "lawn" on the dish in which "holes" or plaques (free of bacteria) are formed wherever a phage is present. The characteristic phenotype of the plaques and their numbers make phage biology accessible to quantitative analysis.

Seymour Benzer perfected genetic screening procedures in order to achieve a molecular level of resolution. He took advantage of the finding that the *r*II mutants that produce a phenotype of large plaques with rough borders on *E. coli* bacteria of stock B were lethal on bacteria of stock K (formally denoted K12(λ)). This enabled him to maintain *r*II stocks on B bacteria and screen for extremely low frequencies of recombination within the *r*II gene on K bacteria. Further analysis found some mutations which did not recombine with two or more others that did recombine *inter se*. Benzer concluded that formally these were "deletions." Such deletions were used to divide the gene into sections, in each of which the *r*II "point"-mutations could be located by their *not* recombining with specific deletion(s) but recombining with each other. The topographic map of the *r*II region included over 300 sites (see Figure 11.1).

Benzer's analysis showed that viral genes could be mapped by adapting the principles of the Morgan–Sturtevant inter-gene linkage analysis to the phage-dimension. As with Drosophila pseudoalleles, what appeared to be one gene were actually two adjacent complementation units, or functionally related entities. However, unlike experiments with Drosophila, in the phage-screening tests only the trans-constellation could be studied. It was deduced that the *r*II gene was composed of two

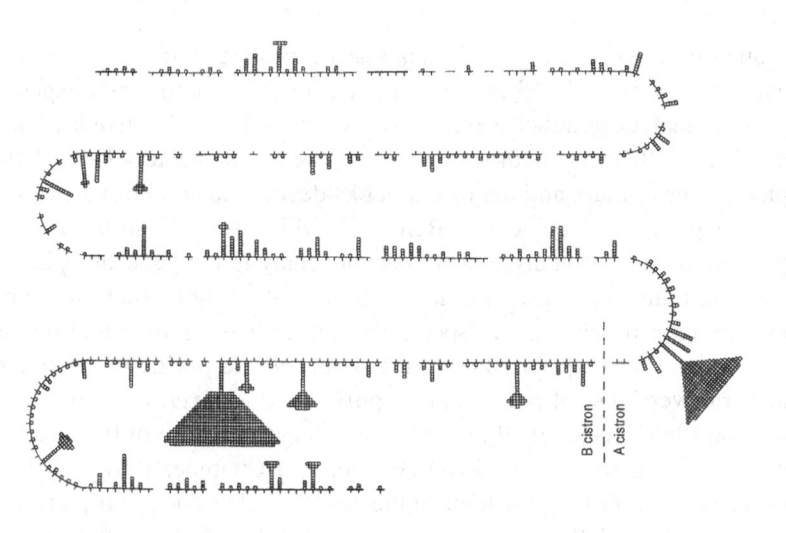

Figure 11.1. Map of the *r*II locus of the T4 bacteriophage: two cistrons, A and B, each contain many recons. Mutons in recons differ in their specific mutation rates (after Benzer, 1957).

trans-complementary entities, or *cistrons*. In each cistron, many sites of recombination – *recons* – were identified at which mutations occurred at specific frequencies, suggesting that each recon may be composed of several sites that may mutate – *mutons*. Calculating the recombination frequencies between the outermost mutations on each cistron (6 units for cistron A and 4 for cistron B) of the total length of the phage's recombination map (700 units) and imposing this on the total nucleotide number of T4 (2×10^5) enabled researchers to convert genetic analysis units into estimated DNA nucleotide numbers (1,700 for cistron A and about 1,100 for cistron B). The great variability of the mutation rates per recon indicated that these sites could be further resolved (or that the mutation rate per site depended on some "position effects" with neighboring sites).

Although the proteinaceous product of the *r*II cistrons was not known for several years and it took some time for experiments to produce cis-combinations of pairs of *r*IIs and demonstrate the shortening of recombination frequencies of the "deletions" (compare the evidence provided by Mohr for the nature of deletions in Drosophila, Chapter 6), Benzer's analysis of the *r*II gene provided a pivotal instrument for *genetic analysis* at the *molecular level*. Benzer's heuristics was (and remained also when he turned to behavioral genetics) consistently "classic genetic" reductionist, and with the *r*II analysis he pushed it to its limits. His experiments were the ultimate fulfillment of the experimental expectations of analytic geneticists (see Holmes, 2000, 116–117). Salvador Luria considered Benzer's heuristics a "key system" in the topology of the phage genetic map, and major textbooks devoted lengthy discussions to "the experimental evidence" Benzer provided in the transition from phenomenological analysis to molecular analysis for gene mapping in particular and genetics in general. Suffice it to note that the early experiments on mutagenic specificity, on classifying of mutations as transitions (of one pyrimidine to the other, or one purine to the other) and transversions (of pyrimidine to purine and *vice versa*) (Benzer and Freese, 1958; Freese, 1959), on understanding of the nature of the effect of mutations (missense mutations) (Benzer and Champe, 1961, 1962), as well as on establishing the logic of the genetic code as one of triplets that are read sequentially from given starting points (Crick, Barnett, Brenner, and Watts-Tobin, 1961), all took advantage of the *r*II genetic analysis. The correspondence of the topology of the mutation sites in the tryptophan-synthetase gene of *E. coli* and that of the replaced amino acids in the polypeptide (Helinski and Yanofsky, 1962) was a further triumph derived from extending the reductionist methodology of genetic analysis to the molecular level.

Bacteria and viruses, or *prokaryotes* – organisms lacking a membrane-bound, structurally discrete nucleus and other subcellular compartments – became efficient tools for genetic analysis that appeared for some time to completely take over the role of traditional *eukaryotic* organisms – those that have a nucleus bounded by a membrane, within which most of the cellular DNA is organized, such as Drosophila, corn, mice, and Neurospora. Genetic analysis became molecular.

12

Molecular "cytogenetics"

Starting with the achievements of microscopists in the 1880s, attention was directed at the cell's nucleus and its chromosomes as the carriers of the genetic material, to the exclusion of the role of the cytoplasm and all its components (the cell membrane was considered inherently irrelevant). Nuclei were composed of nucleo-proteins. Based on the repetitive tetranucleotide structure hypothesis, first proposed by Hermann Steudel in 1906 and developed by Phoebus Levene in 1931 (Deichmann, 2004), the prevailing conception was that nucleic acids comprised stoichiometric aggregates of nucleotide tetrads. Proteins, being the material of enzymes, were celebrated at the first half of the twentieth century as the essence of life. They seemed to be able to provide the diversity that would be expected of the material of heredity thanks to the large number of known amino acids involved in their composition. Toward the end of the 1940s Fred Singer directly demonstrated the diversity of polypeptide chains, and Wendell Stanley's initial erroneous claim in 1935 that crystallized tobacco mosaic virus was a pure self-replicating protein (a study for which Stanley was awarded the 1946 Nobel Prize) further upheld the centrality of proteins in the deliberations on the nature of the hereditary material (see Deichmann, 2004; Lederberg, 1994). These studies were foreign to genetic analysis and its methodology (see Chapter 1).

Two kinds of nucleic acids were conceived, named after their main source at the time, alkali-stable "thymus" nucleic acid, and alkali-labile "yeast" nucleic acid; corresponding to today's DNA and RNA, respectively. Evidence proving the central role of nucleic acids was accumulating during the late 1920s and got important experimental support during the 1930s in studies carried out mainly by physicians and microbiologists whose agenda was primarily to reduce pathogenesis to physico-chemical terms. In 1928 the phenomenon of bacterial transformation was discovered by Fred Griffith. Mice injected with nonpathogenic pneumo-cocci derived from type II bacteria together with heat-killed type III bacteria succumbed to infection and died of type III virulent bacteria. Griffith concluded that a component of the killed type III bacteria transformed the non-virulent living bacteria into virulent ones. The transformation was reproduced in the early 1930s in Oswald Avery's laboratory and opened the way for purifying the transforming factor to a degree unheard of among biologists (Avery, MacLeod, and McCarty, 1944). The conclusion "that a nucleic acid of the deoxyribose type is the fundamental unit of the transforming principle of Pneumococcus Type III" did not relate the transforming principle explicitly to a gene and had little impact on geneticists and cytogeneticists. Edmund B. Wilson, who in 1896 had asserted that there is "considerable ground for the hypothesis that in a chemical sense this substance [nuclein, a compound composed of nucleic acid and albumin] is the most essential nuclear element handed on from cell to cell," changed his mind in 1925 when Feulgen's staining method for DNA failed to stain chromatin in meiosis: "The nucleic acid component comes and goes in different stages of cell-activity, and it is the oxyphilic [protein] component that seems to form the essential structural basis of the nuclear organization" (see Deichmann, 2004; Wilson, 1924, 633). As noted by David Stadler, "Biologists at that time didn't know much chemistry and didn't think about the implications of a limit to variability in the primary sequence of proteins. We were much more impressed by their *biological* properties" (Stadler, 1997, 865).

However, this was not true for all biologists. Alfred Mirsky, Avery's colleague and a pioneer of molecular biology, adhered to the dogma of the inadequacy of nucleic acids as the exclusive carriers of genetic information and suggested that the less than one percent protein impurity in Avery's preparations could be the real genetic material. J. B. S. Haldane suggested in 1937 that nucleic acids and proteins might provide mutual templates upon which the proteinaceous genetic information was reassembled dur-ing cell replication. Moreover, Jack Schultz of Morgan's fly-group con-cluded that, rather than the indirect genetic analysis through mutagenesis as advocated by Muller the nature of the gene should be analyzed directly, and that "much new data is necessary before we can exclude the possibility of specificities in the nucleic acids themselves" (Deichmann, 2004, 218). Schultz collaborated with Torbjörn Casperson, who developed micro-photometric methods that differentiated between proteins, ribonucleic acids, and deoxyribonucleic acids in the cell according to their absorption of ultraviolet light of different specific wavelengths (Casperson and

Schultz, 1938). Casperson and Jean Brachet demonstrated that whereas RNA was present in both the cytoplasm and the nucleus and was quantitatively related to the cell's proteins and to the metabolic activity of the cell, DNA was concentrated in the nucleus and located in the chromosomes: the absorption pattern of ultraviolet light at different wavelengths in metaphase chromosomes overlapped precisely the pattern typical of DNA.

As put by Delbrück, geneticists "were very much aware of what Oswald Avery was doing – and also of his discovery, that this transformation worked if you took a preparation which ostensibly contained DNA." But there was skepticism:

> Either DNA was not a stupid molecule, or – the thing that did the transformation was not the DNA. . . . So then the dilemma became – can DNA carry specificity, or are we really not dealing with specificity at all but with a third possibility, some very special case where all the specificity is already there but just needs a stupid kind of molecule to switch it from one kind of production to another. . . . for example, you can switch a tree from non-flowering to flowering . . . and if you do it by manipulating the daylight-and-dark cycle then you are not putting in a molecule with information but just throwing a physiological switch. . . . So, while it was true that Avery's discovery . . . led to the conviction that there had to be enough specificity in the DNA – you really did not know what to do with it.
>
> Judson (1979, 59–60)

Inspired by Avery's experiments, Erwin Chargaff and his colleagues introduced in the late 1940s the new paper-chromatography technique to clean DNA and characterize its base composition. They soon refuted the Levene tetranucleotide theory by showing on the one hand, differences in DNA composition between species and on the other hand, that there was a characteristic proportional relationship between the nucleotide bases of all DNA molecules: the percentage of adenine equals that of thymine and that of guanine equals that of cytosine; no such constant relationships were observed with RNAs. This presented the clue for how to get out of the paradox.

But, perhaps not surprisingly, notwithstanding the impressive advances in biochemical analysis, it was a biological, "hybridization" experiment rather than a biochemical experiment that finally convinced biologists of the fundamental role played by DNA as the genetic material. This was the case although the biological experiment was far less clear-cut than Avery *et al.*'s evidence of the transformation capacity of purified DNA. Alfred Hershey and Martha Chase infected bacteria with bacteriophages that were labeled either with radioactive sulfur that is localized in proteins or with radioactive phosphorus that is localized in nucleic acids. Whereas the radioactive phosphorus was significantly transferred to generations of progeny phages, only relatively little of the sulfur was found in the immediate progeny. "The experiments reported in this paper show that one of the first steps in the growth of T2 is the release from its protein coat of the nucleic acid of the virus particle, after which the bulk of the sulfur-containing protein has no further function" (Hershey and Chase, 1952, 39). It is obvious from their experimental procedure that Hershey and Chase considered seriously all possible alternatives to the role of DNA in virus infection: they themselves needed a lot of convincing to accept DNA, rather than the proteins, as the carriers of the hereditary information. Their preliminary experiments meant that bacterial intracellular DNA derived from phage "is not merely DNA in solution, but is part of an organized structure at all times" and that the sensitization of phage DNA to DNase at phages' adsorption to bacteria "might mean that adsorption is followed by the ejection of the phage DNA from its protective coat" (Hershey and Chase, 1952, 45). The experiments that followed confirmed "that the bulk of the phage sulfur remains at the cell surface during infection, and takes no part in the multiplication of intracellular phage. The bulk of the phage DNA, on the other hand, enters the cell soon after adsorption of phage to bacteria." In their discussion they state:

> We have shown that when a particle of bacteriophage T2 attaches to a bacterial cell, most of the phage DNA enters the cell, and a residue containing at least 80 per cent of the sulfur-containing protein of the phage remains at the cell surface. This residue consists of the material forming the protective membrane of the resting phage particle, and it plays no role in infection after the attachment of phage to bacterium.
>
> Hershey and Chase (1952, 54)

Of the remaining 20 percent of sulfur-containing protein, they found that little or none of it was incorporated into the progeny of the infecting particle. Phosphorus and adenine derived from the DNA of the infecting particle, on the other hand, were transferred to the phage progeny to a considerable extent. Thus they inferred that whereas protein has no function in phage multiplication, DNA has some function.

> Our experiments show clearly that a physical separation of the phage T2 into genetic and non-genetic parts is possible. A corresponding functional

separation is seen in the partial independence of phenotype and genotype in the same phage.

Hershey and Chase (1952, 54)

Against this background, Watson and Crick's 1953 model of the structure of DNA as a double helix of two anti-parallel uniform sugar-phosphate backbones held together by weak hydrogen bonds between complementary nucleotides immediately appealed to the geneticists' quest for material hereditary transmission by hybridization. As Watson and Crick were quick to point out, the model of specific pairing immediately suggested a mechanism of auto-replication (Watson and Crick, 1953b). Indeed, in their second paper Watson and Crick more specifically referred to the "Genetical implications of the structure of deoxyribonucleic acid,"

> Many lines of evidence indicate that it is the carrier of part of (if not all) the genetic specificity of the chromosomes and thus of the gene itself . . . Until now, however, no evidence has been presented to show how it might carry out the essential operation required of a genetic material, that of exact self-duplication.

Watson and Crick (1953a, 964).

Watson and Crick specified how their "recently proposed structure for the salt of deoxyribonucleic acid, which, if correct, immediately suggests" the properties that Muller allocated to genes as the atoms of heredity:

> [1] The first feature of our structure which is of biological interest is that it consists not of one chain but of two. . . .
> [2] The other biologically important feature is the manner in which the two chains are held together. This is done by hydrogen bonds between the bases, . . . The important point is that only certain pairs of bases will fit into the structure. . . .
> [3] It follows that in a long molecule many different permutations are possible, and it therefore seems likely that the precise sequence of the bases is the code which carries the genetical information. . . . Thus one chain is, as it were, the complement of the other, and it is this feature which suggests how the deoxyribonucleic acid molecule might duplicate itself. . . .our model for deoxyribonucleic acid is, in effect, a *pair* of templates, each of which is complementary to the other. . . .
> [4] Our model suggests possible explanations for a number of other phenomena. For example, spontaneous mutation may be due to a base occasionally occurring in its less likely tautomeric forms.

Watson and Crick (1953a, 965 and 966)

The model indicated not only a mechanism of self-replication but also, once it became clear that the three-dimensional structure needed for proteins' function is immanent in the polypeptides' primary sequence (Anfinsen and Redfield, 1956), that a code of linear nucleotide sequence may convey information to the linear amino-acid sequence that is essential to their assembly into functional specific three-dimensional constructs. Likewise, it allowed changes in sequence without losing the properties of self-replication, as needed for inherited mutations.

Notwithstanding, the model posed some serious difficulties at the physico-chemical level. Watson and Crick mentioned a major one: "Since the two chains in our model are intertwined, it is essential for them to untwist if they are to separate... Although it is difficult at the moment to see how these processes occur without everything getting tangled, we do not feel that this objection will be insuperable" (Watson and Crick, 1953a, 966). A prominent skeptic was, however, Max Delbrück, who put special emphasis on the difficulty of the unwinding of the double-helix structure at replication. This difficulty exposed, perhaps better than any other problem, the tension between the wish to come forward with elegant models for biological phenomena, and the need to reduce the biological phenomena to stringent physico-chemical terms. The experiment of Meselson and Stahl (1958b), which demonstrated the semi-conservative nature of the replication of the DNA molecules, firmly established the model of Watson and Crick and at the same time elegantly demonstrated the power of the basic conception and methodology of hybridization of the Mendelian science of genetics even in the era of molecular analysis, although it too did not prove that the semi-conserved entities of replication were each single strands of the double-helix model, neither did it resolve the uncoiling conundrum.

> To test the Watson-Crick hypothesis, one need only "label" the DNA of an organism, allow it to reproduce in a non-labeling medium, and then determine the distribution of parental label among progeny DNA molecules.
> Meselson and Stahl (1958a, 9)

Meselson and Stahl noted that hypotheses for the mechanism of DNA replication "differ in the predictions they make concerning the distribution among progeny molecules of atoms derived from parental molecules."

> We anticipated that a label which imparts to the DNA molecule increased density might permit an analysis of this distribution by sedimentation

techniques. To this end, a method was developed for the detection of small density differences among macromolecules. By use of this method, we have observed the distribution of the heavy nitrogen isotope N^{15} among molecules of DNA following the transfer of a uniformly N^{15}-labeled, exponentially growing bacterial population to a growth medium containing ordinary nitrogen isotope N^{14}.

Meselson and Stahl (1958b, 671–672)

In essence, *E. coli* bacteria were fed for several generations with nutrients that contained a heavy isotope of nitrogen (N^{15}). When all the nitrogen of the bacteria DNA was of the heavy isotope, the bacteria were allowed to replicate (synchronously) once, twice, and three times in a medium with only light DNA precursors. Samples of DNA from given cycles of replication were run in the ultracentrifuge (in cesium chloride solutions of different concentrations, which had buoyancies such that DNA molecules would equilibrate with the CsCl, giving bands somewhere in the middle of the tube's length). One generation of replication produced hybrid molecules that gave a band located half-way between that of the heavy DNA and the light DNA. Another cycle of replication produced equal amounts of two buoyancies, the hybrid buoyancy and the light-only buoyancy. Subsequent cycles further diluted the band of the hybrid buoyancy.

The results permitted the following conclusions:

1. *The nitrogen of the DNA molecule is divided equally between two subunits which remain intact through many generations. . . .* That is, the subunits are conserved.
2. *Following replication, each daughter molecule has received one parental subunit. . . .*
3. *The replicative act results in a molecular doubling. . . .*

The results . . . are in exact accord with the expectations of the Watson-Crick model for DNA duplication. However, it must be emphasized that it has not been shown that the molecular subunits found in the present experiment are single polynucleotide chains or even that the DNA molecules studied here correspond to single DNA molecules possessing the structure proposed by Watson and Crick.

Meselson and Stahl (1958b, 676–678)

Of the three alternative models, the conservative, the semi-conservative, and the distributive, DNA replication in *E. coli* evidently followed the semi-conservative pattern.

The experiment was thus not only "the most beautiful experiment in biology" (Holmes, 2001), but also the most fundamental experiment to link molecular genetic research to the established science of genetics: hybridization was not merely a research method; it was a fundamental property of hereditary matter. Furthermore, the fact that DNA of the bacteria replicated *as a unit*, semi-conservatively, rather than distributively, strongly indicated that all the DNA of these cells acts as one continuous molecule. This, of course, made the difficulty of unwinding even more serious, but the elegance of the experimental evidence made it easier for most investigators to ignore it. The problem was, of course much more serious for eukaryotic chromosomes that were orders of magnitude longer.

Meselson and Stahl extended the experiments to the DNA of organisms with more DNA per chromosome, like salmon sperm. Their findings, however, were equivocal (Meselson and Stahl, 1958b, 679–681). When the physical structure of the bacterial chromosome as a circular structure was elucidated, Stahl suggested that the eukaryotic chromosome might be constructed of bacterial-type DNA circles attached to each other by protein connectors (Stahl, 1961). This would partially alleviate the unwinding difficulty. However, autographic studies of tritiated thymidine (thymidine-H^3) pulse labeled chromosomes of plant root cells enabled Herbert Taylor to perform experiments analogous to those of Meselson and Stahl in plant chromosomes (Taylor, Woods, and Hughes, 1957). Chromosomes were labeled with thymidine-H^3 during the interphase when they were duplicating (later denoted the S-phase). Colchicine was applied to destroy the mitotic spindle and thus prevent sister-chromatids from moving apart.

> The equal distribution of the label at the first division after a single duplication in thymidine-H^3 tells us very little by itself. . . . But at the next division, after one duplication free of the labeled precursors . . . the labeled chromosomes regularly produced one labeled chromatid and one completely free of label. Since the labeled anaphase chromosomes, which have duplicated once in thymidine-H^3, conserved this DNA and passed it on to only one of their daughter chromosomes, the original unlabeled DNA must likewise have been conserved and passed on to the other daughter. Therefore, the chromosomes must have been composed of two subunits of DNA.
> Taylor (1959, 64)

This indicated that chromatids, like bacterial chromosomes, were double structures that replicated semi-conservatively, and Taylor concluded that

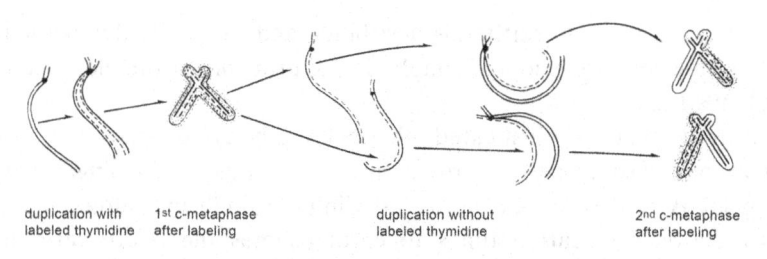

duplication with   1st c-metaphase   duplication without   2nd c-metaphase
labeled thymidine   after labeling   labeled thymidine   after labeling

Figure 12.1. Scheme of replication of a chromosome of *Vicia faba* in the presence of a radioactively labeled medium. Each line represents one subunit of the chromosome or chromatid. Broken lines indicate labeled subunits; solid lines indicate unlabeled subunits. C-metaphase indicates metaphase in the presence of colchicine that prevents separation of centromeres (Taylor *et al.*, 1957).

chromosomes are composed of "two and only two functional sub-units of DNA" (Taylor, 1959, 64). But, of course, there is no evidence to indicate that each of the double entities was one strand of complementary DNA strands (Figure 12.1).

> The DNA of *E. coli* and the chromosomes of *Vicia* are each composed of two sub-units which are assorted into progeny units according to the same set of rules. If we assume that the DNA of *Vicia* would behave in a DNA transfer experiment exactly as does that of *coli*, we can deduce the following relationship between DNA molecules and chromosomes: the two subunits of each DNA molecule segregate with different sub-units of the chromosome. The establishment of this rule tempts one to construct models of chromosomes with specified *structural* relationships between chromosomal and molecular sub-units.
>
> Meselon and Stahl (1958a, 11) and Holmes (2001, 390)

Each chromatid, one could speculate, was composed of two (or several) double helices that were distributed semi-conservatively.

> For the sake of definiteness, it could be supposed that DNA in *coli* . . . exists as a *pair* of Watson-Crick molecules associated laterally or, perhaps, end-to-end, and that it is these *intact* Watson-Crick molecules which are the sub-units in the N^{15} transfer experiments. . . . An end-to-end association might be expected to escape detection even upon direct observation in the electron microscope. So far as we can see, however, such models can be accepted only at the expense of abandoning the reasonable and detailed molecular mechanism for the specific synthesis of new DNA offered by the Watson-Crick hypothesis.
>
> Meselson and Stahl (1958a, 10–11).

199

There was no way to settle this possibility, and the single double-helix DNA molecule was adopted simply on the principle of the elegance of Occam's Razor.[1]

It took more sophisticated physical methods to show that the eukaryotic chromosomes were essentially composed of a single, very long DNA sequence. Kavenoff and Zimm applied measurements of visco-elastic retardation times to chromosomes that were carefully extracted. They determined the molecular size of the largest molecule for different species of Drosophila (with different karyotypic arrangements of their chromosomes), as well as for a *D. melanogaster* translocation (that increased the size of one chromosome significantly) and a pericentric inversion (that changed the location of the centromere without changing the size of the chromosome). Their results indicated that chromosomes were indeed single DNA molecules that stretched the chromosome from end to end to include centromeres and other cytologically observed structures (Kavenoff and Zimm, 1973).

The difficulty of unraveling the long double-helix structure at replication was overcome when it was shown that replication initiated at numerous sites along the chromosome, each unraveling, followed by synthesis, was extended in *both* directions in the form of a "replication-bubble" that eventually coalesced with its neighbor bubbles (Huberman and Riggs, 1968).

The replicating enzyme, DNA-polymerase, needs an RNA-primer, and synthesis is always in one direction (the $5' \rightarrow 3'$ direction). At the origins of replication at each replication bubble, synthesis is straightforward on one strand (the $3' \rightarrow 5'$ of the template) in each direction, and it unravels on the other strand of the template in short backward spells of "Okazaki fragments" so as to allow the enzyme's $5' \rightarrow 3'$ synthesis (Okazaki, Okazaki, Sakabe, Sugimoto, and Sugino, 1968).

What were expressed in molecular terms were not only the organization of the chromosomes and their replication, but also the mutations and mechanisms of gene expression that comprised a dominant volume

---

[1] An additional observation of Taylor's experiments was that a label switched between chromatids indicated some kind of chromatid recombination. It was an open question whether these label exchanges were markers of normal interchromatid mitotic recombination events, or the effect of the radioactive isotope-induced breakage and rejoining.

of the molecular research effort. A special issue was that of recombination. As noticed earlier (Chapter 5), the chiasmatype mechanical stress model was replaced by enzymatic models based on pairing of DNA-strands from two non-sister chromatids. Staggered breaks in one strand of each chromatid provided the basis for hybridization at the molecular level.

13

Recombination molecularized

> Most geneticists believe in crossing over, that is to say
> breakage of two chromosomes at homologous points followed
> by exchange. This process has generally been thought of in
> mechanical terms. It is at least equally fruitful to think of it in
> chemical terms.
>
> Haldane (1954, 110)

When recombination was observed in bacteria, Lederberg re-evoked the
old idea of copy-choice that Belling (1928) and Freese (1957) had suggested, proposing that it could also be extended to eukaryotic chromosomes in order to overcome abnormalities such as gene conversion (see
Chapter 6): If daughter chromatids were formed by conservative replication, then a copy-choice mechanism might switch synthesis from one
chromatid to another, and if the synthesis of the two daughter chromatids from each parental chromosome was not quite synchronized, both
might be copied over a short interval from the same parent, thereby
giving rise to a 3:1 ratio at a heterozygous site. Replicas can switch from
copying off one chromosome to copying off the other repeatedly in short
regions of effective pairing, and the switching need not be reciprocal,
resulting in both gene conversion and negative interference (Holliday,
1964; Whitehouse, 1965, 317). However, as noted earlier (Chapter 5),
there were serious experimental inconsistencies with any copy-choice
model of eukaryotic chromosome replication: The model predicted
recombination between only the two replicated daughter chromatids,
ignoring the fact that all possible pairs of the four chromatids may be
involved in crossing-over events; chromosome replication occurred at
the S-phase of the cell cycle, prior to meiosis and to its chromosome pairing (Swift, 1950); and this replication of the DNA was

semi-conservative (Taylor, 1959). Finally, it was demonstrated that a burst of a small amount of DNA-replication occurred at zygotene when exchange pairing presumably occurred (Stern and Hotta, 1973) suggesting the involvement of some DNA-repair mechanism.[1]

As noted by Holliday, "one advantage of the copy choice hypothesis is that it makes several quite specific predictions," and it has become evident that most of these predictions are not supported by experimental evidence. Yet gene conversion was shown to be related to recombination events and it appeared that conversion was impossible without some genetic replication. And "genetic results can only be explained if the switch takes place in such a way that no deletion or duplication of the genetic material occurs." This being so, recombination by torsion rupture and reunion must be "disregarded on the grounds that it could not be a process precise enough to explain the genetic data" (Holliday, 1964, 283).

A model based on breakage and rejoining that challenged the accepted chiasmatype hypothesis of recombination was proposed by H. L. K. Whitehouse. He suggested that the increased torsion of the paired homologs at meiotic prophase led to the mechanical rupture of only *one* strand of the double helix of the two chromatids. Following this,

> the reconstruction of DNA molecules...took place not by end-to-end association of broken chains but by a lateral association of complementary segments from homologous regions, to give *hybrid DNA*. End-to-end joining would appear readily to allow deficiencies or duplications of nucleotides to arise, whereas lateral association would automatically give great precision to the recombination process.
>
> Whitehouse (1965, 318)

Whitehouse referred to Marmur and Lane's (1960) demonstration that "complementary nucleotide chains of DNA will associate spontaneously *in vitro* under suitable conditions to give the duplex molecule. It therefore seemed likely that re-assembly of broken molecules might occur *in vivo* in a similar way" (Whitehouse, 1965, 318). By this Whitehouse was a pioneer who extended the genetic notion of experimental hybridization to the molecular level and introduced a revised outlook on crossing over from a pure mechanical to a molecular process that is enzymatically

---

[1] Although semi-conservative replication was found to be the rule, conservative replication may occur by the "rolling circle" replication mechanism. This mechanism is found in some phages as well as in the ribosomal-RNA amplifying mechanism in the oocytes of the toad *Xenopus*.

controlled: "If the initial breakage was in homologous regions of complementary chains, one from each molecule, crossing-over could occur by a succession of steps involving chain separation, synthesis of new chain segments, and base pairing of complementary segments from the two molecules" (Whitehouse, 1965, 318). "The trick was to make recombination involve a heteroduplex, in which a marked segment of one chain carries information from one chromosome while the corresponding segment of the other chain carries information from its homolog" (Stahl, 1994, 241).

Holliday went one crucial step further than Whitehouse and suggested specificity in the initial breakage process, rather than random mechanical breakage, namely, that the initial breakage was in chains of the *same polarity*. He too "uses the complementarity of the two strands of DNA in order to explain specific chromatid pairing at the molecular level. After genetic replication and general genetic pairing at certain points DNA molecules from opposite homologous chromatids unravel to form single strands, and these then anneal or coil up with the complementary strand from the other chromatid. Thus specific or effective pairing over short regions could occur" (Holliday, 1964, 284). These may be resolved in one of two ways:

> If one of the pairs of non-complementary strands is involved in this process, then the half chromatid chiasma is converted into a whole chromatid chiasma, thus recombining outside markers . . . If the other pair is involved then no chiasma is formed and there is no recombination of outside markers. In both cases there is a region of 'hybrid' DNA . . . if the annealed region spans a point of heterozygosity – a mutant site – then mispairing of bases will occur at this site.
>
> Holliday (1964, 285)

Assuming that such mispairing is unstable, one or both such bases may get involved in reactions that restore a complementary pair within the helix by normal hydrogen bonding. Holliday made the point that "it is most reasonable to suppose that such exchange reactions would be enzyme mediated," like that of one of the DNA repair mechanisms. Depending on which direction these repairs occurred, 2 : 2, "1 : 3 or 3 : 1 ratios for particular alleles could be achieved in the absence of any DNA synthesis" (Holliday, 1964, 286). The observation that about half of the tetrads of fungi that manifest gene conversions were also recombinants for markers flanking the site of conversion, and that the remaining gene-conversion tetrads were parental with respect to the outside markers, provoked the notion of a structurally symmetric four-chain intermediate. It could resolve to give crossover and non-crossover chromatids with equal probability. The proposed model "explains why conversion and crossing-over are correlated, why when both events do occur only two of the four chromatids are involved, and why conversion can occur without crossing-over" (Holliday, 1964, 286). Enzymes that cause junctions to slide were found in bacteria, and enzymes that are capable of resolving Holliday junctions (as the four-chain intermediates were now called) *in vitro* were found in T4 phage, in *E. coli*, and elsewhere. Mutants lacking these enzymes are frequently recombination deficient.

As noted by Stahl, "Holliday's junction has been a cornerstone of recombination models since its introduction. Consequently, it has been a focus for biochemical investigations, as well," and his model "was the lightning rod for 30 years of research, and its central assumptions, though modified, have survived every strike" (Stahl, 1994, 246). Molecular biologists sometimes believed that they had re-invented genetics: Calling attention to ignoring Whitehouse's contribution, he noted: "I suspect that the overthrowing of the chiasmatype model meant less to a phage geneticist (like me) than to a Drosophila geneticist (like you). From 1953 on, we sought to explain recombination in terms of DNA molecules, and we were ignorant of the classical past." On the other hand, Max Delbrück "scorned a suggestion that the relevant correction enzymes might remove stretches of DNA of appreciable and variable length, preserving intragenic mapability" on sound physico-chemical considerations, and "dictated against the invention of an enzyme just because genetic phenomenology called for it" (Stahl, 1994, 246). Delbrück was wrong, but as Stahl wrote: "Max D. once proclaimed that he expected to succeed in understanding biological replication because he was not encumbered by the baggage of classical genetics."[2] Whether it was out of ignorance or intentionally, a point was made to conceive of molecular genetics as independent of "classical genetics."

Holliday, on the other hand, was the first to stress: "Clearly the mechanism of conversion and crossing-over that is proposed is a general rather than a specific one," although "there are strong indications that whatever the basic mechanism is operating, the details of this mechanism may not be the same in different organisms" (Holliday, 1964, 288). And the genetic theory of the phenomenon of recombination by sheer mechanical

---

[2] Stahl–Falk e-mail correspondence, November 2006.

tension of twisting chromosomes has given way to a molecular mechanism based on the homologous annealing properties of DNA strands.

In the initial Holliday model no DNA synthesis was needed, except for the ligation of the broken ends and the repairing of the mismatches. However, soon it became necessary to postulate some DNA synthesis at the site of recombination. Hastings and Whitehouse (1964) postulated strand breakage at only specific sites on the chromatids followed by unraveling of the DNA strands and synthesis of the complementary strands filling the gaps, in order to account for the polarity in recombination. The necessary support for such models was provided by experimental evidence for a pulse of DNA synthesis during pachytene (Hotta, Ito, and Stern, 1966). But genetic analysis asked also specifically for breakage-induced enzymatic DNA-repair mechanisms.

By 1949 Kelner (1949) showed that the lethal effect of ultraviolet light on the conidia of *Streptomyces griseus* could be reversed by subsequent treatment with visible light. This introduced the phenomenon of mutagenesis repair mechanisms. An enzyme system was described that was capable of bringing about in various organisms "repair" of DNA damage caused by ultraviolet. *In vitro* experiments showed that irradiating solutions of thymine with ultraviolet led to the formation of thymine-dimers. And *in situ* experiments showed that the observed photoreactivation was due to an enzymatically induced split of the dimers. Setlow and Carrier (1964) and Boyce and Howard-Flanders (1964) showed that an ultraviolet-resistant strain of *E. coli* had the ability to remove the thymine dimers from its DNA, whereas an ultraviolet-sensitive strain lacked this capacity, because of a mutation in a specific gene. In due time many more DNA-repair mechanisms were discovered and the genes involved in the enzymatic processes that such mechanisms demanded were uncovered.

Although recombination turned out to be a more complicated process than Whitehouse and Holliday had initially described, the model was attractive: It was an enzymatically guided process of breakage, unraveling double strands, complementary synthesis of stretches of DNA-sequences, heteroduplex-configurations, and resolving these configurations by enzymatic DNA-repair mechanisms. More specifically (see Figure 13.1): (i) the junction could be cut in two ways (d): one resolved without crossing over of flanking sites; the other resolved by crossing over of flanking sites, at equal probability. (ii) It allowed a two-step process: two strands of the same polarity were cut first (a); if and only if they were truly homolog, permanent partners, were they swapped (b). If

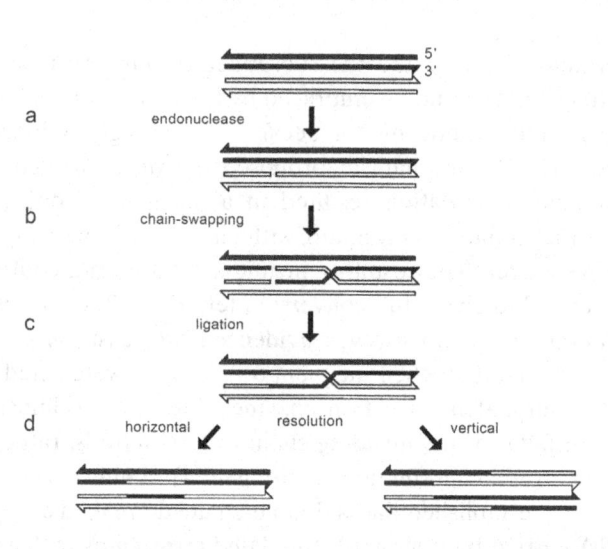

Figure 13.1. The Holliday model for enzymatically guided recombination between two chromatids, each composed of a double helix molecule of DNA (Stahl, 1994).

they were not, each could retreat to the old partner. (iii) A heteroduplex was produced on both chromatids involved (c). (iv) The heteroduplex was equally extensive on both strands. This last requirement predicted symmetric mismatch correction that would result in 3 : 1 conversions as often as 1 : 3 conversions. However, it became apparent that Holliday's model was too symmetric to deal with the experimental data. This could be overcome by relaxing the requirement that the initiating cuts be precisely isolocated: once chain swapping had been effected, appropriate enzymes could trim or fill as necessary (Stahl, 1994).

Meselson and Radding (1975) modified the model. In their model, a recombinogenic single chain was displaced from a chromatid by the action of polymerase operating in the chain displacement mode. This chain invaded the homolog (exploiting the as yet to be discovered "strand invasion" activity of RecA protein), displacing the resident chain of like polarity. Nuclease activity was postulated to remove this displaced chain, and the genetically asymmetric but structurally symmetric Holliday junction was restored. A marker in this region would show half conversion (segregate 5 : 3) if it were not mismatch-corrected. Diffusion-driven or enzyme-driven sliding of the junction away from the point of initiation would result in segments of reciprocal (symmetric) heteroduplex DNA. By appropriate adjustment of the relative durations of the initial asymmetric phase and the subsequent symmetric phase a wide range of fungal data could be embraced by the model (Stahl, 1994, 242).

Obviously the genetic model became increasingly a biochemical-enzymatic one. The mechanism proposed by Meselson and Radding for recombination initiation resulted in a net gain of one (simplex) copy from the initiating chromatid with the loss of one simplex copy from the responder. The resulting incipient 5 : 3 tetrad, could be mismatch-corrected to give a full conversion tetrad (6 : 2) or to restore the Mendelian ratio of 4 : 4. However, evidence from yeast (*Saccharomyces cerevisiae*) showed that when incipient 5 : 3 tetrads were acted upon by presumptive mismatch-correction enzymes, they were (almost) always converted in favor of the invading chain to 6 : 2 tetrads, rather than in favor of the invaded chromatid (restoring 4 : 4) (Stahl, 1994).

When it was established that a double-chained break in a fragment of yeast DNA carried by a plasmid stimulated crossing-over that incorporated the plasmid into the chromosome, the mechanism of recombination was revolutionized once again (Orr-Weaver, Szostak, and Rothstein, 1981). The recombinogenicity of a double-chained break, and the demonstration that it stimulated DNA synthesis that fully incorporated the plasmid in the chromosome, was equivalent to a full conversion without mismatch correction: the Holliday junction model was modified accordingly so that the information for repairing each chain was derived directly from the intact homolog. However, in the new model (i) initiation was no longer symmetric, (ii) both chains of a duplex were cut to initiate recombination, and (iii) heteroduplex DNA was relegated to a minor role and conversion occurred without a requirement for mismatch correction (Stahl, 1994). As it turned out, meiotic initiators of recombination were sites for spontaneous meiosis-specific double-chain breaks. Although this model settled many experimental results, some data still needed further analysis of this DNA-strand hybridization model of pairing by invasion (or aggression) of DNA strands and complementary synthesis of missing segments or of correcting mispairings.

Recombination became a matter of induced breakage at sensitive sites of the chromosomal DNA sequence, and repair mechanisms became a discipline in which not much of the rationale of genetic analysis remained, except for the notion of DNA-strands' specific hybridization properties.

VI

Deducing genes from traits, inducing
traits from genes

Mendel's decision to select such unit characters that would provide insight into the laws of hereditary transmission, whatever their physiological or developmental properties may be, was crucial in the establishment of genetics as an independent discipline of *inheritance* (Falk, 1995). But when de Vries in 1900 recognized Mendel's work it was because he believed it to contribute to his hypothesis of *Intracelluläre Pangenesis*, a bottom-up determinist theory of organisms' *development* and *function*, as part of the strategic goal of his *Die Entstehung der Arten durch Mutation*. Indeed, the "unit character" was as much a notion of development as one of inheritance. As conceived by Bateson, unit characters were *defined* by properties that segregate according to Mendelian rules, and he tried to explicate their role in the physiology and development of the organisms accordingly (see Chapter 3).

Thus, from its inauguration the discipline of genetic research was burdened or blessed with the dialectic tension between the methodological reductionism of hybridists, deducing genes from traits, and the conceptual determinism of morphogenists, inducing traits from genes.

Although geneticists never lost interest in problems of physiology and development (Gilbert, 1998), they did adopt, as a rule, a bottom-up perspective of individual traits rather than a top-down perspective of organisms in their context. However, they needed some empirical handle to examine speculations such as the Weismann–de Vries hypothesis of specific nuclear particles being differentially exported to the cytoplasm of somatic cells. De Vries suggested dominance: why and how did it happen that although two alternative factors are transmitted equally in the germ-line, as a rule only one factor is expressed, and in so far as a factor varies, it is the established "wild type" variant that is usually the dominant allelomorph.

Johannsen's discrimination between genotype and phenotype appeared to settle the dispute of hybridists and morphogenists. Hybridists conceived of traits as convenient markers for the study of the mechanisms of inheritance, whereas morphogenists studied the physiological, developmental and behavioral phenotypic consequences of given genotypic set-ups. Richard Woltereck attempted to stress the always present role of environment in shaping phenotypes by introducing the concept of the norm of reactions (NoR) (see Chapter 4). He suggested that there might not be a way to predict the NoR of one strain from that of another strain to the same array of environmental conditions (Woltereck, 1909), but that suggestion was largely thought to be a defeatist "Lamarckian" position (Harwood, 1996). Contrary to Woltereck, the pathologist Oskar Vogt and geneticist Nikolai Timoféeff-Ressovsky took an extreme genocentrist determinist position, according to which genes have specific effects, though for "physiological" reasons not all carriers of specific genetic constitutions display the phenotype of the genetically determined trait (penetrance), and it may not be expressed to the same extent in all individuals showing the trait (expressivity) (Timoféeff-Ressovsky and Timoféeff-Ressovsky, 1926; Vogt, 1926; see also Falk, 2000a; Sarkar, 1999). Few of the studies directed at specific physiological or biochemical mechanisms were concerned with how genetics and environment interacted in the formation or the function of a trait, and most genetic analysis in the decades to come was increasingly "genocentric," stressing the determinist role of genes in the induction of traits.

The establishment of the Watson–Crick model of the material basis of inheritance was the triumph of this genetic determinism. Extending phenomenological genetics to methodologies at the biochemical level opened genetic analysis not only to the study of the molecular organization and mechanics of DNA, but even more so to research on the physiology of gene involvement in cellular functions and organismic development. This culminated in the Central Dogma of Heredity and the uncovering of the genetic code, which indicated that "genetic information" was transmitted one-way from DNA by transcription to RNA and from RNA by translation to proteins (Crick, 1966). Even the intervention of the environment was genetically regulated (Jacob and Monod, 1961), to the extent that when reverse-transcription was discovered by Temin and Mizutani (1970) and by Baltimore (1970), it was rejected for a long time because it appeared to refute Crick's Central Dogma (see Crick, 1970).

14

How do genes do it?

For Mendel the function of hereditary factors was a secondary issue. When he encountered the phenomena of *dominating* and *recessivity*[1] he referred to them simply as facts that must be considered when dealing with the transmission of the factors:

> Although the intermediate form of some of the more striking traits . . . is indeed nearly always seen, in other cases one of the two parental traits is so predominant that it is difficult, or quite impossible, to detect the other in the hybrid. . . . This is of great importance to the *definition and classification* of the forms in which the offspring of hybrids appear.
>
> Mendel, in Stern and Sherwood (1966, 9, italics added)

Mendel further emphasized that although "some of the more striking traits" show intermediate forms in hybrids, he deliberately selected for his experiments such traits that could be discerned qualitatively, so that he would be able to classify his data into binary categories (Falk, 2001a).

Not so for de Vries, for whom dominance was the clue to understanding function and development. He introduced his 1900 paper *Das Spaltungsgesetz der Bastarde* in terms of function rather than transmission:

> According to pangenesis the total character of a plant is built up of distinct units. These so-called elements of the species, or its elementary characters, are conceived of as tied to bearers of matter, a special form of material bearer corresponding to each individual character. Like chemical molecules, these elements *have no transitional stages* between them.
>
> de Vries, in Stern and Sherwood (1966, 107, italics added)

Thus, whereas Mendel deduced *Faktoren* of transmission from the segregation of traits in hybridization experiments, de Vries induced the

---

[1] Mendel used the terms *dominirende* and *recessive Merkmake* (Mendel, 1866, 11).

development of traits from the *Faktoren* of Mendel's hypothesis. Indeed, the most basic model for the function of the Mendelian factors was that of Bateson's Presence and Absence Hypothesis (see Chapter 3). This is an all-or-none version of the notion of unit characters and may be conceived as a "one unit character – one function" hypothesis of segregating *Faktoren* action: where the factor was present there was an action, where it was absent, nothing was produced; hence dominance and recessivity were simply reflections of a kind of *Faktor*-effect identity. That Bateson may have had enzymes in mind can be inferred from his relations with Archibald Garrod, with whom he discussed as early as 1901 the rare human condition of alkaptonuria as an inborn error of metabolism (Bateson and Saunders, 1902, 133–134). Bateson's co-worker, C. C. Hurst, was more explicit in referring to the chemical nature-function of the hereditary unit character, calling the difference between red and yellow flowers "a distinct chemical reaction" (Hurst, 1906, 117). The conception was strictly determinist: each unit character had a *specific* effect or lost it. Characters too were the results of factors acting specifically "on the basis" of the function of other factors, or as independent steps in the expression of physiological or morphological pathways (see Schwartz, 2000, 2000b). Mendel's pea varieties were not yellow *versus* green, but rather varieties of yellow and non-yellow. In reality, the Mendelian contrasting pair, yellow and green, should be regarded as the "presence and absence of yellow on a basis of green" (Hurst, 1906, 119).

The tools to examine such hypotheses were limited, however, considering "there can be no doubt that the *visible* Mendelian characters are always secondary, and but little doubt that they are all dependent at some stage of analysis upon chemical relations" (Shull, 1909, 415). With factors being deduced, analysis of gene function at the phenomenological level is only a proxy to the more fundamental, physico-chemical level of processes. Bateson and his colleagues recognized two difficulties in adopting the Presence and Absence Hypothesis of unit characters: one empirical, namely, it is impossible to observe *dominant* traits of "absence": white is the absence of pigment, yet white fur of rodents may be dominant (over grey fur); the other conceptual, namely, it is impossible to envision evolution on the basis of new factors that were absence-alternatives to present dominant alleles. These difficulties could be overcome once the disjunction between the phenotypic unit characters and the genotypic factor was established.

Three kinds of evidence were brought up to refute the Presence and Absence Hypothesis at the genotypic level.

a. *Multiple alleles.* No more than two alleles were possible if the hypothesis related to the factor and its absence.
b. Reverse mutations. No reverse mutation to the wild-type allele is possible if the mutant allele is an "absence."
c. *Change of phenotype.* Correns had shown already in 1903 that dominance and recessivity relations of allelomorphs might change even within the same plant. It was, however, mainly William Castle who pointed out that recessive mutants might change their phenotype from "absence" to "presence" as a function of change in environmental conditions. The classical example is that of the Himalayan rabbit, whose fur color depends on the temperature at which it develops. Obviously an "absent" allele cannot become "present" with change of the environment. Likewise, Morgan and his associates called attention to the horned condition in sheep hybrids that was dominant in one sex and recessive in the other.

Although genetic analysis of alleged multiple allelism and reverse mutations continued well into the 1950s these were arguments referring to the nature of the gene, not to its function, and they too lost their power once it was shown that the structural unit of function was *not* a uniform, indivisible atomic entity. However, the empirical fact that the majority of newly induced mutations are recessive, i.e., that the effect of the "wild-type" alleles is, as a rule, completely (or nearly completely) dominant over the effect of the mutant allele, together with the finding that as a rule, the phenotype provided by the "wild-type" allele is much more stable over environmental fluctuations (such as temperature-shifts, see Plunkett, 1932), was so overwhelming that it was imperative to refer to dominance as a qualitative property of genes. Dominance was a clue to the function of genes and/or to the adaptive power of natural selection (see Chapter 19).

Morgan consistently insisted that each gene might affect many characters and that not all alleles affected the characters to the same extent. Consequently, he thought "it unwise to commit ourselves any longer to a view that a recessive character is necessarily the result of a loss from the germ-cell." He replaced the model of dominance and recessivity as *presence and absence of the genes* by a model of dominance as *a property of the genes*. This preserved a basic one-to-one relationship between the gene and its primary *product* (to distinguish from its *effect*!), according to which "there need be no loss, but only a change in configuration with a corresponding change in the end product in which the changed part plays a role, along with the other parts of the cell" (Morgan, 1913a, 11).

A more direct reference to the biochemical level was Troland's (1917) attempt to explain genetic autocatalysis in enzymatic terms, "due to an attraction between the solid crystal and the molecules in solution ... If he is right, each different portion of the gene structure must – like a crystal – attract to itself from the cytoplasm materials of similar kind" (Muller, 1922). Although Troland's attention was directed primarily at the problem of autocatalysis, this notion of enzyme-like action of the genes on their cytoplasmic surrounding was the basis of Muller's thinking about gene action.

Studies of variations of temperature on "phenogenesis" led Plunkett early on to suggest "a general theory ... based on orthodox concepts of physical chemistry and general concepts of genetics" of the processes involved in the phenotypic effects of genes (Plunkett, 1932):

> The usual difference ... between wild-type and mutant characters ... are ascribed, on the basis of this theory, to differences in the distance of the developmental process from its asymptote at the time it ends: processes controlled by wild-type genes usually closely approach their asymptotes, while those modified by mutant genes may be terminated by the effects of other developmental processes while still very incomplete. This conception is applicable ... also to other usual differences between wild-type and mutant characters, in respect to variability in general and in respect to dominance, which is merely a special case of interaction of factors affecting the same phenotypic character. This peculiarity in the physico-chemical kinetics of "wild-type" developmental processes is a result of natural selection, which tends, in general, to favor a genotype which produces a relatively uniform phenotype.
>
> Plunkett (1932, 159–160)

This functional model of dominance as a *by-product* of a more general selection for the stability of traits, was further elaborated by J. B. S. Haldane (1930) and H. J. Muller (1932). The studies of Stern (1929, 1960a) with mutations of *bobbed* and those of Muller with *white* (1932, 1950a) suggested that at least for some genes of Drosophila, the difference between the wild-type and mutant alleles was in the *amount* of the product: three doses of the mutation (in an otherwise diploid) did not give a more extreme aberrant phenotype than two doses, but rather one approaching that of the wild-type. This indicated that the mutations were "hypomorphs" – to use Muller's terminology – i.e., produced the same product as the wild-type alleles, only less so. Haldane and Muller, following Shull (1909, 415) and Wright (1929b, 278), suggested that the amount of product produced by alleles of a gene varied asymptotically.

According to Haldane, wild-type alleles were selected to be in the relatively far asymptotic segment of the curve, so as to provide for stability in phenotypes facing environmental fluctuations. Muller similarly extended the model to selection acting for alleles that were beyond the saturation value, as a homeostatic device that ensured independence of the "normal" phenotype from genetic or environmental haphazard. Muller concluded that "the effect of these normal genes must be regarded as like that of a quantitatively greater amount of mutant genes" but that with increasing efficiency of the gene "there ensues a progressively lesser rise" in phenotypic effect, until the gene-dosage/phenotype curve reaches saturation, obeying the "law of diminishing returns" (Muller, 1950a, 179–180). According to Muller, "there is no *a priori* reason why characters in general should represent the maximum biochemical possibilities. . . . The clue to an interpretation . . . is provided by the fact that mutant genes are on the whole much more variable in their expression than normal genes" (Muller, 1950a, 180–181). The phenotype of mutants, whose product is in the linearly increasing part of the gene-activity/ phenotypic-effect curve, may vary considerably with any fluctuation in external conditions (including the effects of other genes). For mutant alleles, "any agency which operates, during their activity in determining the character, in such a way as, in effect, somewhat to diminish or increase this activity, will result in a visible alteration of the character, for it will have an effect like that of changing the dose" (Muller, 1950a, 181) (see Figure 14.1). On the other hand, the phenotype obtained in the presence of wild-type alleles is remarkably stable under a wide range of environmental and genetic conditions. The wild-type alleles are better buffered than the mutant alleles. This could have been achieved either by natural selection for such alleles whose activity is in the asymptotic (horizontal) leg of the gene-activity/phenotypic-effect curve (as suggested by Haldane, 1930), or by modifiers of existing alleles (as suggested by Muller, 1932). Within limits, the further the genotypic activity is on the horizontal part of the curve, the more stable is the phenotypic effect.

It should, however, be noted that Muller was aware that the analogy to a dose/response curve was only that: *a phenomenological analogy*: denoting an allele "hypomorphic"

> does not imply that the mutant gene's action is qualitatively or in a biochemical sense exactly the same as that of the normal allele, only less. . . .
> What is implied is only that the mutant gene works in such a way as to produce a *final effect* similar to that of the normal, but a lesser effect, like

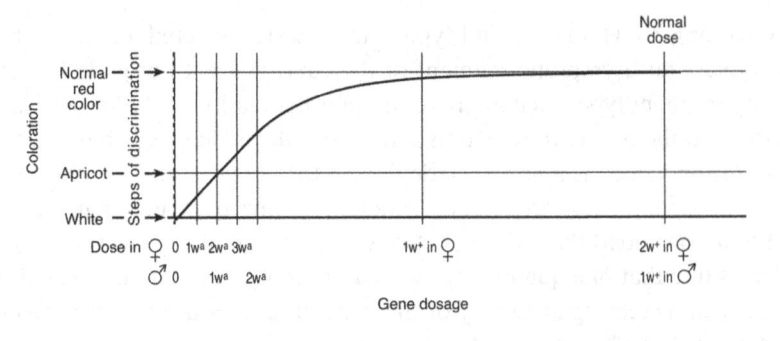

Figure 14.1. The relation between eye-color and gene dosage or activity at the *white* gene of *Drosophila*. w – white eye allele; w^a – white-apricot allele; w^+ – wild-type allele of the eye-color gene (after Muller, 1950a).

> that which would presumably have been brought about if the concentration or the activity of the normal gene itself (or of its products) had been reduced. It is in this sense that the mutant gene is less active.
>
> Muller (1950a, 177)

The dose/response analogy could work well also for "amorph" mutations, alleles that do not produce any of the normal effect, but hardly for the "hypermorph" ones, and not at all for the "neomorph" or "antimorph" mutations, those that produce more, a new, or an antagonistic effect to that of the normal, respectively.

A variety of terms have been introduced into genetic analytic discourse that presumably referred to the action of genes: penetrance and expressivity, pleiotropy, epistasis, and phenocopy were probably the most popular ones. They were introduced, however, not as explanations of gene actions, but rather as formal devices to overcome inconsistencies in the determinist "genocentric" conceptions of genetics in the 1920s to 1940s (Falk, 2000a; Sarkar, 1999). However, once introduced, these terms became *instrumentally* important in designing experiments to examine their validity and impact: the term epistasis has been assigned by Bateson to "characters which have to be, as it were, lifted off in order to permit the lower or *hypostatic* character to appear" (Bateson, 1907, 653). It refers to the interference of one gene with the phenotypic expression of another *nonallelic* gene (or genes), so that the phenotype is determined effectively by the former and not the latter when both genes occur together in the genotype (Rieger, Michaelis, and Green, 1976, 188). Testing for epistasis has been maintained in the present-day molecular and genomic analysis as a test for genes acting in the same or in

216

different functional or developmental pathways: if the effect of A and B is similar to that of A, then A is epistatic over B and both affect different steps of the same pathway, whereas when the effect of A and B is more extreme (e.g., additive) than that of A and B alone, they affect different pathways (Falk, 2001b). Similarly, phenocopy, a nonhereditary, phenotypic modification (caused by specific environmental conditions) that mimics a similar phenotype caused by a gene mutation was elaborated by Goldschmidt (1938b) into an efficient tool for the analysis of gene action: he assumed that the duplication of mutant effects by environmental agents applied at specific times might be used to investigate when and how the mutant gene changed the rates of a specific process (Sinnott, Dunn, and Dobzhansky, 1950, 407–408). In many mutants of Drosophila it was possible "to reproduce the phenotypic likeness of many mutants ... by treating larvae of a definite stage with temperature shocks, and this in a quite regular manner" (Goldschmidt, 1938b, 7). This turned out to be an effective method for determining the "sensitive period" of genes' action and of the rate of the processes (e.g., ether vapors induced bithorax phenocopies in adult Drosophila flies, when applied to early embryos), although attempts to identify specific gene functions by using reagents that would identify specific reactions had to wait until genetic analysis of prokaryotes and cell cultures allowed such devices.

Once it turned out that many phenotypic traits could unequivocally be related to the same genetic unit (as determined by complementation tests as well as by linkage analysis) the question was whether *pleiotropism*, the production by one particular mutant gene of apparently unrelated multiple (or multifold) effects at the phenotypic level (Rieger *et al.*, 1976), was "primary" or "secondary." Did the gene primarily induce various products or did genes produce only one specific product that interacted down the functional or developmental pathway with other products? Since there was no handle to analyze the primary gene product, the genetic analysis relied on examining whether the multiple effects could be converged to one specific effect. Of special interest were lethal factors, i.e., gene mutations whose presence in the genotype blocks normal development or a function that results in death (before sexual maturity):

> Development, as we know it, is ... a process of extraordinary precision. ... Each step is dependent upon the normal appearance of the preceding one, and the normal result depends ... upon the orderly sequence of events in quality, quantity, time and space. ... Most of the changes of individual processes that might be produced will throw out of gear the combined

system, and the result will be destructive. But certain processes may be changed without deleterious consequences; and if this is done by a genetic change, we call it a mutation.

Goldschmidt (1938b, 3)

Two examples of such pleiotropic lethal mutations are a complex "pedigree of causes" in the rat that converged to an anomaly of cartilage differentiation (Gruneberg, 1938), and the plethora of symptoms of the lethal sickle-cell disease in humans, which could all be traced to the β-component of the hemoglobin molecule (Neel and Schull, 1954), and eventually to the substitution of the sixth amino acid from glutamic acid to valine, due to a single base substitution in the DNA sequence (Ingram, 1957).

Lethals provided a rather abundant and quite well defined category for investigators, especially for Drosophila workers, whether in detecting mutagenic effects (see Muller, 1927a and Chapter 8), or the abundance of concealed genetic variability in natural populations (see Muller, 1950b), or in the detailed analysis of the physiological and developmental pattern of specific lethal genes (see Hadorn, 1961 and Chapter 16).

Plants provided especially adequate materials for studying the function of specific genes long before the introduction of systematic analysis of genetic control of metabolic pathways in Neurospora and other fungi and prokaryotes. Flower pigments are almost all of two types, plastid pigments and water-soluble pigments, anthocyanins and anthoxanthins. Haldane and his colleagues describe, beside genes, the recessive alleles which completely or partly block formation of one kind of pigment or another, and in some cases substitute one pigment for another, or even mutations that affect petal pH. Of special interest were genes in which change can produce local chemical differentiation (see, e.g., Haldane, 1954, chapter V, 53–62).

In 1922, when Muller set out on his project to characterize the genetical properties of the gene since "[t]he chemical composition of genes, and the formulæ of their reactions, remain as yet unknown," he made it a point to stress that,

> We do know, for example, that in certain cases a given pair of genes will determine the existence of a particular enzyme (concerned with pigment production), that another pair of genes will determine whether or not a certain agglutinin shall exist in the blood, a third pair will determine whether homogentisic acid is secreted into the urine ("alkaptonuria"), and so forth. But it would be absurd, in the third case, to conclude that on this

account the gene itself consists of homogentisic acid, or any related substance, and it would be similarly absurd, therefore, to regard cases of the
former kind as giving any evidence that the gene *is* an enzyme, or an
agglutinin-like body. The reactions whereby the genes produce their
ultimate effects are too complex for such inferences.

Muller (1922, 32–33)

However, one of the most basic facts of genetic analysis has been that the
function of the gene controlling the typical development of hereditary
traits cannot be studied directly but may only be extrapolated from the
knowledge of a mutated gene. Thus, the existence of the normal, "wild-
type" gene has only been inferred from the existence of a mutant allele
showing Mendelian behavior (Goldschmidt, 1938b).

With the establishment of a strong determinist, bottom-up genetics,
interactions were of secondary interest beyond – or above – that of the
function of each gene as a discrete entity that had a unique and specific
product. From the organismic top-down geneticists' vantage point
interactions were immanent in the basic organization of genetic material
as a component of the cell as a whole, or even in the organism as such;
whereas discrete genes were only instrumental devices.

Sturtevant's discovery that "two genes lying in the same chromosome
are more effective on development than are the same two genes when
they lie in different chromosomes" (Sturtevant, 1925, 147) reminded us
of the significance of *position effects* (see Chapters 8 and 9). The position
and number of copies of the genetic material *per se* may have phenotypic
consequences. Or as put by Goldschmidt: "Two adjacent Bar genes
reinforce each other's action, as compared with two opposed Bar genes.
The position of the gene has an influence upon its action" (Goldschmidt,
1938b, 304). In modern molecular genetic terms this would probably
correspond to a CNV, copy number variant, known to be involved in
many human diseases (Cohen, 2007).

While there was no way to directly study the pathway from genes to
traits as long as there was no well-defined physico-chemical description
of the matter of heredity; nevertheless, genetic analysis laid down the
foundations on which biochemical analysis built the molecular pathway
once deoxyribonucleic acid was recognized to be that matter and its
structure was suggested by the model of Watson and Crick.

15

The path from DNA to protein

In their short paper of April 25, 1953, Watson and Crick introduced their model of the structure of DNA. Although this is a presentation of a physico-chemical model, and the authors make "the usual chemical assumptions," they emphasize that "The structure has novel features which are of considerable biological interest."

> The structure has two helical chains each coiled round the same axis. . . . The novel feature of the structure is the manner in which the two chains are held together by purine and pyrimidine bases. . . . They are joined together in pairs . . . adenine (purine) with thymine (pyrimidine), and guanine (purine) with cytosine (pyrimidine). . . . The sequence of bases on a single chain does not appear to be restricted in any way. However, if only specific pairs of bases can be formed, it follows that if the sequence of bases on one chain is given, then the sequence on the other chain is automatically determined.
>
> Watson and Crick (1953b)

The authors notice that the model appears to be "roughly compatible with the experimental data, but it must be regarded as unproven until it has been checked against more exact results." Nevertheless, in the final paragraph they state that it "has not escaped our notice that the specific pairing we have postulated immediately suggests a possible copying mechanism for the genetic material" (Watson and Crick, 1953b).

In their second paper, a month later, Watson and Crick specify five features of their structure that are "of biological interest," namely:

1. "It consists not of one chain, but of two."
2. The "manner in which the two chains are held together" by hydrogen bonds between the bases. "The important point is that only certain pairs of bases will fit into the structure."

3. "[A]ny sequence of pairs of bases can fit into the structure. It follows that in a long molecule many different permutations are possible, and it therefore seems likely that the precise sequence of the bases is the code which carries the genetical information."
4. "[S]pontaneous mutations may be due to a base occasionally occurring in one of its less likely tautomeric forms."
5. "[T]he pairing between homologous chromosomes at meiosis may depend on pairing between specific bases" (Watson and Crick, 1953a, 965–966).

Despite a number of uncertainties they feel that their "proposed structure for deoxyribonucleic acid may help to solve one of the fundamental biological problems – replication." The hypothesis they are suggesting is "that the template is the pattern of bases formed by one chain of the deoxyribonucleic acid and that the gene contains a complementary pair of such templates" (Watson and Crick, 1953a, 966). Obviously their physico-chemical model is inspired by the biological problem, but more important, it is related directly to the problems of genetic research as formulated by Muller in his manifesto-paper, thirty years earlier.

> The most distinctive characteristic of each...gene – is the property of self-propagation....This action fulfills the chemist's definition of "autocatalysis"; it is what the physiologist would call "growth"; and when it passes through more than one generation it becomes "heredity." . . .
>
> The fact that the genes have this autocatalytic power is in itself sufficiently striking...and it is difficult to understand by what strange coincidence of chemistry a gene can happen to have just that very special series of physico-chemical effects upon its surroundings which produces – of all possible end-products – just this particular one, which is identical with its own complex structure. But the most remarkable feature of the situation is...the fact that, when the structure of the gene becomes changed, through some "chance variation," the catalytic property of the gene may become correspondingly changed, in such a way as to leave it still *auto*catalytic. . . .
>
> What sort of structure must the gene posses to permit it to mutate in this way?...it is evident that it must depend upon some general feature of gene construction...which gives each one a *general* autocatalytic power – a "carte blanche" – to build material of whatever specific sort it itself happens to be composed of.
>
> Muller (1922, 33–34)

Like Muller, Watson and Crick explicitly refer to the properties of "the gene itself." They also refer to the gene's "heterocatalytic" properties, the path from the gene to its cellular product and function, a subject that occupied much of the attention of molecular biologists over the decade that followed (see, e.g., Rheinberger, 1997).[1] Template models of autocatalysis – and heterocatalysis too – must be juxtaposed with models in which the entities function as catalysts rather than as templates. As pointed out by Haldane, "there has been an immense amount of speculation as to the method by which chromosomes and genes are reproduced." The two simplest hypotheses as to the copying process are as follows:

1. The model is spread out in a one-dimensional chain or a two-dimensional sheet...On this chain or sheet another precisely similar chain or sheet is laid down by a process analogous to crystallization...

2. The genes and the whole chromosome are "copied" into a completely different structure, the relation being, perhaps, like that of antigen and antibody. This "template" or "negative" is again copied, giving *two* new positives (Haldane, 1954, 107–108).

Haldane elaborated on one of the models that might take place throughout the nuclear cycle. "Most of the copies float off, and are primarily gene products, chemically similar to the gene. Only once in a nuclear cycle is a whole chromosome copied, the gene copies adhering to it" (Haldane, 1954, 108). Haldane's book was based on lectures given in 1950 and 1952 aimed at biochemists and was "intended to summarize some of the main facts concerning a branch of science which is growing so rapidly that, had the book been up to date when it was written, it would have been out of date at the time of publication" (Haldane, 1954, 5). Thus it is significant that gene replication and gene function were conceived as processes of template-copy or transcription of information rather than processes of catalytic induction, and that such a conception prevailed in genetic analytic deliberations. Nevertheless, Haldane concluded that "there is good reason to suggest that the genes are direct catalysts" (Haldane, 1954, 112ff.).

Once the Watson–Crick model of DNA structure as a double helix of a complementary sequence of nucleotide bases was adopted, the idea of template synthesis of proteins was obvious. Since the primary sequence

---

[1] It is interesting that in all three formative papers published in 1953 on the molecular structure of DNA (Watson and Crick, 1953a, 1953b, 1953c), Watson is the senior author, although both by age and by expertise in physical chemistry he was the junior author. Evidently the biological aspects and speculative consequences dominated the formulation of the physico-chemical model of the double-helix of the DNA molecule.

of amino acids uniquely determined the secondary and tertiary structures of the active proteins in the cytoplasm (Anfinsen and Redfield, 1956), it was enough for the hereditary system to provide information on the primary sequence of amino acids to obtain proteins' functional specificity. Genetic analyses of *Aspergillus* and of the T4 bacteriophage proved the gene, or *cistron* as the functional unit was called, to be composed of many sites of mutation. Genetic analysis provided also the notion of templates for transfer of sequential information from DNA to proteins. Although the proteinaceous product of the T4 *r*II gene was not known at the time, it was possible to correlate the topology of mutations in the genetic linkage map of *Escherichia coli* with that of the amino acids of specific peptides, including cases in which two closely linked mutations affected the same amino acid, suggesting that more than one nucleotide in the sequence codes a specific amino acid (Yanofsky, Carlton, Guest, Helinski, and Henning, 1964).

Once the concept of transcription of "information" from the DNA sequence and its translation into polypeptide sequences was conceived, much of the experimental tasks was at the biochemical level. But this emergence of new "epistemic things" (Rheinberger, 1997) was very much directed, or at least accompanied, by genetic analysis. To mention just three issues: the analysis of genetic regulation, the elucidation of the genetic code, and the formulation of the Central Dogma.

GENETIC REGULATION

Regulation of gene activation has bothered geneticists ever since it became obvious that the genotype of somatic cells does not differ from cells of one tissue to another, not even from germ-line cells, yet differentiation occurs (Falk, in press). Jacob and Monod simplified the problem of differentiation and reduced it to one of regulation at the cellular level, namely the adaptive regulation of enzyme production by bacterial cells exposed to varying environmental inputs. Induced gene mutations or loss of the adaptive property introduced "regulation of enzyme production" as markers and handles for genetic analysis: "Regulatory genes" constrain and regulate the response pattern of the bacterial cell, whereas "structural genes" "account for the multiplicity, specificity and genetic stability of protein structure, and it implies that such structures are not controlled by environmental conditions of agents" (Jacob and Monod, 1961, 318). The synthesis of the enzyme β-galactosidase in *E. coli* is controlled by the presence of the sugar lactose in the bacterial medium. Practically no enzyme is present in cells if their carbon source is glucose. However, within minutes of transfer to lactose the enzyme appears, and it disappears within minutes of its elimination from the medium. Mutations in the bacterial gene z affect the amino-acid sequence of the enzyme β-galactosidase; two other adjacent genes involve the structure of a "lactose-permease" (y) and an "acetylase" (a). But mutations in another gene (i), identified as a distinct complementation unit and located by recombination tests somewhat "upstream" to the three linked structural genes on the bacterial chromosome, affected the cells' adaptive property. Mutations at i turned the cells constitutive to all three enzymes' synthesis. Genetic analysis of heterozygotes (such as i^+z^-/Fi^-z^+, F being an episome that carried a short segment of the chromosome, thus turning the cells into a merozygote [a zygote which is diploid only for part of its genetic material and haploid for the remainder] and partially heterozygote) proved the i^+ to be a regulator gene producing a cytoplasmic product repressing the enzymes' synthesis. The regulator gene and its function are both *highly specific*, since mutations of i gene do not affect any other system, and *pleiotropic* since both galactosidase and acetylase are affected simultaneously and quantitatively to the same extent, by such mutations (Jacob and Monod, 1961, 334). The gene is a repressor since in the recessive i mutants regulation was lost: the three genes, z, y, and a produced their specific enzymes constitutively; i^s mutations were constitutive and dominant. Jacob and Monod concluded that the gene product of the i^+ "wild-type" allele forms a stereo-specific combination with lactose, thus relieving the cell from repression in the presence of the sugar, allowing the synthesis of the enzymes. But it must be assumed further that when not interacting with lactose the repressor acts on some site that coordinately affects the synthesis of all three enzymes:

> This controlling element we shall call the "*operator*." We should perhaps call attention to the fact that, once the existence of a specific repressor is considered as established, the existence of an operator element as defined above follows necessarily. Our problem, therefore, is not whether an operator exists, but where (and how) it intervenes in the system of information transfer.
>
> Jacob and Monod (1961, 341)

Another locus o, mutations of which coordinately affected all three enzymes, was identified adjacent to the locus of z. o^c constitutive mutations were assumed to result from loss of sensitivity of the operator to the repressor, and as predicted, they were dominant in o^c/o^+ merozygotes. Regarding molecular considerations, it was assumed that the interacting repressor molecule would be the RNA transcript of the repressor gene, but later it turned out to be a protein (Müller-Hill, 1996).

Jacob and Monod's analysis and model of the genetic control of cellular regulation opened a wide door to genetic and biochemical mechanisms of gene functions, in the best tradition of reductive heuristics. But, as Sydney Brenner commented: "The paradigm does not tell us how to make a mouse but only how to make a switch" (Brenner, Dove, Herskowitz, and Thomas, 1990, 485; and see Chapters 13 and 18).

THE GENETIC CODE

On July 8, 1953, three months after the presentation of the DNA model, an excited George Gamow dispatched a letter to Watson and Crick, pointing out that "[i]f your point of view is correct each organism will be characterized by a long number written in quadrucal system with figures 1, 2, 3, 4 standing for different bases" (Kay, 2000, 131), and in October of that year he published his proposal for the "coding problem" as the "possible relation between deoxyribonucleic acid and protein structure" (Gamow, 1954). Gamow's "diamond code" was an overlapping triplet code and it was based on strictly DNA–protein interactions, of "key-and-lock" relations between various amino acids and the rhomb-shaped "holes" formed by various nucleotides in the DNA chain (Kay, 2000, 136). This was obviously *not* an analytical genetic problem; for example, it was shown by Brenner (1957) that the combinations of pairs of amino acids recovered from proteins were too numerous for the constrained number possible by an overlapping DNA code. But genetic analysis soon became crucial in resolving the problem of reading the code. Whereas deciphering the code words for the twenty amino acids was done by biochemical methods of protein synthesis by adding polynucleotides to cell-free systems, genetic analysis of amino acid replacements in mutations and reverse-mutations at the tryptophan-synthetase (*trpA*) gene served to confirm *in vivo* the suggested codons for the amino acids (Helinski and Yanofsky, 1962).

It was, however, the genetic analysis of recombinants of *r*II mutations in T4 that elucidated the principles of coding the sequence of amino acids of peptides in the DNA: reading the code with no comma, one triplet after another sequentially from a fixed starting point. Crick and his associates (1961) reasoned that acridine-induced mutations were of a nature different from mutations induced by nucleotide analogs (like 2-aminopurine, 5-bromouracil): most of them were complete function-loss mutations, and although they might be reverted by acridine, they were not reverted by the nucleotide-analog mutagens. It was assumed that, contrary to the nucleotide-analog mutagens that induced base-substitution mutations, acridines by their physico-chemical properties intercalated in the nucleotide chain and thus induced deletions or insertions of nucleotides in the sequence at replication. Acridine-induced *r*II mutation (denoted "+"-mutations) may be reverted by acridine-induced mutations (denoted "–"-mutations). Moreover, +-mutations were sometimes reverted by –-mutations of independent origin located nearby (and vice versa, –-mutations could be reverted by independent adjacent +-mutations) but no reversion of mutations of the same type (+ or –) could revert each other's phenotype. However, combining three mutations of the *same* symbol often resulted in reversion: Crick *et al.* took advantage of the finding by Benzer that a deletion extending from cistron A into cistron B (as shown by recombination analysis) disabled cistron A but not cistron B (as shown by complementation tests). The acridine mutations and reversions were induced in cistron A and their effect was tested in cistron B. Many of the reversions obtained by combining acridine mutants were not full-fledged reversions but showed reduced efficiency (Crick *et al.*, 1961).

These conclusions from genetic analysis of acridine-induced "frame shift" *r*II mutations were confirmed by Streisinger and his colleagues. By following the change in the amino-acid sequences in the lyzozyme protein of the T4 bacteriophage, they found that a sequence of lysine-serine-proline-serine-leucine-aspargine-alanine was replaced in an acridine-induced revertant of an acridine-induced mutation by the sequence lysine-valine-histidine-histidine-leucine-methionine-alanine, thus confirming both the nature of frame-shift mutations and the assignment of amino acids to triplets of nucleotides (Streisinger *et al.*, 1967).

THE CENTRAL DOGMA

... the most significant thing about proteins is that they can do almost anything.... It is at first sight paradoxical that it is probably easier for an organism to produce a new protein than to produce a new small molecule, since to produce a new small molecule one or more new proteins will be required in any case to catalyse the reaction.

The path from DNA to protein

I shall argue that the main function of the genetic material is to control (not necessarily directly) the synthesis of proteins. There is little direct evidence to support this, but to my mind the psychological drive behind this hypothesis is at the moment independent of such evidence. Once the central and unique role of proteins is admitted there seems little point in genes doing anything else.

Crick (1958, 138–139)

In fact an understanding of the exact relationship between genetic material, its cellular products (proteins), and its consequent effects, only began to develop once molecular studies could be combined with genetic analysis of how biological traits are transmitted.

Strickberger (1976, 67–68)

So what molecular biology has done . . . is to prove beyond any doubt but in a totally new way the complete independence of the genetic information from events occurring outside or even inside the cell – to prove by the very structure of the genetic code and the way it is transcribed that no information from outside, of any kind, can ever penetrate the inheritable genetic message. . . . This was what Francis called the Central Dogma: no information goes from protein to DNA. And he had to call it a dogma at the time, because at the time so little was yet known.

Monod, in Judson (1979, 217)

Once it became clear, from genetic analysis more than from direct biochemical evidence, that the main function of DNA was to control protein synthesis, a central issue became how the "information" residing in the sequence of the DNA nucleotides is transmitted to the sequence of amino acids of the polypeptide molecules (Crick, 1958). There is no doubt that the "information" metaphor, borrowed from mathematical cybernetic theory, played an important role in formalizing the concepts of transformation of the linear sequence of nucleotides into the linear sequence of amino acids. It must, however, be kept in mind that the metaphor was used in a context far from that of information theory, which dealt with the probabilistic reliability of transmission of signs (see Part V). The information metaphor of genetic analysis is related to transmission of *semantic* information, as obtained by the methodology of hybridization of living organisms or DNA and/or RNA nucleotide sequences. The metaphor of genetic transfer is a direct extension of the Weismann–de Vries notions of one-way traffic of nuclear determinants from the nucleus to the cell as the mechanism of differentiation. It is not related to Shannon's conception of information. Crick's powerful *credo*

227

presented the Central Dogma five years after the publication of the model of DNA structure and three years after the extension of the resolving power of genetic analysis to molecular dimensions, in terms of the determinist conceptions of preformationist reductionism at the beginning of the century (see, e.g., Loeb, 1912). The saying "What is true for *E. coli* is true for the elephant", sometimes also attributed to Monod (Judson, 1979, 613; Weinberg, 2007, 1), may mark the peak of reductionist genetic analysis.

Increasingly sophisticated biophysical and biochemical methodologies of experimentation provided evidence for the genetic model that culminated in Crick's Central Dogma. Thus Volkin and Astrachan demonstrated the existence of a small fraction of the total bacterial RNA that had a highly specific activity in correlation to genetic function (Volkin and Astrachan, 1957). Pardee, Monod, and Jacob embarked on a series of experiments, known as the "PaJaMa" experiments (Pardee, Jacob, and Monod, 1959), investigating the genetic control of lactose metabolism by hybridizations of lac$^+$ and lac$^-$ bacteria, in what Lily Kay called "an example par excellence of scientific trading zones as exchange sites for different theoretical commitments and material cultures" (Kay, 2000, 212). The experiments provided the basis for an intermediate messenger-RNA and for its negative regulation in the conversion of DNA-information to a polypeptide sequence.

Crick translated all the phenomenological genetic findings into a consistent physico-chemical model. Summarizing the experimental evidence, he concluded that both DNA and RNA "carry some specificity for protein synthesis": amino acids, on their way to becoming protein, have been shown to pass rapidly through microsomal cytoplasmic particles, of which most of the protein and RNA is metabolically inert. Evidence indicates that the first step in protein synthesis involves enzymes that appear to form activated amino acids that are transferred to the microsomal particles. "The soluble RNA also appears to be involved in this process" (Crick, 1958, 151–152). This led Crick to formulate his ideas on cytoplasmic protein synthesis as an "outline sketch," based on two principles, which he called the Sequence Hypothesis, which is essentially a biophysical hypothesis, and the Central Dogma, which is essentially a genetic hypothesis. Crick repeatedly emphasized the speculative nature of these. He "found them to be of great help in getting to grips with these very complex problems," pointing out, however, that although "[i]t is an instructive exercise to attempt to build a useful theory without using [these principles] . . .one generally ends in the wilderness" (Crick, 1958, 152).

– *The Sequence Hypothesis* in its simplest form assumes that "the specificity of a piece of nucleic acid is expressed solely by the sequence of its bases, and that this sequence is a (simple) code for the amino-acid sequence of a particular protein." The hypothesis unites several pairs of generalizations: (1) The central biochemical importance of proteins and the dominating biological role of genes; (2) The covalent linearity of protein molecules and the genetic linearity within the functional gene; and (3) The simplicity of the composition of protein molecules and the simplicity of the nucleic acids.

– *The Central Dogma* states that "once 'information' has passed into protein *it cannot get out again.* . . . Information means here the *precise* determination of sequence, either of bases in the nucleic acid or of amino-acid residues in the protein" (Crick, 1958, 152–153).

Granted that molecular genetic analysis indicates that the RNA of the microsomal particles is the template for protein synthesis, Crick rejected a Gamow-version of a code of steric correspondence between grooves in the DNA double helix and amino acids and reverted instead to the "hybridability" property of nucleic acids: "One would expect, therefore, that whatever went on to the template in a *specific* way did so by forming hydrogen bonds." Thus, he suggests "a natural hypothesis that the amino acid is carried to the template by an 'adaptor' molecule, and that the adaptor is the part which actually fits on to the RNA. In its simplest form one would require twenty adaptors, one for each amino acid" (Crick, 1958, 155). Although, "what sort of molecules such adaptors might be is anybody's guess," Crick mentions "one possibility which seems inherently more likely than any other – that they might contain nucleotides." This is so because it would "enable them to join on to the RNA template by the same 'pairing' of bases as found in DNA" (Crick, 1958, 155).

The principle that goes through the whole hypothesis of protein synthesis is the specific "hybridability" of nucleotide-pairs:

It will be seen that we have arrived at the idea of common intermediates without using the direct experimental evidence in their favour; but there is one important qualification, namely that the nucleotide part of the intermediates must be specific for each amino acid, at least to some

229

extent. . . . Thus one is led to suppose that after the activating step . . . some other more specific step is needed before the amino acid can reach the template.

Crick (1958, 156)

Crick examined different hypotheses, but in all these speculations, the leading notion was the quality of nucleotide base-pairing, or hybridability.

> It is an essential feature of these ideas that there should be *at least two types of RNA in the cytoplasm*. The first which we may call 'template RNA' is located inside the microsomal particles. It is probably synthesized in the nucleus under the direction of DNA, and carries the information for sequentialization. . . . The other postulated type of RNA, which we may call 'metabolic RNA' . . . its sequence is determined by base-pairing with the template RNA. . . . it becomes 'soluble RNA' and is constantly being broken down to form the common intermediates with the amino acids.
>
> Crick (1958, 157–158)

In conclusion, two things particularly strike Crick:

> first, the existence of general ideas covering wide aspects of the problem. It is remarkable that one can formulate principles such as the Sequence Hypothesis and the Central Dogma, which explain many striking facts and yet for which proof is completely lacking. . . . Second, the extremely active state of the subject experimentally both on the genetical side and the biochemical side.
>
> Crick (1958, 160–161)

It is, however, important to note that Crick explicitly stated that although "once 'information' has passed into protein *it cannot get out* . . . the transfer of information from nucleic acid to nucleic acid, or from nucleic acid to protein may be possible, but transfer from protein to protein, or from protein to nucleic acid is impossible" (Crick, 1958, 153). Reductionist geneticists ignored this caveat and rejected in the name of the Dogma findings such as those of Temin and of Baltimore of reverse transcription from RNA to DNA (Baltimore, 1970; Temin and Mizutani, 1970), until Crick called attention to the fact that he *did not* exclude the possibility of "reverse transcription" from RNA to DNA (Crick, 1970). It was only the transfer of "information" from proteins to nucleic acids that the Dogma rejected.

16

Genes in the service of development

Concerning the matter of functioning of genes during development, I have contrasted . . . two possible views, and suggested a third. . . . It is known that the protoplasm of different parts of the egg is somewhat different, and that the differences become more conspicuous as the cleavage proceeds . . . From the protoplasm are derived the materials for the growth of the chromatin and for the substances manufactured by the genes. The initial differences in the protoplasmic regions may be supposed to affect the activity of the genes. The genes will then in turn affect the protoplasm, which will start a new series of reciprocal reactions. In this way we can picture to ourselves the gradual elaboration of differentiation of various regions of the embryo.

Morgan (1934a, 9–10)

In his classical paper of 1913 on the linear arrangement of sex-linked factors in Drosophila, Sturtevant described the phenotypes of the wings in combinations of genotypes: "The normal wing is RM. The rM wing is known as miniature, the Rm as rudimentary, and the rm as rudimentary-miniature" (Sturtevant, 1913b). The miniature phenotype is produced by a mutation in the rudimentary gene and vice versa, the rudimentary phenotype is produced by a mutation in the miniature gene.

This terminology, unlike the terminology we use today, had been introduced by Morgan: it reflected his sensitivity to the epistemological significance of the notation describing experimental results. Morgan had been highly critical of any theory which referred adult traits to hereditary particles, which to him were "nothing more than an up-dated version of preformism" (Allen, 1984, 723–724). Although Morgan's

231

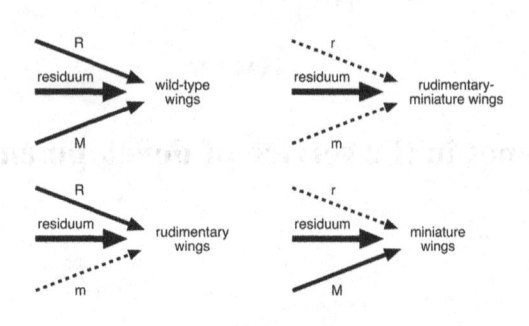

Figure 16.1. Schematic description of Morgan's conception of "wild-type" and mutant wing production as reflected in his mutation nomenclature.

opposition to a particulate notion of heredity receded once he conceived of the distinction between phenotype and genotype, his interest in genes for mapping development continued. Morgan introduced the mutants that affected the development of Drosophila's wing-shape as follows: "If M stand for Miniature, the miniature wing will occur when *M* is present, and *R* is absent (*r*), i.e., *rM*. If R is the symbol for Rudimentary, the rudimentary wing will be *Rm*" (Morgan, 1912, 324–325). *R* is needed to convert the miniature phenotype to normal, and similarly, *M* is needed to convert the rudimentary phenotype into normal. Such a nomenclature of mutants signified that it was not a single gene but rather the whole genome of an organism that was involved in the development of traits. *There are no genes for traits*; once one of the genes of an organism mutated (was deleted) it could no longer contribute to the trait, and the phenotype was determined by those genes that remained intact (the *residuum*) (Falk and Schwartz, 1993). True to the Mendelian conception, Morgan treated each gene as controlling a *specific* developmental step, but each such developmental step was *integrated* with the others (see Figure 16.1). The notation for the mutations clearly reflected Morgan's developmental hypothesis:

> It may seem, on first thought, that no wings at all should appear with *M* and *R* absent; but such an interpretation would rest on a false conception, as I take it, of Mendelian factors; for, the absence of *R* and of *M* does not mean that all factors for wing are lost – there may be hundreds of factors that enter into the production of wings – but only that when a certain factor, *R*, is lost from the complex, a miniature wing is produced by the remainder; and when the factor *M* is lost from the complex of wing-factors, a rudimentary wing is produced by the remainder. When both *R* and *M* are absent the remaining factors are still capable of forming as much of the

wing as is shown by the rudimentary-miniature wing. In fact, this last type of wing bears the same relation to miniature wing that ordinary rudimentary bears to long wing.

Morgan (1912, 331)

By using such a nomenclature Morgan hoped to counter the Weismannian factorial concept that identified "each character with a special determinant of that character" and according to which development "is a process of sorting out of the determiners of the germ-plasm into different regions of the body" (Morgan, 1913a, 10).

Unfortunately, Morgan soon realized that extending his notation based on multiple developmental steps beyond the simple case of two factors acting independently on a character, actually led to an impossible situation with regard to the use of factors: "a double meaning was attached to [notation like that of] *P*, for it stood both for the *P*-factor . . . and also for the residuum as a whole" (Morgan, 1913a, 9). Consider a third mutation, say *Vg* (*vestigial*), affecting the wing in the same way that *R* and *M* do. *Vg* exhibits a distinct mutant wing phenotype in combination with each of the other two mutations. By specifying the effect of *Vg* rather than including its effect in the genotypic residuum of the two previously specified genes, the residuum also changes. This would imply profound rephrasing of the reference to *all* three genes. The three single-mutation phenotypes should be now described as miniature-rudimentary for the *M R vg* genotype, rudimentary-vestigial for the *m R Vg* genotype, and miniature-vestigial for the *M r Vg* genotype. The specification of any other gene of the developmental residuum would demand further revision of the description of all the previous genes involved (Falk and Schwartz, 1993).

Morgan soon realized that it was not only the nomenclature, but also empirical considerations that required him to abandon efforts to reconcile his embryological top-down conceptions with Mendelian bottom-up research. It was "with much reluctance" that he suggested a change of his nomenclature. "The change is not one of any theoretical importance, but a practical necessity for all cases of this kind" (Morgan, 1913a, 14). He conceded that the name of a new character stands merely as its symbol. "[T]hus *P* stands for the pink factor and small *p* stands for the correlative factor of the pink-eyed fly. Whether small *p* represents the loss of the *P* factor, or a change in that factor when the pink eye appears, is immaterial" (Morgan, 1913a, 13). When Castle argued that Morgan did not go far enough, suggesting that Morgan should have further simplified his nomenclature so that "it commits us to no physiological theory, but simply states the facts" (Castle, 1913b, 181), Morgan noted that such a Baconian ideal was a price that we could hardly afford, because it jeopardized the basic information about the *genetic* hypothesis that we wished to transmit by the adopted nomenclature (Morgan, 1913c). In retrospect, Castle may have had a point: the new nomenclature adopted by Morgan and his students was not "simply stating the facts." It was strongly implying exactly what Morgan wished to avoid, namely a "genocentric" notion of genes being *for* traits. Geneticists continued to include facts, if not hypotheses about the role of genes in the development of traits, in a nomenclature that denoted dominant mutations with capital letters and recessives with small letters. It was only when much of genetic research turned to haploid organisms, from Neurospora to bacteriophages that the nomenclature became more "neutral."

THE ROLE OF GENES IN DEVELOPMENT

Whereas gene function continued to play a central role among most European students of genetics who were anchored in the organismic-developmental research tradition, for American scholars evolution was the point of departure. Accordingly, their interest was mainly in the mechanics of inheritance, but they did not miss any opportunity to study the involvement of specific genes in development (see Harwood, 1987, 1993).

The intensive research in Drosophila and maize made these organisms especially conducive to the design of genuine genetic methods of analysis of development, but interest in mammals, especially mice and rats, did not trail far behind. Besides elucidating the nature of the effect of various genes or gene-combinations, such studies also contributed to insights into general principles of gene organization and function. In maize, an abnormally high frequency of mutations in genes that affect the development of the seeds' pericarp pigmentation (Emerson, 1914) culminated in Barbara McClintock's model of regulatory genes "jumping" to different chromosome sites due to breakage and rejoining cycles (McClintock, 1951). In the rat, Hans Gruneberg studied the effects of specific genes on complex phenotypes and concluded, even prior to Beadle and Tatum's one gene – one enzyme hypothesis, that each gene has only one primary product that may react or interact in many secondary and tertiary processes: pleiotropism is secondary (Gruneberg, 1938). Also Sturtevant, in his search for genetic manipulations that would affect specific developmental functions, studied Himalayan rabbits in which the fur pigmentation depended on the skin temperature at hair-growth (Sturtevant, 1913a), and the Drosophila Bar-eye phenotype, the intensity of which was due to the position effect of a short segment of its X-chromosome (Sturtevant, 1925). Salome Gluecksohn-Waelsch, on the other hand, directly studied the embryological effects of single mutations, which elucidated interactions of basic embryological induction processes in skeleton and urogenital differentiation in the mouse (Gilbert, 1991b).

As early as 1923, Sturtevant called attention to the finding by Boycott and Diver "on the inheritance of dextral and sinistral coiling in the snail *Limnaea* ... of 'maternal' inheritance that is nevertheless dependent upon the chromosomes." He ends his short note commenting: "Further data on the case of *Limnaea* will be awaited with interest, for it seems likely that we shall have here a model case of the Mendelian inheritance of an extremely 'fundamental' character, and a character that is impressed on the egg by the mother" (Sturtevant, 1923). Such a maternal effect had been interpreted for many years, together with some other exceptions, like pigmentation of larval skin and eyes in the moth *Ephestia kühniella*, as a phenomenon of "predetermination": the gene acts on the eggs in the ovary, predetermining the direction of an early cell division and thus the type of asymmetry and the pattern of the future individual (Sinnott, Dunn, and Dobzhansky, 1958, 361–362). However, all this was bottom-up, reductionist heuristic genetic thinking. Thinking like embryologists, the top-down or systems' perspective, presented by Morgan in 1932, was missing. Although it was clear that such cases provided entry points into genetic analysis of development, little had been done until half a century later, when Christiane Nüsslein-Volhard took up another conspicuous maternal effect in the development of the Drosophila embryo (the "bicaudal" mutation of Bull, 1966) and extended it by systematically looking for maternal effect mutations in order to study the earliest steps of genetic control of Drosophila embryogenesis (Nüsslein-Volhard, 1977, and Chapter 20).

Still, genetic analysis provided increasingly effective new *tools* for research of old developmental problems. One of the first characters that were subjected to genetic analysis was sex determination in Drosophila. As described in Chapter 7, sex determination in Drosophila was found to depend on X-chromosome/autosome ratio. Sturtevant realized that genetic markers could be extremely helpful markers of cell-fate, replacing the rather limited method of following the cells' fate by injecting vital dyes (which would become quickly diluted in sequential cell divisions).

Frequent chromosome non-disjunction in early cleavage divisions of the embryo was found to occur in a stock of *Drosophila simulans*. When euploid XX zygotes lost one X-chromosome in an early mitotic division, roughly half of the cell nuclei of the fly's body were XX, i.e., female, and half were XO, i.e., male (Sturtevant, 1929): gynandromorph flies developed. Starting out with heterozygotes for an X-chromosome's recessive mutation, like *y* for yellow cuticle, produced some gynandromorphs in which male parts were yellow and female sections were wild-type (when the non-*y* marked chromosome was the one lost). This indicated the following. First, the effect of *y* is cell-autonomous since there was a clear demarcation line between the female wild-type and male yellow parts (and not a checkered pattern), even though the course of demarcation varied from one gynandromorph to another. Second, the nuclei of early mitotic divisions remained compact clonal clusters; and, third, in the early cleavage divisions nuclei were non-determinate with respect to organ differentiation. Any section or organ of the fly could be yellow or wild-type in different gynandromorphs. Furthermore, using the eye color mutation *vermilion* (*v*), at another X-chromosome-linked gene, Sturtevant found that, fourth, contrary to *y*, the function of *v* was non-autonomous. Even when one eye was male and the other female (according to the *yellow* marker), both eyes were either vermilion or wild-type. Sturtevant located a region at the thorax (which turned out to be that of the ring-gland) to be the *focus*, the site of the fly where the genotype determined whether the phenotype of both eyes would be wild-type or vermilion. Sturtevant thus put genetic analysis at the service of the classical "fate mapping" of embryologists.

These methods were extended by Hotta and Benzer (1972) in order to construct fate maps of the whole embryo and especially to locate the functional foci of behavioral patterns. Basically the correlation of the marker phenotype – the pigmentation of the cuticle, or the twisting of bristles, marking X-linked recessive mutations – of any pair of properties, for example, the antennae and the proboscis, the foreleg and the shaking phenotype, would indicate their common developmental origins. The proportion of discordance in the four-cell matrix of the phenotypes of the paired properties is represented by units of "distance" on the fate map: the further apart the sites in the zygote where cleavage nuclei must be at the time of determination (in the embryo), the higher the probability of discordance of nuclear markers. Benzer called these units *Sturts* in honor of Sturtevant, who initiated this genetic analytical mapping (sixty years after Sturtevant had introduced linkage maps, the distances of which were measured in *centi-Morgans*). The maps that were constructed corresponded remarkably well with those of "classical" embryologists' methods, such as the injection of vital stains and the inactivation of specific nuclei and cells by localized heating or pencil laser beams (Benzer, 1973).

To the extent that genetic analysis was directed at the role of genes in development this was often serendipitous. This changed in the early 1930s, when Beadle and Ephrussi set out to deliberately combine the research methods of experimental embryology with those of genetic analysis (Beadle and Ephrussi, 1936, and Chapter 11). They studied the fates of reciprocal transplantations of imaginal discs among larvae of Drosophila of various eye-color mutants. The red-brownish eye color of wild-type Drosophila flies is due to two pigments: ommochromes and pteridines. Their research project provided insight into the specificity of genes in enzyme production and their place in the developmental pathways (e.g., *v* affecting an earlier step than *cn* in the synthesis of ommochrome); it did not build up, however, into a developmental research project.

A similar attempt to harness genetic analysis to developmental biology was carried out in Kühn's laboratory in Germany, studying pigment formation in the moth *Ephestia kühniella*. It is open to speculation whether the limited successes of Kühn's associations with the pioneer of hormone biochemistry Adolf Buttenandt were due to the complexity of the system they worked with, or to the difficulties imposed in war-time in Germany of the 1940s. But as noted by Laubichler and Rheinberger, Kühn, himself a student of Weismann, may be conceived as one who "connects Weismann and the late 19th century program in developmental genetics with modern molecular developmental genetics and the discovery of the *homeobox*, both intellectually and in the form of direct lineage of prominent scientists" (Laubichler and Rheinberger, 2004, 108). Directly and indirectly he launched the careers and the research programs of developmental genetics in both the United States – Ernst Caspari was his student – and in Europe – "the young Ernst Hadorn, who was a student of Fritz Baltzer, became acquainted with Kühn as well as his ideas" (Laubichler and Rheinberger, 2004, 109).

Hadorn's interest was in the developmental physiology of lethal factors (Hadorn, 1961). He showed that lethals were a heterogeneous group of phase-specific mutations whose physiologies could be studied individually. He soon applied the embryologists' methods of organ transplantations to the transplantation of imaginal discs of Drosophila larvae. Imaginal discs were cultured in the abdomen of adult flies by serial transfers from one fly to another. Thus Hadorn introduced an "*in vivo* tissue culture" that could be maintained indefinitely (for many years it has been impossible to design media for *in vitro* tissue culture of Drosophila) (Nöthiger, 2002). This opened the path for genetic analysis of development at the cellular and eventually the molecular level.

The developmental capacity of discs was examined at will by implanting cultured discs (or parts of discs) into larvae about to pupate; the disc's developmental fate could be inspected in the abdomen of the emerging adult. Hadorn and his students showed that specific regions of the discs were determined to become specific elements of the adult organ and that this determination of the cells of the imaginal discs was maintained over many generations. Occasionally, however, "transdetermination" to other specificities occurred, such as an eye disc that developed into an antenna (see, e.g., Gehring and Nöthiger, 1973). These were obviously too frequent to be due to gene mutations (Hadorn, 1968). Furthermore, the direction of these transdeterminations was not random. This provoked speculation about the genetic control of epigenetic processes (see, e.g., Kauffman, 1973). Indeed, Hadorn's student Walter Gehring demonstrated by clonal cell analysis (utilizing a genetic marker induction by somatic recombination – see below) that transdetermination was *not* a clonal event but rather an epigenetic regulative event, induced in groups of cells that happened to be in the right place at the right time (Gehring, 1967).

The tools for extending genetic analysis to the molecular level awaited application. This happened when reductionist genetics reached a hiatus and the imminent crises were lurking around the corner (see Stent, 1969). Alan Garen, a successful molecular biologist, recalled:

> By 1965 it was evident that the reductionist approach of molecular biology to complex biological phenomena was proving remarkably effective, at least as applied to bacteria and their viruses. The basic mechanisms underlying the genetic control of cellular processes seemed sufficiently well understood for Jacques Monod to proclaim that what was true for *Escherichia coli* is true for elephants. . . . Nevertheless, the obstacles facing anyone hoping to achieve a comparable understanding of developmental processes in higher organisms were daunting. . . . What was needed as a first step into the field was a conceptual framework for transforming such enormous phenomenological complexity into experimentally manageable problems.
>
> Garen (1992, 5)

Garen "had the good fortune of attending a lecture at Yale by Ernst Hadorn, a pioneer in reestablishing the vital link between genetics and development that had been forged earlier by Thomas Hunt Morgan but surprisingly neglected afterwards" (Garen, 1992, 5). Garen adopted Hadorn's conception of lethal mutations and screened, in association with Shearn, for lethal mutations that affect imaginal discs' development in the larvae (the working hypothesis being that larval development is largely independent of the development of the imaginal discs; the latter may be conceived as parasitic to the larvae). Indeed, in roughly half the lethal mutants, imaginal discs were either missing or defective. By his collaboration with Hadorn and his student Walter Gehring, Garen brought the feasibility of extending genetic analysis to complex systems and their development to the community of molecular biologists facing the challenge of the elephant rather than that of *E. coli*. As we shall see, Hadorn and his students were also instrumental in turning attention to the genetic analysis of the development of Drosophila's larval and embryological stages as major tools of the molecular level control of development and differentiation (see Gehring, 1998).

CLONES AND COMPARTMENTS

The Sturtevant–Benzer notion of genetic fate-mapping of Drosophila was further extended by taking advantage of induced genetic changes, such as mitotic recombination, in successively later embryonic and larval somatic cells. Mitotic recombination was discovered by Curt Stern after irradiating Drosophila larvae heterozygous for conspicuous somatic markers with mild doses of X-rays. "Spots" of the hidden recessive markers could be detected in the adult fly. The frequency of such spots in heterozygotes for *yellow* (*y*) pigmentation of the cuticle and *singed* (*sn*) bristles, was too high to be attributed to induced mutations (Stern, 1936). By examining trans heterozygotes such as *y* +/ + *sn* Stern observed "twin-spots" – a yellow cuticle spot next to a bristle spot (whereas in the *cis* construct there were no twin spots, and spots were either yellow or, less frequently, yellow and singed) (Figure 16.2). Stern concluded that such spots were the consequence of induced mitotic crossing over that yielded homozygosis for the genetic markers. Support for this interpretation was that their frequency reflected their relative distance from the centromere: the further away, the more probable a recombination event. Stern utilized these localized spots of mosaicism to study the

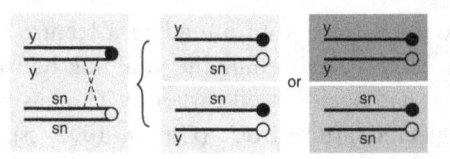

Figure 16.2. "Yellow" (upper right) and "singed" (lower right) twin spot of neighboring cells produced by somatic recombination in a cell of a "wild-type" fly (left).

determination of the pattern of bristle number and spacing on the flies' mesothorax (Stern, 1955).

The experimental manipulability of the induction of genetic change in somatic cells became a powerful tool in the hands of Antonio Garcia-Bellido (who was trained in Hadorn's lab in Zurich; see Lewis, 1998) and associates, not only for cell-marking but also for the study of genes that regulate differentiation (Garcia-Bellido, 1998). Garcia-Bellido and Merriam induced somatic recombination in larvae of different ages (Garcia-Bellido and Merriam, 1969), and followed the dynamics of the number of spots and their size in adult flies (Garcia-Bellido and Merriam, 1971). The size of spots decreased and their number increased linearly with the age of induction. More significantly, whereas early-induced spots extended freely from one organ to the other, the later the spots were induced, the more limited was their extension from one organ to the other (corrected for decreasing spot size). Thus, spots extended freely from one segment to the other in gynandromorphs when induced in the first cleavage divisions, but were limited to adult segments when induced in earliest larvae. Whereas early-induced spots extended freely from wing to leg in the mesothorax, later on they were limited to either leg or wing, and so forth. Genetically controlled differentiation proceeded in a hierarchically ordered sequence of increasing specification of *compartments* (Garcia-Bellido, Lawrence, and Morata, 1979; Garcia-Bellido and Merriam, 1971; Garcia-Bellido, Rippoll, and Morata, 1973).

Efforts were concentrated on the mesothorax and especially on the fly's wings that provided large flat surfaces. Sites of early-induced spots (in different animals) often partly overlapped. "This means that even though the final structure of each wing is the same, the cells that produce it do so in different ways in different wings" (Garcia-Bellido *et al.*, 1979, 104). To overcome the decreasing size of induced spots with age of induction, Garcia-Bellido designed a method of inducing somatic crossing over in heterozygous for *Minute* mutations (in addition to the spot marker mutations). Minute flies are slow in development. Non-Minute induced spots grow considerably faster than the surrounding tissues, as indicated by following the attached cell-markers. The possibility of producing fast growing clones further stressed the indeterminacy of the location of spots. On the other hand these were effective in determining the borders of compartments, especially in the later steps of compartmentalization when induced spots would otherwise be too small to reliably demarcate borders. Huge non-Minute spots on a Minute background indicated unequivocally the precise location of compartments' border-lines. Furthermore, even the largest clones never completely filled a compartment (as determined by comparisons of overlapping clones in different flies), indicating that like transdetermination, compartmentalizations were epigenetic events that affected groups of cells; compartments were established by "polyclones," i.e., not based on common cell-ancestry, but rather on cells being at the right time in the right place (Crick and Lawrence, 1975).

The mutant allele of the gene *engrailed* was characterized by an additional sex-comb on the first leg pair of *Drosophila melanogaster* males. Garcia-Bellido showed that in the mutant, instead of the normal distinct anterior and posterior patterns of the leg, the posterior pattern was a mirror image of the anterior pattern. This was also the case in the wing (and other segmental organs, though less conspicuously to the untrained observer): in the mutant flies the determination of the posterior compartment of the segment maintained its "anterior" pattern: *en* was a selector gene for turning anterior compartments into posterior ones. Cell clones that respected the (virtual) anterior-posterior compartment border in wild-type flies expanded into the otherwise alien compartment in *en* flies (Garcia-Bellido *et al.*, 1979; Garcia-Bellido and Santamaria, 1972). Again, like transdetermination, these were shown to be epigenetic events that affect groups of cells, not necessarily clonally related, "polyclones" that happen to be at the right place at the right moment (Crick and Lawrence, 1975).

Engrailed is a homeotic mutation, altering one organ of a homologous series from its own characteristic form to that of some other member of the series: in engrailed flies the posterior compartment appears to be transformed into an anterior one, though in reality, the cells of the posterior compartment failed to be transformed into the structures typical of the posterior compartment. Genetic analysis implicated *engrailed* to be a specific *selector*-gene in the control of a determination event. Also other homeotic phenotypes, like antennapedia, or proboscipedia seemed to be due to a failed "binary switch" caused by a mutation in "selector" genes. But other homeotic genes were apparently more complex than binary "selectors" or switches. For many years Ed Lewis followed the mutations in the complex locus of *bithorax* on the third chromosome of Drosophila. Although a series of mutations had been isolated in this locus, each defined as a binary wild-type mutant phenotype, they all interacted as pseudoalleles, indicating a higher level of integration. Once enough genetic analysis of the phenomenological level of cell differentiation was obtained (Lewis, 1978), molecular methodologies could be applied to carry on (see Chapters 19 and 20).

Contrary to the selector-genes' directed epigenetic changes that characterize compartmentalization, genetic analysis detected that "marked" individual cells and all their clonal progeny play a significant role in the epidemiology of cancer. Many tumors start out as a result of mitotic recombination or other chromosomal events in cells heterozygous for a mutation at a tumor suppressor gene. This was first proposed by Alfred Knudson in 1971 for retinoblastoma, where essentially two groups of patients were discerned, those in which tumor manifestation was sporadic and those with a familial history. The probability for further tumors remained low in those whose tumor manifestation was sporadic. In patients with the familial disposition for the development of a tumor, it appeared to be inherited as a dominant allele of a mutation in the gene *Rb*. In these persons the probability for further tumors increased linearly with age. Knudson suggested that a tumor-suppressor gene was involved. At the cellular level the disposition was due to a recessive factor (for the changing meaning of "dominance" with the expansion of research to haploid organisms, see Chapter 19). For a cell to become tumorous it must be homozygous (or hemizygous) for the recessive allele of the wild-type retinoblastoma-suppressor allele. In familial manifestations of retinoblastoma, the recessive allele is transmitted (in the gametes) to members of the family. Those who are heterozygous for the recessive allele of *Rb* are at risk: any event in one of their cells that "exposes" the recessive allele, i.e., removes the tumor-suppressing capacity of the wild-type allele, promotes the probability of the cell becoming tumorous. This model was confirmed once linked markers to *Rb* (and later also the gene *Rb* itself) were sequenced. Mitotic recombination brings about such an effect, as do other mechanisms for the loss of the tumor-suppressor allele, such as deletions of the wild-type chromosome segment. The phenomenon was eventually termed LOH, loss of heterozygosity, and is found also in other tumor-suppressing genes (Weinberg, 2007, 214ff.).

By the mid-1960s genetic analysis was ready methodologically to expand to the molecular level of analysis and conceptually to return to a system's approach, top-down analysis. It had developed the tools to carry this out.

VII

What is true for *E. coli* is not true for the elephant

Thus eventually one may hope to have the whole of biology "explained" in terms of the level below it, and so on right down to the atomic level. And it is the realization that our knowledge on the atomic level is secure which has led to the great influx of physicists and chemists into biology.

Crick (1966, 14)

... the analysis of the hierarchy of living things shows that to reduce this hierarchy to ultimate particles is to wipe out our very sight of it. Such analysis proves this ideal to be both false and destructive.

Polanyi (1968, 1312)

By 1960, with Crick's Central Dogma and the statement that "once the central and unique role of proteins is admitted there seems little point in genes doing anything else" (Crick, 1958, 139), genetics' determinism reached its peak. Jacob and Monod's analysis of "genetic regulatory mechanisms in the synthesis of proteins" further provided the logical extension of Crick's physico-chemical reduction of genetics. In their opening sentences they state that "the 'structural gene' accounts for the multiplicity, specificity and genetic stability of protein structures, and it implies that such structures are not controlled by environmental conditions or agents" (Jacob and Monod, 1961, 318). But then, throughout their study they take us through "genes" that do not qualify for that model. This starts with "primary products of the i^+ gene," the repressor which "may be a polyribonucleotide" (Jacob and Monod, 1961, 333; this turned out to be wrong), and ends with the site of the repressor's interaction which they called "operator," that coordinately regulates the heterocatalytic activity of the "operon" – a battery of structural genes.

> Structural genes obey the one-gene one-protein principle, while regulator
> genes may affect the syntheses of several different proteins ... a *single*
> operator gene may control the expression of *several adjacent structural*
> *genes*, that is to say, by demonstration of the *operon* as a co-ordinated unit
> of gene expression.
>
> Jacob and Monod (1961, 353)

The spectacular achievements of the decade following the introduction
of bacterial (and bacteriophage) genetic analysis and the presentation of
the Watson and Crick model of DNA, led several scientists at the front
line to conclude that the fundamental concepts had been established and
that from now on only the task of filling in the missing details remained.
Gunther Stent was probably the most explicit in expressing this spirit in
his "The Coming of the Golden Age: A View of the End of Progress"
(Stent, 1968, 1969). Stent had been impressed by Niels Bohr's hope to
discover unknown laws of *physics* in biological systems. However, when
it turned out that the only "laws of biology" beyond the laws of physics
and chemistry were the historical constraints to the evolution of life on
earth and their idiosyncrasies, Stent shifted his research to what he
believed to be more promising fields. So did Crick. Others changed their
focus to alternative areas within the sphere of genetic research, like
Benzer who turned to the study of the genetic basis of behavioral traits in
Drosophila, or Brenner who set out to design a new model organism,
Caenorhabditis elegans, for the remaining problems of developmental
genetics.

A crucial step towards the new era of molecular genetic analysis was
introducing another technique into the experimental toolkit: hybridization at the molecular level. The strands of the double helix of DNA were
separated and then allowed to reanneal with strands from other sources
or with RNA strands. This technique, first developed *in vitro*, but later
extended *in situ* (one of the most popular present-day methods of genetic
research is FISH, or Fluorescent *In Situ* Hybridization), *de facto* revolutionized the experimental and consequently the conceptual approach
to analyzing genes in development as well as in evolutionary history,
and shifted attention from genes to genomes. Molecular polynucleotide
hybridization became pivotal to the study of the structural organization
of genetic matter. With the introduction of techniques such as the PCR
(Polymerase Chain Reaction), molecular hybridization heralded an era
of genetic engineering epitomized by the Human Genome Project and
the new disciplines of genomics.

One of the first "unexpected" experimental results that challenged the reductionist concept came in 1968 from the US Terrestrial Magnetism Laboratory. A significant fraction of dissociated mouse DNA reannealed (hybridized) much faster than that of bacteria. This suggested that even though the amount of mouse DNA was orders of magnitude larger than bacterial DNA, it was not much more complex. This result challenged an axiomatic assumption of reductionist genetics that repetitive DNA sequences could contribute little to the transmission of genetic "information." Such repetitive sequences were considered by many to be "Junk DNA." Others were more lenient in ascribing to DNA, including the repetitive sequences, a role in such functions as chromosome-household functions and regulation of the genome and cellular integrity.

The determinist genetic conception of the gene as an open reading frame (ORF) DNA sequence that is transcribed to a corresponding messenger-RNA (mRNA), which is translated on the ribosomes into sequences of amino-acid polypeptides, was challenged further when it turned out that many transcribed sequences were subjected to various processing routes before becoming translatable mRNAs. Elaborate splicing mechanisms, including alternative splicing, intervene between the transcribed nuclear RNA and the translated mRNA, and impose top-down constraints on the functional–developmental specifications of a cell, a tissue, and even a whole organism on the bottom-up notion of "genetic information transfer." By the time such mechanisms as overlapping reading frames from a given segment of DNA, anti-sense trans-splicing, or alternative trans-splicing, also from different chromosomes, were described, no one was shocked, not even when evidence for modifications of the sequences of the transcribed RNA before translation was revealed.

In a recent review of "What is a gene?" Helen Pearson noted:

> The more expert scientists become in molecular genetics, the less easy it is to be sure about what, if anything, a gene actually is. . . . Instead of discrete genes dutifully mass-producing identical RNA transcripts, a teaming mass of transcription converts many segments of the genome into multiple RNA ribbons of different lengths. These ribbons can be generated from both strands of DNA, rather than from just one as was conventionally thought. Some of these transcripts come from regions of DNA previously identified as holding protein-coding genes. But many do not.
>
> Pearson (2006, 399)

Indeed, the term "gene" became again a generic term, an intervening variable, rather than a hypothetical construct (Falk, 2000b).

Genetic analysis as a reductionist methodology, now primarily molecular and relying increasingly on modern complexity analyses and on computer methods, turned to the analysis of development and evolution and to the combination of these (evo-devo). Constraints of development and of evolutionary history together with those of gene-expression, turned genetic conceptions top-down and genetics became a less distinct discipline of Life Sciences: there is hardly a discussion in the Life Sciences that does not involve reference to "genes." If there is still something that distinguishes genetics as a discipline it is its hybridist research methods.

17

Extending hybridization to molecules

It is clear that the correlation between the structure of deoxyribonucleic acid (DNA) and its function as a genetic determinant could be greatly increased if a means could be found of separating and reforming the two complementary strands.

Marmur and Lane (1960, 453)

When methods for *in vitro* hybridization of polynucleotide molecules were developed by Doty and Marmur and their colleagues (Marmur and Lane, 1960), it opened new vistas for genetic research. These new methods soon transformed and extended much of genetic research to genetic analysis of *hybridization at the polynucleotide level*. Such notions had been anticipated by Crick regarding the genetics and taxonomy of proteins (1958, 142): "Biologists should realize that before long we shall have a subject which might be called 'protein taxonomy' – the study of the amino acid sequences of the proteins of an organism and the comparison of them between species." The beginnings of these analyses, developmental (pathological) on the one hand and taxonomic on the other, were already evident from the studies of changes of single amino acids in hemoglobin molecules (Ingram, 1957, 1963).

Hoyer, Bolton, and McCarthy isolated RNA complementary to DNA, based on the demonstration that specific cellular RNA could be hybridized with heat-denatured DNA from the same cells. The hybrids were isolated by cesium chloride density gradient centrifugation (Bolton and McCarthy, 1962). These researchers examined primarily bacterial species relationships, "where there exists only the faintest paleontological record and the simplest of all ontogenetic processes." Taxonomic relationships were demonstrated by DNA:DNA interactions and differentiation of active and inactive genes by comparing messenger RNA pools from cells of different fates competing for sites of hybridization with the species' DNA. These studies demonstrated the sensitivity of *in vitro* hybridization methods for detecting differences in genes' functions and evolutionary pathways in bacteria (McCarthy and Bolton, 1963). Further studies of site-competition hybridizations of populations of RNA messages present in differentiated cells of higher organisms appeared to neatly confirm earlier conclusions of genetic and cytogenetic analyses that whereas all tissues carry the same genetic material, different pools of genes are functional in different tissues (McCarthy and Hoyer, 1964).

REDUNDANT DNA

In 1968, Roy Britten and David Kohne introduced their paper in *Science* with the "remarkable fact" that separated complementary strands of *purified* DNA derived from organisms ranging from the bacteriophage MS2 to the calf recognize each other, and under appropriate conditions specifically reassociate (Britten and Kohne, 1968, 529). This was remarkable, as "it had been expected that it would be very difficult to observe the reassociation of the DNA of vertebrates and other higher organisms." Because of "the enormous dilution of individual nucleotide sequences in the large quantity of DNA in each cell" it was expected that months would be required for the completion at practical concentrations. This paradox led to the conclusion "that some nucleotide sequences were frequently repeated in the DNA of vertebrates" (Britten and Kohne, 1968, 529). Roughly 10 percent of the DNA of the mouse reassociated extremely rapidly and was shown to consist of a million copies of a short nucleotide sequence; still more of the DNA was repetitive, though to a lesser extent. When it was shown that repeated nucleotide sequences are the rule in eukaryotes the notion of the one-to-one relationship of DNA and genetic material was threatened, even if allowance was made for some unknown "regulatory genes" beyond those discovered by Jacob and Monod. The term "Junk DNA" soon entered deliberations. Gradually the reductionist notion of a necessary one-to-one relationship of discrete sequences of DNA and genes was overcome, and DNA was conceived in the wider context of cellular metabolism.

The association of polynucleotide molecules is measured by passing samples of single- and double-stranded molecules through hydroxyapatite

250

columns, and measuring the amount of double-stranded molecules adhering to the column. Reassociation of a pair of complementary sequences results from their collision. The product of the initial DNA concentration (C_o) and the time of incubation (t) is the controlling parameter for estimating the completion of the reaction. Thus, "the DNA of each organism may be characterized by the value of $C_o t$ at which the reassociation reaction is half completed under controlled conditions" (Britten and Kohne, 1968, 530). The $C_o t$ factor in the experiments was found to range over at least eight orders of magnitude.

Buoyancy-ultracentrifugation of eukaryotic DNA in CsCl (and also other carriers like $HgCl_2$), also revealed the highly repetitive DNA sequences by forming humps, or "satellites" (satDNA) on the roughly normal buoyancy-distribution curve.

For the mouse two fractions of DNA were isolated. One fraction – the mouse "satellite DNA" representing 10 percent of its DNA – reassociates more rapidly than that of the smallest virus, while 60 percent of the remaining DNA reassociated 500 times more slowly than bacterial DNA. In calf thymus DNA, too, a clear separation could be discerned between DNA which reassociated very rapidly and that which reassociated very slowly. Also much of salmon DNA is made up of repetitive sequences of different degrees of repetition. "Whenever a region of sequence homology occurs between two fragments of the genome, it is likely to continue for at least several hundred nucleotides. It does not usually continue perfectly, however" (Britten and Kohne, 1968, 535). The strong reduction in the thermal stability of many of the reassociated DNA preparations indicated, however, actual dissimilarities in the sequences of the reassociated DNA, and provided a measure of the precision of the repetitious DNA preparations. Whereas many types of eukaryotic organisms, from protozoa through invertebrates, to vertebrates, and to plants, were found to contain repetitive sequences of DNA, species of prokaryotes (and yeast) appeared not to contain such repeats (Britten and Kohne, 1968, 533).

All forty chromosomes of the mouse are acrocentric, i.e., they are one-armed, the centromere being located at their tip. *In situ* hybridization of isolated satDNA with cytological preparations of mouse chromosomes overwhelmingly located them in the proximal region of the chromosome (Pardue and Gall, 1970). Those segments adjacent to the centromeres are heterochromatic, known to be relatively poor in genes (Chapter 6). The *in situ* technique of polynucleotide hybridization was further adapted to become directly observable in the electron microscope, thus providing a

Percent of mammalian DNA content

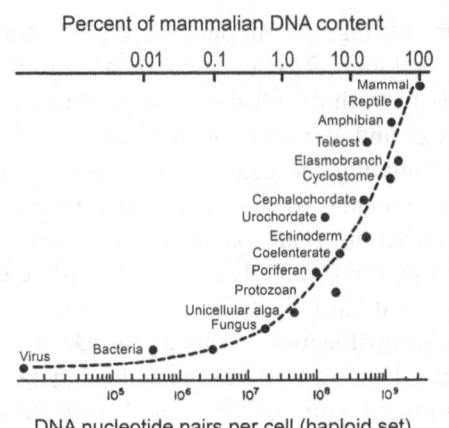

DNA nucleotide pairs per cell (haploid set)

Figure 17.1. Minimum amount of DNA per cell in various systematic classes (Britten and Davidson, 1969).

cytogenetic analytic tool at the molecular level analogous to that of the cytological method of observing homologies in dipteran polytene chromosomes: Labeled DNA and RNA strands were hybridized *in situ* and followed microscopically to detect homologies and rearrangements, such as duplications, deletions, inversions and translocations.

Whereas the presence of highly repetitive DNA sequences was one of the earliest observations when researchers tried to extend what they learned from *E. coli* to elephants, soon other surprises turned up.

Broadly speaking, genome size is expected to increase with the grade of organization of eukaryotes, but the basic needs of cellular household structure and function common to all organisms would presumably indicate largely similar genomic dimensions. The wide range of sizes often observed even among closely related creatures was puzzling. Tabulating the minimum amount of DNA observable at each grade of organization shows that the major steps of evolution between viruses and higher chordates involved an increase in nucleotide pairs per cell ranging over five orders of magnitude (Britten and Davidson, 1969) (see Figure 17.1).

An amphibian, *Amphiuma*, is the organism known to have the largest genome, having 8×10^{10} nucleotide pairs (about 25 times the size of the human genome). Only 20 percent of its DNA was recovered in the slowly reassociated fraction. It would appear that much of the eukaryotic DNA must be redundant. This discrepancy became known as the C-value paradox.

Repugnant to Britten and Kohne was "that about half of the DNA of higher organisms is trivial or permanently inert (in an evolutionary time scale)," and they suggested an integrative conception: The genetic material is not a "collection of different and unrelated genes," but rather a consequence of evolutionary processes and of functional needs:

> A large part is made up of families of sequences in which the similarity must be attributed to common origin. . .There is no doubt that a great increase in DNA content has occurred during the evolution of certain species . . . [That] the low average melting temperature of strands of DNA from different species are reassociated . . . indicates that the members of families of repeated sequences in the DNA of a species slowly change in nucleotide sequence.
>
> Britten and Kohne (1968, 537–538)

In addition, there are "certain minor classes of DNA," such as ribosomal genes, that for functional reasons consist of many hundreds or thousands of copies. Although Britten and Kohne suggest that multiple, nearly exact copies of a gene could provide higher rates of synthesis of structural proteins, they note that the very large number suggests a regulatory role, and as proposed by Crick, "the repetitive segments of the DNA play their role structurally through interaction of these particular nucleotide sequences with the proteins of the chromosome" (Britten and Kohne, 1968, 540, note 49).

Molecular genetic analysis called for reconsideration of cellular function and evolutionary assumptions. But the proposed models did not explain the sequences repeated a hundred, thousand, or a million times (even when allowing for some degree of imprecision of repeats), nor that of the C-value paradox.

Four years prior to Eldredge and Gould's introduction of the heretic "Punctuated equilibrium" as an alternative to phyletic gradualism in evolution (Eldredge and Gould, 1972), Britten and Kohne had already laid the ground for it.

> However, the significance of the very large number might be less direct. It might, for example, raise to a useful level the probability of some rare event such as the translocation of certain DNA sequence fragments into adjacent locations in the genome.
>
> Saltatory replications of genes or gene fragments occurring at infrequent intervals during geologic history might have profound and perhaps delayed results on the course of evolution.
>
> Britten and Kohne (1968, 539)

They quote G. G. Simpson's *This View of Life* (1964, 164): "The history of life is decidedly non-random." Striking phenomena of evolutionary trends are "copiously documented by fossils."

> These phenomena are far from universal; they are not "laws" of evolution; but they are so common and so thoroughly established by concrete evidence that they demand a definite, effective directional force among the evolutionary process. They rule out any theory of pure random evolution such as the rather naïve mutationism that had considerable support earlier in the twentieth century.
>
> Simpson (1964, 164)

For more than sixty years genetic research had been a dialectic confrontation of Mendel's thesis of methodological reductionism with de Vries's antithesis of conceptual determinism. The paper by Britten and Kohne anticipated the emergence of a possible synthesis that would combine a more profound, molecular methodological reductionism with a new genomic conceptual integrationism.

INTEGRATED NETWORKS

Within a year Roy Britten, together with the molecular embryologist Eric Davidson, published an extensive paper: "Gene regulation for higher cells: A theory," subtitled: "New facts regarding the organization of the genome provide clues to the nature of gene regulation" (Britten and Davidson, 1969). In it the authors laid out a new integrative working program for the presumably estranged disciplines of embryology and genetics: A framework of top-down concepts, without giving up the methodological bottom-up approach:

> Cell differentiation is based almost certainly on the regulation of gene activity, so that for each state of differentiation a certain set of genes is active in transcription and other genes are inactive.
>
> Britten and Davidson (1969, 349)

This was not a new speculation, and evidence to support it had been accumulating. "Little is known, however, of the molecular mechanisms by which gene expression is controlled in differentiated cells" (Britten and Davidson, 1969, 349). Systematic bottom-up experimental evidence suggested that at a minimum, the following features for the system should be considered:

1. Change in the state of differentiation is often mediated by simple external signals.
2. A given state of differentiation tends to require the integrated activation of many noncontiguous genes.
3. Many genomic sequences which are transcribed in the nuclei of higher cell types appear to be absent from cytoplasmic RNAs.
4. The genome of higher cell types is extremely large, compared to that of prokaryotes.
5. This genome is enriched by a large fraction of repetitive nucleotide sequences which are scattered throughout the genome.
6. Repetitive sequences are transcribed in differentiated cells according to cell-specific patterns.

Regulatory processes occur at all levels of biological organization, but the authors' theory is restricted to the processes of cell regulation at the level of genomic transcription. Their model suggests that a single initiating event sets into action regulatory mechanisms affecting numerous genes. Significantly, a *gene* is described *functionally*, rather than structurally, in generic terms, as "a region of the genome with a narrowly definable or elementary function. It need not contain information for specifying the primary structure of a protein" (Britten and Davidson, 1969, 349–350). No more genes *per se*; there are *Producer genes*, *Receptor genes*, *Integrator genes*, and *Sensor genes* that act in the context of batteries of genes.

Britten and Davidson recalled that "most of the known biosynthetic pathways are already represented in unicellular organisms." Thus, "it seems unlikely that the 30-fold increase from poriferan [sponge] to mammal can be attributed to a 30-fold increase in the number of producer genes."

> Quite possibly, the principal difference between a poriferan and a mammal could lie in the degree of integrated cellular activity, and thus in a vastly increased complexity of regulation rather than a vastly increased number of producer genes. Much of the DNA accumulating in the genome toward the upper end of the curve . . . might then have a regulative function. . . . It is likely that an ever-increasing library of different combinations of groups of producer genes is needed as more complex organisms evolve.
>
> Britten and Davidson (1969, 352)

Conceptually, many of the ideas of researchers on genetic regulation and its complexity as a differentiation system were directed at the sequence-specific hybridizability properties of DNA and RNA poly-nucleotides. As noted earlier, Jacob and Monod originally attributed the function of a transcribed repressor RNA molecule to directly interacting by competitive hybridization with the operator DNA sequence of the *lac*-operon (see Chapter 15, and Müller-Hill, 1996 for the history of the recovery of the lac-repressor). Attention was directed to the role of "sequence recognition between the special chromosomal RNA's and the DNA [that] specifies the pattern of gene activity." It was noted that "the chromosomal RNA's have the property of binding, in what is apparently a sequence-specific way, to double-stranded native DNA" (Britten and Davidson, 1969, 354). Although it turned out that the repressor molecule was a protein, the significance of this line of investigation as an experimental test of the idea of activator RNAs is obvious.

Britten and Davidson point out that one distinguishing characteristic of such regulatory loci would be their "pleiotropic effects on the activity of a number of producer genes" and the pattern of their integration. They suggest that "a number of good cases of this genre actually exist, particularly for drosophila and maize," and point at *Notch*, mutations of which display a very large variety of developmental abnormalities, all affecting early embryonic organization. This is consistent with what would be expected of mutations in integrator gene sets (Britten and Davidson, 1969, 354). Eventually, *Notch* became one of the most profoundly studied ubiquitous genes involved in cell–cell signaling (see, e.g., Artavanis-Tsakonas, Rand, and Lake, 1999). In addition, the authors noted that McClintock and others had demonstrated the "presence of other control sites adjacent in the genome to the same producer genes as those controlled by the distant regulatory elements," as exemplified by the Ac-Ds system (Britten and Davidson, 1969, 355).

No less significant for the paper's role in re-establishing the integration of genetic and developmental research, not to speak of evolution and development, is the authors' attention to the "Studies with drosophila imaginal disk cell determination and transdetermination carried out by Hadorn and his associates" which demonstrated "the existence of an apparatus in the genome for specifying integrated patterns of activity in the various cell types derived from the disk cells" (see Chapter 16):

> In experimental imaginal disk systems, highly exact specification of the patterns of producer gene activity is heritable through many cell divisions and is separated in time from producer gene function per se (that is, manifest dedifferentiation).
>
> Britten and Davidson (1969, 355)

A SOURCE OF INNOVATION

Britten and Davidson emphasize that "the apparently universal occurrence of large quantities of sequence repetition in the genomes of higher organisms" strongly suggests that they have an important current function. But this present-day function for the repeated DNA sequences must be considered "in addition to their possible evolutionary role as the raw material for creation of novel producer gene sequences." Repeated sequence families are observed with degrees of similarity varying from perfect matching to matching of perhaps only two-thirds of the nucleotides. Many of the repeated sequences are intimately interspersed with nonrepeated sequences throughout the length of the genome. As the authors note, evolution is often an increase in complexity and an acquisition of functional specificity, and this is precisely the pattern required when repeated sequences are usually or often regulatory in function (Britten and Davidson, 1969, 355).

> At higher grades of organization, evolution might indeed be considered principally in terms of changes in the regulatory systems. It is therefore a requirement of a theory of genetic regulation that it supply a means of visualizing the process of evolution.
>
> Britten and Davidson (1969, 356)

They further mention that:

> [T]he genome of an organism can accommodate new and even useless or dangerous segments of DNA sequence such as might result from a saltatory replication. Initially these sequences would not be transcribed . . . Only by inclusion in integrated producer gene batteries . . . would their usefulness as producer genes be tested. . . . The new families of repeated sequences might well be utilized to form integrator and receptor gene sets specifying novel batteries of producer genes. Thus saltatory replications can be considered the source of new regulatory DNA.
>
> Britten and Davidson (1969, 356)

In other words, Britten and Davidson note the possibility that imprecisely repeated sequences may reflect also the evolution of producer genes – an idea long advocated by Ohno (1970) that soon became central to considerations of evolution in an increasing body of paralogous genes.

Britten and Davidson end their paper by pointing out the interesting and surprising predictions of their model. These properties "suggest that *both the rate and the direction of evolution . . . may be subject to control by*

257

natural selection" (Britten and Davidson, 1969, 356, italics in original). Genetic research transcends its reductionist conception phase and advances toward the integrative phase, which increasingly applies the high-resolution methods of molecular biology to developmental and evolutionary aspects of living organisms and to the combination of the two, now called evo-devo.

18

Overcoming the dogma

In December 1992 Bertil Daneholt presented the Nobel laureates in physiology and medicine for that year to the assembly of the Karolinska Institute in Stockholm:

> In the middle of the last century, the Austrian monk Gregor Mendel conducted his famous breeding experiments with the garden pea. . . . To Mendel a gene was an abstract concept, which he used to interpret his breeding experiments. He had no idea of the physical properties of genes.
>
> Only in the mid-1940s could it be established that in terms of chemistry, genetic material is composed of the nucleic acid DNA. About ten years later the double helical structure of DNA was revealed. Ever since then, progress within the field of molecular biology has been very rapid . . .
>
> Initially, genetic material was studied mainly in simple organisms, particularly in bacteria and bacterial viruses. It was shown that a gene occurs in the form of a single continuous segment of the long, thread-like DNA, and it was generally assumed that the genes in all organisms looked this way. Therefore, it was a scientific sensation when this year's Nobel Laureates, Richard Roberts and Phillip Sharp, in 1977, independently of each other, observed that a gene in higher organisms could be present in the genetic material as several distinct and separate segments. Such a gene resembles a mosaic. . . . It soon became apparent that most genes in higher organisms, including ourselves, exhibited this mosaic structure.
>
> Roberts' and Sharp's discovery opened up a new perspective on evolution, that is, on how simple organisms develop into more complex ones. Earlier it was believed that genes evolve mainly through the accumulation of small discrete changes in the genetic material. But their mosaic gene structure also permits higher organisms to restructure genes in another, more efficient way. This is because during the course of evolution, gene segments – the individual pieces of the mosaic – are regrouped in the genetic material, which creates new mosaic patterns and hence new genes.

This reshuffling process presumably explains the rapid evolution of higher organisms.

http://nobelprize.org/nobel_prizes/medicine/laureates/
1993/presentation-speech.html)

Daneholt's short review of the history of genetic thinking, as he saw it, represents the frame of mind of scientists of the molecular era: reductive genetic analysis was successfully translated into terms of chemistry, to the extent that they are surprised and conceive it a "scientific sensation" when the apparently straightforward bottom-up progression is disturbed by "a gene in higher organisms [that] could be present in the genetic material as several distinct and separate segments," resembling a mosaic of coding and non-coding sequences. This "makes no sense" in terms of the logic of reductive physics and chemistry, and it is "blamed" on biology, the lawfulness of which is top-down historical constraints. Living organisms have a history of evolution and cannot be reduced to Cartesian machines anymore.

INTERVENING SEQUENCES

The Central Dogma left no room for cellular modifications in the path from DNA to polypeptide: circumstances that initiate or suppress the regulatory systems might vary greatly, but once transcription is started from a given site it would go on along the open reading frame (ORF) and eventually be translated, without the active involvement of cellular mechanisms in the semantics of the transmitted information. The polypeptide phenotype is determined by the polynucleotide genotype.

Mutations are changes that modify the transcribed sequence at the locus of a gene. From the perspective of the central dogma these changes might affect transcription as much as translation in this determinist path. Accordingly, one could conceive of the folded, active protein as comprising the *primary phenotype*, and conceive of DNA, the transcribed and translated mRNA, and, as a matter of fact, the translated polypeptide chain, as equally representing the *genotype*. To the extent that one observed pleiotropism, it was secondary not only to the nucleotide sequence of the DNA and the RNA, but also to the polypeptide.

These notions of genes as discrete material entities lost ground as biochemical inductive evidence increased: towards the 1970s, Jim Darnell and his colleagues observed that the nuclear RNA molecules of higher eukaryotes were on average much larger and more complex than mRNA molecules. They "began a lengthy series of hybridization and hybridization competition experiments ... designed to test their sequence relatedness" (Darnell, 1999, 1687). Unstable heterogeneous nuclear RNA molecules (hnRNA) were found to precede mRNA molecules, and consequently the transcribed RNA must be processed prior to being translated into the polypeptide sequences. Stated differently, in eukaryotes the reductionist conception of the Central Dogma of the gene as a discrete entity of DNA that provides all the essential functional information for a protein up to its immanent three- (and four-) dimensional folding, must be abandoned. At least in eukaryotes, the mRNA nucleotide sequences, as well as the translated polypeptide, are the products of a processed transcript of the DNA sequence, and are as much phenotypic markers as is eye color, or the enzymes producing eye color. Once this is conceived it is obvious that these molecules are all the phenotypic expression of an abstract gene concept, the primary material expression of which happens to be (usually) a sequence of DNA nucleotides. The nucleotide sequence of DNA happens to be the material link of the machinery that provides the cell's function. It is the primary phenotype of the Mendelian–Johannsenian notion of genotype.[1]

Phillip Sharp and Richard Roberts complicated the elegant picture of the Central Dogma in a way that opened up new and powerful possibilities. They independently found that an individual gene doesn't have to consist of a single, continuous coding stretch of DNA, but can instead be made up of several DNA segments separated by presumably "irrelevant" DNA. A given gene was found to be a mosaic of coding exons and intervening introns. Other researchers would soon discover that such a discontinuous gene structure is in fact the most common gene configuration in higher organisms, encompassing from one discontinuity in most yeast genes to hundreds in humans, and whereas exons are typically of the order of 150, introns can be as long as 800,000 nucleotides. The coding sequences comprise often less than 10 percent of the gene's total length. It was only necessary that a precise molecular mechanism splice the introns out during RNA-processing. The most common splicing mechanism was one dependent on a pair of nucleotides (GU) at the initiation 5′ splice site and another pair of nucleotides (AG)

---

[1] The genome, as the term is understood today, provides the phenotypic, molecular information of the cell's genotype.

at the 3' end splice site (as well as a nucleotide – A – in the intron, indicating the branch point site for the excised intron-lariat-structure produced at splicing) and a complex splicosome, to guarantee proper splicing. Considering that transcription proceeds at a rate of forty nucleotides per second, these intervening segments that are spliced out can considerably affect the rate of availability of transcripts and may provide a primitive regulatory mechanism.

However, processing the "information" of the DNA on its way to the protein was like opening a Pandora's Box: once it was opened, there was no way to close it. Once there was an efficient mechanism for splicing flanking exons, it did not take much to find a way to join non-following exons: in alternative splicing, exons may be skipped and a given exon joined to one further downstream. Going still further, two exons carried on different RNA molecules may be trans-spliced (common in the nematode *Caenorhabditis elegans* and in protozoan trypanosomes). Alternative splicing was found to be a very important process: in humans it is estimated that 60 percent of the genes are spliced in alternative ways to generate more than one protein per gene. Some genes encode hundreds of alternative versions. The most spectacular example of multiple versions appears to be the *DSCAM* gene of Drosophila: of its twenty-four exons, four have alternative versions: exon no. 4 with twelve alternatives, exon no. 6 with forty-eight alternatives, exon no. 9 with thirty-three alternatives, and exon no. 17 with two alternatives, providing for 38,016 alternative versions of a 78kb mRNA. All are involved in an axon-guidance receptor that is responsible for directing nerve cones to their proper target. The gene *cSlo* of the chicken is active in the hair cells of the inner ear and provides codes for 576 alternative versions of a protein that has a role in determining the sound frequency to which the ear cells respond. The gene for the T-antigens of the SV40 virus encodes two protein products – the large T-antigen and the small t-antigen – the result of alternative splicing. Both forms of the T-antigen are made in all cells infected by the virus, but the ratio of the two forms produced differs depending on the splicing protein.

Splicing and alternative splicing seems to be a most significant epigenetic regulatory process in the development and functioning of eukaryotic organisms. It may also be one of the most ancient processes: accumulating evidence suggests that it was lost in the evolution of prokaryotes (rather than gained in that of eukaryotes) probably for the gain in reproduction expenses in organisms where functional complexity may not be very demanding.

Once the transcript was processed, more possibilities were detected, such as overlapping reading frames of the same strand or of complementary strands. A mutation in the overlapping sequence would affect both traits, and trans-splicing would locate the gene in two loci, sometimes even on different chromosomes. Whereas the earlier mentioned examples could be conceived as confirming the claims for secondary pleiotropism, the later examples must be considered genuine cases of primary pleiotropisms, namely a DNA sequence affecting immanently more than one RNA transcript.

But how do we define the spliceable gene in relation to the Mendelian *Faktor*, Muller's discrete gene, or Benzer's cistron? What does it mean that a trait is a marker for a gene when this gene can produce more than one version, sometimes very many versions indeed? For many cases we may go back to Bateson's definition of unit character as that which segregates as a Mendelian unit in hybridization experiments (see Chapter 3). Even if in some of the above-mentioned cases the alternative mRNAs, although coded for different proteins, may be viewed together on a higher, cellular, or organismic level, as one trait marking one gene: an axon-guidance receptor protein, a protein tuning inner ear cells to sound variations, and antigenicity, respectively; in other cases this does not hold. Further evidence for top-down regulation is the prevalence of backup mechanisms, as indicated by the fact that many "knockout mutations" do not end in lethality. Yet, this is not the case for many others.

In a recent attempt to clarify the meaning researchers allot to the concept of the gene, Paul Griffiths and Karola Stotz compiled thirteen different empirical cases in which transcripts were modified in various ways, including: translational recoding as a result of ribosomal slippage, overlapping genes without shared coding sequences, an exon repetition (homotype trans-splicing), an anti-sense trans-splicing, an alternative anti-sense trans-splicing, co-transcriptional splicing between two ORFs, alternative trans-splicing from different chromosomes, anti-sense overlapping genes, and more. Questionnaires with the details of these cases were presented to professional researchers. For each case they were asked if they would describe it as one in which one or more than one gene was involved in generating the final transcript, and how appropriate it would be to ascribe the descriptions of the cases to one or to more genes. Not surprisingly, the answers from some 500 respondents were quite confused (see insert in Griffiths and Stotz, 2006; and also insert in Pearson, 2006, 401). It appears that the dialectic confrontation between the gene as a hypothetical construct and the gene as an intervening variable was resolved in favor of the latter: as emphasized repeatedly in recent chapters, the gene assumed the role of a generic term (see Falk, 2000b).

INTEGRATING THE GENOME

Once conceptions started to move toward a genomic, top-down perspective, notions such as "superfluous" or "junk" DNA that the Central Dogma era assigned to non-coding sequences gave way to a fresh approach, calling explicitly for a revised definition of the gene in the genomic era (Snyder and Gerstein, 2003). Gibbs (2003) quoting Mattick (2003) suggested that "the failure to recognize the full implications of . . . the possibility that the intervening noncoding sequences may be transmitting parallel information in the form of RNA molecules – may well go down as one of the biggest mistakes in the history of molecular biology."

The Human Genome Project that produced a sequence that was annotated for potential genes and other genomic features was crucial, yet it was only one step in the "whole genome" effort to understand the concepts of heredity in an organismic, or even more accurately, in a life-as-a-system, top-down approach that emerged in the 1970s.

The ENCODE approach (**En**cyclopedia **of** **D**NA **E**lements) is to map a wide variety of sequence elements and to elucidate the status of each element at each stage of an organism's development and function. The pilot project limited attention to 1 percent (~30Mb) of the human genome, including forty-four representative regions. Yet, it was found that nearly the entire genome may be represented in primary transcripts that extensively overlap and include many non-protein-coding regions. Although the idea of "a network of transcripts" had been suggested before, "data from the ENCODE project provide firmer footing . . . for this challenge to the concept of lone transcript units." Indeed, "a richer view of the connection between chromatin structure, regulation of transcription, and replication has emerged from integrating these data sets" (Weinstock, 2007).

The extent of the difficulty of differentiating between methodological reductionism and determinist conceptions in molecular genetic research is revealed again in Gerstein and his associates' paper "What is a gene, post ENCODE?" (Gerstein *et al.*, 2007). They admit that the classical view of a gene as a discrete element in the genome has been shaken by

ENCODE; yet, their own review of the evolution of operational definitions of a gene over the past century – from the abstract elements of heredity of Mendel and Morgan to the present-day ORFs enumerated in the sequence databanks, including their elevation of integrating interactions to the level of an "interactome" (Gerstein, Lan, and Jansen, 2002) – shows that as early as the 1970s the discrepancy between "nominal genes" and functional entities in the context of an integrated genome became central to genetic analysis (see Falk, 2004; Griffiths and Stotz, 2006). The crucial finding appears to have been the discrepancy between the smaller number of protein-coding genes that sequence annotation could identify (~21,000), and the ENCODE indications that to produce all the detected RNAs, >90 percent of genomic sequences appear to be transcribed as nuclear primary transcripts. This highlighted the number and complexity of the RNA transcripts that the genome produces (Gerstein *et al.*, 2007; Gingeras, 2007).

> The discrepancy between our previous protein-centric view of the gene and one that is revealed by the extensive transcriptional activity of the genome prompts us to reconsider now what a gene is.
>
> Gerstein *et al.* (2007, 669)

The current findings provide a computational metaphor for the genome complexity "in terms of subroutines in a huge operating system," which leads to proposing a tentative update to the definition of a gene: "A gene is a union of genomic sequences encoding a coherent set of potentially overlapping functional products" (Gerstein *et al.*, 2007, 669). A better reversion to the Mendelian–Johannsenian notion of the gene is hard to imagine! The gene is the something *defined by its unity of function* that is provided by matter (DNA sequences). As Gerstein and his colleagues note, their "definition sidesteps the complexities of regulation and transcription by removing the former altogether from the definition and arguing that final, functional gene products (rather than intermediate transcripts) should be used to group together entities associated with a single gene. It also manifests how integral the concept of biological function is in defining genes" (Gerstein *et al.*, 2007, 669).

> ... insofar as the nucleotides of the genome are put together into a code that is executed through the process of transcription and translation, the genome can be thought of as an operating system for a living being. Genes are then individual subroutines in this overall system that are repetitively called in the process of transcription.
>
> Gerstein *et al.* (2007, 671)

NON-CODING DNA

Once even molecular biologists overcame referring to a gene as "a genomic region producing a polyadenylated mRNA that encodes a protein," the way was opened to reconsider the function of non-coding DNA sequences.

The emergence of a large collection of unannotated transcripts with apparently little protein-coding capacity often overlapping protein-coding genes further blurred the physical boundaries (Gingeras, 2007). In spite of developments in molecular hybridization methodologies, including "large scale studies of cDNA cloning and interrogation of genome tiling arrays," most non-coding RNA (ncRNA) "genes" are hard to identify by genetic analysis. Some, with strong evolutionary and/or structural constraints, can be identified computationally through RNA folding and coevolution analysis. Yet, the role of non-coding RNAs has established the conception of the genome as an integrated system on a new level (Mattick, 2005).

Non-coding RNA may account for up to 98 percent of the transcriptional output of the human genome. Of the well-characterized non-coding transcripts with known functions the best known are of course ribosomal RNAs, transfer RNAs, small nuclear RNAs such as those involved in splicing, as well as components of RNases. However, much recent attention has been directed at the abundant micro- and small-RNAs which participate in the RNA-interference (RNAi) pathway. These have regulatory functions at transcriptional and post-transcriptional levels, based on nucleotide sequence-pairing properties that seem as if they were taken directly from a handbook of methods of genetic analysis.

The gene-silencing mechanism known as RNAi is thought to defend cells from viruses and "jumping genes": it hybridizes with single-stranded RNA to form molecules of double-stranded RNA (dsRNA). The enzyme Dicer cuts up the offending double-stranded RNA, creating fragments 21–25 nucleotides long, called small interfering RNAs (siRNAs). These fragments bind to copies of the original RNA, targeting them for destruction. A linked phenomenon is that of microRNAs (miRNAs) of about 22 nucleotides long, first identified in the nematode *Cae-norhabditis elegans*. It is involved in gene regulation in a wide variety of organisms, by binding to messenger RNAs and preventing their translation to proteins. Small RNAs appear to be associated with various epigenetic phenomena in plants as well as in animals (Dennis, 2002b).

It has been suggested that miRNA interactions with the network of protein-coding genes evolved to buffer stochastic perturbations and thereby confer robustness on developmental genetic programs (Hornstein and Shomron, 2006).

These mechanisms have been exploited by biologists and have become a popular method for gene-silencing studies by RNA-sequence hybridizations (Dennis, 2002a).

Although some invoke the old "Junk DNA" notions, believing that transcription is a messy business, that it is cheap to make RNA and that cells can tolerate high levels of "transcriptional slop" it appears likely that non-coding RNAs constitute a critical hidden layer of gene regulation in complex organisms, the understanding of which requires new approaches in functional genomics. The fact that the miRNAs that have been discovered are remarkably similar across different species also indicates that they may have had a role in evolution (Dennis, 2002a).

19

Dominance

The phenomenon of dominance and its meaning accompanied the science of genetics from its initiation. Bateson interpreted it in terms of the Presence and Absence Hypothesis. R. A. Fisher, taking notice of the harmful effects of inbreeding, pointed out that "if we assume that adaptation is the result of selection, the majority of large mutations must be harmful" (Fisher, 1922, 323). Thus a major task of the evolution of dominance would be to equalize the phenotypes of the heterozygotes to that of the homozygote for the wild-type allele (Fisher, 1928). However, doubts were raised about the efficiency of natural selection in carrying out the task (Wright, 1929a). According to Wright's theory of the functional organization of the cell that related the genotype to the phenotype, dominance would be the outcome of interaction:

> On the view that genes act as catalysts and largely through bringing about the production of catalysts of second order, it is easy to show that increase in the activity of a gene should soon lead to a condition in which even doubling of its immediate effect brings about little or no increase in the ultimate effects.
>
> Wright (1929b, 278)

Haldane (1930) and Muller (1932), following Shull (1909, 415) and Wright, suggested that the amount of product produced by the alleles of a gene varied asymptotically, and that the wild-type alleles in diploids have been selected to function in the relatively saturated segment of the asymptotic gene/product curve (see Chapter 14). "[I]t is to the advantage of the organism that most genes shall be very stable, and present-day races are doubtless the products of a long process of selection in that respect as well as in regard to the constancy of the reaction whereby the factors produce the characters" (Muller, 1918, 494).

268

The expansion of genetic research to haploid organisms and the intensification of the analysis of cell metabolism at the molecular level, justified a re-examination of the role of dominance in genetic theory. Dominance and recessivity had been conceived to be the genotypic property assigned to given alleles: there were dominant alleles and there were recessive alleles, irrespective of the specific physiological meaning of this assignment (and irrespective of the fact that given proper environmental conditions the dominant–recessive relationship changed). This rationale became encoded in the genetic terminology introduced for Drosophila (and many other organisms), namely, capital letters for dominant alleles, small letters for recessives (Falk, 2001b).

Kacser and Burns took Wright's physiological model of dominance one step further, suggesting that at that level "dominance" may be an automatic outcome of cell metabolism (Kacser and Burns, 1981; Keightley, 1996). Enzymes do not act in isolation, but are kinetically linked to other enzymes via their substrates and products, comprising an intricate feedback system out of which dominance emerges. "Dominance is neither a 'property' of genes nor a 'property' of characters . . . [rather d]ominance is a shorthand term to describe the *relationship* of the phenotypes" of the two homozygotes and their heterozygote (Kacser and Burns, 1981, 653).

Viewing the organism as "an enzyme system, [which] consists of a large array of specific and saturable catalysts organized into diverging and converging pathways, cycles and spirals all transforming molecular species and resulting in a flow of metabolites," they claim that "the recessivity of mutants is an inevitable consequence of the kinetic properties of enzyme-catalyzed pathways and that no other explanation is required" (Kacser and Burns, 1981, 640–641). Viewed from another level, however, Kacser believes "that the recessivity of most mutants does not require such an explanation." Rather, "[e]nzymes have evolved by natural selection to maximise (or optimise) some *output* of the whole system and hence only indirectly to specify the kinetic parameters in relation to the substrate concentration" (Kacser, 1987; see also Falk, 2001b, 312).

In haploid prokaryotes equalizing the phenotype of a homozygote and a heterozygote seemed of little significance, yet the need for alleles that stabilize phenotypes against possible fluctuations remained. To test the hypotheses of dominance–recessivity relationships, organisms with haploid life cycles and occasional diploidy could be followed in merozygotes (zygotes which are diploid for only part of their genetic material and haploid for the remainder). As a rule, even in organisms in which haploidy prevails wild-type alleles are dominant. In *Chlamydomonas*, an organism that is a "compulsory" haploid, most of a list of fifty-two mutations are recessive to their wild-type alleles (Orr, 1991). This case provides support for the "physiological" explanation of the origin of dominance as an automatic consequence of the kinetics of enzymes in cell systems.

Attempts to reduce dominance to molecular terminology (Schaffner, 1969) and to define "dominance" of a gene (allele) by its capacity to direct the synthesis of an active enzyme, have shown that dominance is a relational property (an allele must be dominant with respect to another) that cannot be formally reduced to molecular terms (Schaffner, 1976, 1993; see also Hull, 1974, and Part V).

As noted by Sarkar, such a reduction of dominance, confined to the relationship between the amino-acid sequence of the primary polypeptide and the nucleotide sequence of the DNA, while sometimes endorsed also by molecular biologists (e.g., Crick, 1958), makes a parody of genetics. Sarkar conceives of dominance as a characteristic of phenomenological genetic analysis that loses its power when relating to enzymes as the products of the molecular mechanisms that translate DNA sequences to polypeptides:

> Genetic rules and explanations are powerful because there are many traits far removed from DNA that can be explained and understood from the presence of particular alleles at a single locus (or at a very few loci). . . .
> When a molecular explanation of dominance emerges – if it does – it must account for dominance at the level of polydactyly or Huntington's disease.
> Thus, whether or not dominance arises is only minimally explained from the properties of individual enzymes. The explanatory weight falls on the *topology* (or architecture) of the network, that is, its connectivity properties . . .
>
> Sarkar (1998, 171–173)

Inevitably, the significance of dominance changed dramatically with the turn of genetic research to studies in haploid organisms. Genetic rules and explanations that were powerful for "final phenotype" traits, far removed from DNA, become trivial when relating to the level of both alleles that are expressed as two polypeptides. To the extent that dominance is understood at the molecular level, it emphasizes that enzymes do not act in isolation (Sarkar, 1998, 172).

Dominance, however, continued to fulfill a crucial role as a *methodological instrument* of the new genetics of haploid genomes. The power of the method was first demonstrated by the utilization of the phenomenon of heterokaryon formation in *Neurospora*, i.e., of cells whose cytoplasm carried multiple haploid nuclei of two parental origins. The phenotype of such heterokaryons simulated dominance–recessivity relations of diploids. The fact that it was argued that this similarity was only "superficial"[1] is another reflection of the dispute of whether dominance was a property of the genes or merely that of a phenotypic interaction at the cytoplasm level.

Pseudoallelism and the breakdown of the complementation test as the arbiter of alleles of a gene further jeopardized the conceptual status of dominance. A precondition for an operational test of allelism was that both mutations were recessive to the wild type.

> What does non-complementarity of recessive mutants ... imply? In a heterozygote $+/m_1$ the normal, dominant structure $(+)$ is capable, at least qualitatively, of carrying out the function in which the mutant, recessive structure (m_1) fails.
>
> Pontecorvo (1958, 41)

Still, dominance and the methodology of complementation had been exploited to its maximum by Benzer (1957) as an operational test, in which bacteria infected by T4 bacteriophages from different strains served to simulate compound heterozygotes, or more properly, heterokaryons, to identify the functional units – *cistrons* – of the bacteriophages.

For some time, therefore, the term "dominance" was referred to in a somewhat inconsistent manner. Symptomatic is the reference to dominance in William Hayes' textbook: "The presence or absence of dominance has probably nothing to do with the gene itself as an entity, but depends on the nature of the particular biochemical process which it controls" (Hayes, 1964, 25). However, when dealing with "genetic expression and its control," dominance served operationally to discriminate between mutants in a structural "gene" and a regulator "gene," rather than between alleles of the same gene (Hayes, 1964, 594ff.). It was in this extension of the concept of the unit of genetics from that of a single *cistron* to a more comprehensive functional regulative entity of the *operon* that Jacob and Monod imputed a strictly descriptive meaning to dominance and recessivity: "The dominance of

---

[1] Such an argument was expressed, for example, in my discussions with the late Larry Sandler.

the inducible over the constitutive allele means that the former corresponds to the active form of the *i* gene" (Jacob and Monod, 1961, 330). A wild-type allele is dominant over a loss-of-product allele when it interacts with its target via a cytoplasm product, even when they are assigned to different genic units; a mutant allele is expected to be dominant when its (or its product's) new conformation affects its capacity to interact (regulatively) with its target. Dominance and recessivity came to describe the functional interactive relations between a source and a target. *Trans-dominance* and *cis-dominance* were first introduced to denote the relation between regulative entities (an *operator* and a *regulator*) belonging to the same operon as a super-entity. This *de facto* acknowledgment of the interchangeability of *dominance* and *epistasis* was extended to describe any interactions between source and target and indicated the final stage in turning the phenomena from the realm of genetic structural organization to the realm of genetic function.

The term epistasis was assigned by Bateson to "characters which have to be, as it were, lifted off in order to permit the lower or *hypostatic* character to appear" (Bateson, 1907, 653). It refers to the interference of one gene with the phenotypic expression of another *nonallelic* gene (or genes), so that the phenotype is determined effectively by the former and not the latter when both genes occur together in the genotype (Rieger *et al.*, 1976, 188). With increasing information on the nature of genes and their function, the conceptual significance of these terms lost weight, although methodologically they continued to play a major empirical role, at least as long as their operational limitations were well defined (Falk, 2001b).

The dependence on genetic analysis of dominance relationships in *cis-trans* tests became largely redundant with the introduction of direct sequencing methods of DNA and RNA and the identification of the relevant proteins directly in terms of their amino-acid sequences. Still, dominance and recessivity are widely used, in such contexts as the genetic basis of human diseases. Surprised at the question, "why are some diseases dominant and others recessive?" Wilkie concludes that the terms are used merely in a descriptive function: "It should . . . be remembered that dominance is not an intrinsic property of a gene or mutant allele, but describes the relationship between the phenotypes of three genotypes" (Wilkie, 1994, 89). However, he too confronts the fact that "most wild-type alleles are dominant over other alleles," yet, he ignores the conceptual framework based on genetic analysis and, perhaps not surprisingly, adopts a molecular-age version of the long overhauled Bateson's Presence and Absence Hypothesis:

> The most likely effects of a random gene mutation are that it will either be neutral (normal phenotype) or inactivating. If the latter, the question is whether the inactivation would be clinically manifest in the hetero-zygote . . . or only in the homozygote.
>
> Wilkie (1994, 90)

It appears that in the case of dominance the shift from diploids to haploids, rather than the molecularization of the terms, had a more profound effect than the increase in the resolving power of genetic analysis. Instead of an analysis of dominance as the functional relationships between two (or more) products of similar repeats of the same entities (alleles) of diploid organisms, dominance became the description of the function and inter-actions expected between the molecular products of such entities. This is most clearly shown by juxtaposing Muller's classification of *amorphs*, *hypomorphs*, *hypermorphs*, *antimorphs*, and *neomorphs* with the molec-ular classification of *haploinsufficiency, ectopic mRNA expression, con-stitutive protein activity, toxic protein*, etc. (see Wilkie, 1994, 90–91 and Figure 2). Many nonsense mutations are said to be dominant negative, rather than recessive, because an interrupted mRNA, if translated, will produce a truncated protein that may be toxic because it disrupts the architecture of multi-subunit complexes (Danchin and He'enaut, 1997). Tumor suppressor genes, like *Rb* and *p53*, are dominant negative because the mutant protein, when combined with the wild-type protein, changes the structural conformation of the complex (Vogelstein and Kinzler, 1992). In the medical literature one may even find that the occurrence of a disease in two or more successive generations is described as "pseudodominant inheritance" (e.g., Ben-Chetrit and Levy, 1998).

Obviously, dominance and recessivity became strictly descriptive terms. In this sense, *Dominance*, or more precisely *dominating*, returned to mean what Mendel originally intended it to mean.

20

Populations evolve, organisms develop

Molecular biology and evolutionary biology are in constant danger of diverging totally, both in the problems with which they are concerned, that is, the "how" as against the "why," and as scientific communities ignorant and disdainful of each other's methods and concepts. The introduction of electrophoresis in evolutionary studies went some way toward impeding that separation and led naturally to an important second stage, the introduction of DNA sequence studies into population genetics.

Lewontin (1991, 661)

In 1966 George Williams made a heroic attempt to maintain the strict, reductionist approach of the New Synthesis of Darwinian evolution in his *Adaptation and Natural Selection: A Critique of Some Current Evolutionary Thought*. For Williams,

[t]he ground rule – or perhaps *doctrine* would be a better term – is that adaptation is a . . . concept that should be used only where it is really necessary. When it must be recognized, it should be attributed to no higher a level of organization than is demanded by the evidence. In explaining adaptation, one should assume the adequacy of the simplest form of natural selection, that of alternative alleles in Mendelian populations, unless the evidence clearly shows that this theory does not suffice.

Williams (1974 [1966], 4–5)

This uncompromising bottom-up doctrine opposed and rejected "certain of the recently advocated qualifications and additions to the theory of natural selection, such as genetic assimilation, group selection, and cumulative progress in adaptive evolution" (Williams, 1974 [1966], 4). With some minor qualifications, for Williams there was no escape from the conclusion that "natural selection . . . can only produce adaptations for the genetic survival of individuals," and he vehemently maintained that any "mechanisms for group benefit is based on misinterpretation, and that the higher levels of selection are impotent" (Williams, 1974 [1966], 7–8). Theories of developmental constraints, such as Waddington's genetic assimilation (Waddington, 1957), and theories of group selection such as those of Wynne-Edwards (Wynne-Edwards, 1962) were accepted – or better said, rejected – by students of the genetic basis of evolution as violating its basic conceptions.

However, the abundant molecular polymorphisms, first shown for the electrophoretic isozymes of natural populations of Drosophila and humans and later found in most eukaryotic species, were inconsistent with Williams' "assumption that the laws of physical science plus natural selection can furnish a complete explanation for any biological phenomenon" (Williams, 1974 [1966], 6–7). The "neutral theory" of evolutionary change was one of the consequences (see Chapter 10). But, whereas the neutral theory basically referred to polymorphism of individual genes, it was the introduction of the concept of "linkage disequilibrium" that indicated most emphatically the return of population genetics to top-down systems, while not abandoning the reductionist methodology of genetic analysis. The criticism of linkage disequilibrium "is really an attack on one of the basic assumptions of population genetics theory, namely that the genotypic array in a random mating population, and evolutionary changes in that array, can be described in terms of gene frequencies at the individual loci" (Franklin and Lewontin, 1970, 707). Reductionist Mendelian methodology would predict independence of the frequencies of genes in populations, each being determined by its specific adaptive value. Thus for the gene A-a, and the gene B-b, the expected frequencies for the four combinations AB, Ab, aB, and ab, would be $f(A)f(B)$, $f(A)f(b)$, $f(a)f(B)$, and $f(a)f(b)$ respectively [$f(A)$, $f(a)$, $f(B)$, and $f(b)$ being the respective allele frequencies of the two genes], even if they are closely linked on the genetic linkage map. However, as it turned out, this is often not the case (see, e.g., Lewontin and Kojima, 1960). Two explanations may be given for such disequilibrium: not enough evolutionary time has elapsed since the establishment of the linkage for equilibrium to be obtained; and there exist functional and developmental interactions between the linked genes such that selection maintains the disequilibrium.

Efforts to compare the evolutionary history of species by comparing their developmental patterns are at least as old as Darwin's theory of the origin of species. Comparisons of embryological developments were one of the main arguments of Darwin himself for the evolution of species, and were used by many biologists to argue for the evolution of species and against it. Probably the most influential among these was Ernst Haeckel with his Biogenetic Law, according to which ontogeny is a condensed version of phylogeny (Gould, 1977). Although the empirical evidence was primarily based on morphological comparisons, it was always implicitly accepted that the comparisons applied also at the physiological, and eventually the biochemical, levels. The very fact that animal experiments were used to test medical theories and techniques was anchored in the assumption of the commonality of their anatomical and physiological properties, though not necessarily on the assumption that these stemmed from evolution from a common ancestor.

With Linnæus and Buffon, the analysis of the relationships between species and their transformation followed two research traditions (Chapter 1). One was that of empirical morphological and physiological investigations of the possible transformation of species, culminating with Darwin's *Origin of Species*. The other was that of the experimental investigation of the transformation of species by hybridization, culminating with Mendel's *Versuche*. In the 1940s when Mayr and Dobzhansky established species as the largest and most inclusive reproductive communities of sexual and cross-fertilizing individuals that share in a common gene pool (Rieger *et al.*, 1976), the hybridists lost their tool for inter-specific comparisons. Lewontin and Hubby (Hubby and Lewontin, 1966; Lewontin and Hubby, 1966) introduced studies of *in vitro* gel-electrophoresis of soluble proteins that were shown to Mendelize *in vivo*, and this opened the way for hybridists' analyses to compete with the morphological-physiological logic of analyzing the evolution of species. "Electrophoresis was also a milestone in that it provides for the first time the possibility of including virtually any organism in the study of evolutionary variation on the basis of a common denominator across species" (Lewontin, 1991, 661). At the same time the barriers that stood in the way of inter-specific hybridization were overcome by *in vitro* DNA:DNA and DNA:RNA methodology (Hoyer, McCarthy, and Bolton, 1964; Marmur and Lane, 1960), and this may have made the electrophoretic comparison of proteins for the establishment of evolutionary relationships secondary.

As noted (see Chapter 17), no evidence presented at the time proved that differences between species in amino-acid sequences of proteins were under the control of Mendelian genes. Still, Crick, referring to the "'family likeness' between the 'same' protein molecules *from different species,"* [like haemoglobin or insulin] predicted that "before long we shall have a subject which might be called 'protein taxonomy' – the study of the amino-acid sequences of the proteins of an organism and the comparison of them between species" (Crick, 1958, 142). Electrophoretic polymorphism immediately provided access to "vast amounts of evolutionary information" hidden away not only within populations, but also in inter-species variability.

After two centuries of essentially different heuristics, the morpho-physiological and the hybridization paths were amalgamated into one. In one fell swoop, the role of genetics in relating the development of organisms to the evolution of populations, or as it became known evo-devo, was established.

BEYOND GENIC ADAPTATION

"Electrophoretic phenotypes ('electromorphs') are discrete differences, almost certain to be a consequence of single gene differences" (Lewontin, 1991, 659). As the concealed variability was much more intensive than the theories of the adaptive hypotheses could explain, the neutral mutation theory was welcomed (see Chapter 10). Lewontin, however, warned against over-enthusiasm for the sweeping conceptions of the new methodology: in the Mendelian system of causal analysis genotype and phenotype appear asymmetrical, thus, an uncritical application of electrophoretic characters to systematics may be misleading (Lewontin, 1992, 138). False convergence may appear in phylogenies because the redundancy of the genetic code may result in the same amino acid in spite of differences in the nucleotide sequences; also, since gel electrophoresis depends upon charge differences, two indistinguishable electromorphs may result from two different amino-acid sequences. A serious underestimate of genetic variation may have resulted. Lewontin warned that "even the richest available static data set on electrophoretic variability lacks the statistical power to discriminate unambiguously between selection and neutrality in large populations with a small amount of migration" (Lewontin, 1991, 660); nevertheless, the neutral mutant theory became so fashionable that for a time it obscured other possibilities, to the extent that geneticists had to struggle for the legitimacy of *non*-neutral theories (Golding, 1994).

Arguably, it was primarily the analysis of linkage disequilibrium that compelled researchers to abandon the perspective of the single gene as the adaptive determinant, and to adopt a genomic perspective of adaptation:

> It appears likely ... that the disequilibrium between pairs of loci is a function primarily of the map length of a chromosome segment and the loss in fitness accompanying homozygosity of this segment. Furthermore, this appears to hold under a wide degree of conditions ...
>
> The model discussed ... provides a possible explanation for the origin and persistence of inversions in natural populations. ... [W]e find considerable organization, and it should be apparent that an inversion encompassing the block of loci would be at a selective advantage. If this were the case, it would not be necessary that the selection coefficients which initially induced the disequilibrium within the set of heterotic loci be maintained – it would only be necessary that some degree of heterosis persist. If inversions do arise that way, it should be noted that coadaptation of the alleles within an inversion is not necessary. Heterosis is the only prerequisite. Co-adaptation may, of course, arise later.
>
> Franklin and Lewontin (1970, 730)

As a matter of fact, deviations from expectations of the neutrality theory provided evidence for the effect of selection. Sequencing a DNA fragment of the fourth chromosome of several lines of *Drosophila melanogaster* and *D. simulans* revealed no, or only little, polymorphism. Since there is practically no recombination in this chromosome, these results indicated that "recent positive selective sweeps ... with extreme hitchhiking" were responsible for these observations (Berry, Ajioka, and Kreitman, 1991). Furthermore, since nucleotide sequences do not have a one-to-one correspondence with amino-acid sequences, the finding of a sequence of the *ADH* (alcohol dehydrogenase) gene locus that is monomorphic for amino acids but polymorphic for synonymous positions at the level of DNA sequence, allows one to conclude that purifying selection has been operating (Kreitman, 1983). Certainly, this extension of genetic analysis to the level of molecular sequencing "is a major methodological break because of the lack of one-to-one correspondence between nucleotide sequences and amino acid sequences" (Lewontin, Paul, Beatty, and Krimbas, 2000, 45).

Whereas arguments of evolution of non-adaptive polymorphism of genic alleles on the one hand and chromosomal- or genome-adaptive values on the other hand, could still be conceived of as extensions of the reductionist conception, the introduction of groups as entities of selection, or of structural constraints in the evolution of species, turned the perspective upside down.

Altruistic traits, however, posed a difficult problem for evolutionary theory as long as fitness was heuristically defined as individual reproductive success. "A behavior is altruistic when it increases the fitness of others and decreases the fitness of the actor" (Sober and Wilson, 1998, 17). Group selection was discredited not by a crucial experiment or even by a new theoretical development, but simply from "the elegance and clarity of Williams's thought in interpreting developments of the previous three decades" (Wilson, 1983; see also Wilson, 1992). Sewall Wright, for example, developed a single-locus model in which one allele codes for a "socially advantageous" character that is "of value to the population, but disadvantageous at any moment to the individual" (Wright, 1945, 416). According to Wright, natural selection is totally insensitive to group benefit, and the relevant allele will always tend towards zero (Wilson, 1983, 164). However, extending the classic genetic analysis and assigning a fitness value to a genotype allows the definition of *inclusive fitness*, namely the genetic contribution to the next generation through both one's own reproduction and the contribution to the reproductive success of other individuals of the same genotype that results from the actions of the individual under consideration. As shown by Hamilton, under such a definition heritable altruistic behavior may evolve if the cost of the altruistic behavior to its displayer (c) is smaller than the benefit to the recipient (b) multiplied by the relatedness (r) of the partners: $c < b \times r$ ("Hamilton's Rule"). Accordingly, altruistic behavior is expected to evolve among close relatives (Hamilton, 1963, 1964a, 1964b). Such kin selection operating on differences in inclusive fitness prompted J. B. S. Haldane to state: "Would I lay down my life to save my brother? No, but I would to save two brothers or eight cousins."

By defining inclusive fitness Hamilton was only following in the footsteps of biologists and philosophers who since Darwin's time had struggled to define the individuals that comprise the units of natural selection and the status of the species in the process. Michael Ghiselin (1971) and David Hull (1976) suggested that, ontologically, species are individuals rather than classes. This would recognize them not only as taxonomic end products of the process of evolution by natural selection, but also as participants in it as such, as units of selection (see the target articles of Ghiselin, 1987 and Mayr, 1987, and the discussion that followed in *Biology and Philosophy* 2(2) and 3(4)). Once the framework of the entities that comprise the "individuals" of natural selection was extended, group selection became a legitimate subject of genetic analysis. David Sloan Wilson and Elliot Sober revived the "super-organism," noting that "individual selection in its strong form is founded on a logical contradiction, in which genes-in-individuals are treated differently than individuals-in-groups or species-in-communities." They brought evidence that superorganisms as units of selection "are more than just a theoretical possibility and actually exist in nature."

> When within-unit selection overwhelms between-unit selection, the unit becomes a collection of organisms without itself having the properties of an organism, in the formal sense of the word. When between-unit selection overwhelms within-unit selection, the unit itself becomes an organism in the formal sense of the word.
>
> Wilson and Sober (1989, 343)

There is no magic or mysticism here; the altruists increase globally, even if they decrease in frequency within each group, "because the two groups contribute different numbers of individuals to the global population." This is an example of the statistical Simpson's paradox (Sober and Wilson, 1998, 23).

The discussion of the evolution of altruistic behavior thus emphasizes the regulative role of the units of selection (Falk, 1988) and signifies the return of the top-down perspective into the consensus of genetic analysis. But arguably the most explicit move in this direction was Gould and Lewontin's paper "The spandrels of San Marco and the Panglossian paradigm: a critique of the adaptationist programme" (Gould and Lewontin, 1979; see also Selzer, 1993). C. H. Waddington called attention to the importance of developmental constraints even as early as the late 1930s. The concept that there are distinct genetically specified self-stabilizing or "buffering" capacities in developing organisms was embedded in Waddington's term "canalization" (Waddington, 1942). However, it was largely ignored at the heyday of the New Synthesis. Now, with the shift of emphasis to the top-down perspective, Gould and Lewontin's paper – provocative as it may have been – was readily adopted by geneticists and evolutionists alike. In essence, it criticized the adaptationist approach that conceived of evolution narrowly as a function of genes' adaptive value and pointed at levels of constraints on the process of evolution, notably developmental constraints: the developmental histories of organisms dictated the constraints of further evolution (as well as the opportunities) to a considerable extent.

Molecular biology followed suit. The introduction of "knock-out" techniques – the complete inactivation of a gene as a model experiment for verifying the function of that gene – often resulted in the embarrassing observation that the knock-out organism did not fulfill expectations; for example, too many "lethals" were not lethal when specifically tested as knock-outs of the relevant genes. This indicated the existence of cellular, or even organismic, backup systems. With the increasing results of DNA sequences, many paralogous sequences, which are highly similar sequences present in the same genome (as opposed to orthologous sequences, which are corresponding, highly similar sequences present in genomes of different species), were identified. At least some of these may function as mutual backup elements in developmental processes (Wilkins, 1997; see also Wilkins, 2002).

Mechanisms that regulate the rate of mutations (and heal early stages of mutations) are also stabilizing mechanisms that provide selective advantages at the level of the genome of populations. An even higher level of "units of selection" may be conceived when the rate of evolution over geological time periods is considered as an adaptive response to global changes in time and space, as specified by Eldredge and Gould in their concept of "punctuated equilibrium" (Eldredge and Gould, 1972).

By the 1970s the genetic theories of evolution overcame the strict conceptual reductionism of the New Synthesis and adopted the top-down view of evolutionary change as a global hierarchical process, in which the stratification of levels of research is largely determined by the subjective needs of the reductionist methodologies.

DEVELOPMENT AND DIFFERENTIATION

Starting in 1951, Edward B. Lewis published a paper on the complex locus of "bithorax" of Drosophila about every ten years, each paper longer than the previous one, and each indicating how hopelessly complex the issue was (Lewis, 1951, 15pp.; Lewis, 1963, 21pp.; Lewis, 1967, 30pp.). Finally, a short paper, five pages long, resolved the issue by brilliant extension of genetic analysis to the processes of developmental control (Lewis, 1978).

The bithorax complex (BXC) mutations, *bithorax* (*bx*), *Ultrabithorax* (*Ubx*), *postbithorax* (*pbx*), *Contrabithorax* (*Cbx*), etc. belong to the class of homeotic mutations, alterations of one organ of a segmental or homologous series from its own characteristic form to that of some other member of the series (Rieger *et al.*, 1976). For example, the mutant *bithorax* transforms the anterior part of the third thoracic segment

281

(carrying a pair of "balancers," the halteres) into a replica of the anterior part of the second, winged segment. Imitation of the bithorax phenotype ("phenocopies") could be induced by treating embryos with ether, indicating that there are crucial stages of normal segmental differentiation of the larvae (and the later metamorphosed imago fly) in the embryo. However, a serious shortage of genetic analysis of Drosophila had been caused by the lack of known larval phenotypic markers to allow analysis of mutations at these early stages; phenomenological analysis concentrated almost exclusively on markers of the adult fly. This is especially remarkable since so much cytological work had been done with the polytenic chromosomes in the larvae. The methodological conception introduced by Hadorn and his associates, and the introduction of new microscopic techniques, enabled Lewis to notice numerous larval markers that were affected by the BXC genes and to extend his genetic analysis to the phenotypes of embryonic and larval stages. This soon proved conducive to new vistas of top-down genetic analysis of the development of systems.

According to Lewis's model, genes that regulate segment-differentiation pathways respect a gene-controlled anterior–posterior gradient in the embryo; the further down the gradient the more genes controlling differentiation of segments are activated (Lewis, 1978). Some mutations affect the gradient (turning all segments in the embryo to posterior or to anterior ones), others shift the response of the gene to the gradient (so that it is activated too late, turning a segment into a more anterior developmental pathway – *bx*, or too early, turning it into a more posterior segment – *Cbx*). This model in the tradition of phenomenological genetics, although later thoroughly modified, heralded a new epoch in the genetic analysis of development. It provided a framework model of differentiation by extending the power of genetic analysis to the effective study of early developmental stages of Drosophila as the model organism. Coinciding with the newly developed molecular "chromosome-walking" method, the bithorax-complex became one of the first to be DNA-sequenced (Bender *et al.*, 1983) and thus extended phenomenological genetic analysis to the molecular level.

The BXC being a complex of homeotic genes was only one of such cascades of regulatory genes. In another homeotic mutation of *Antennapedia* (*Antp* mutations convert the fly's antennae into legs), Walter Gehring and coworkers extended the phenomenological analysis to the molecular level (e.g., Levine, 1988) and identified a DNA sequence of 180 base pairs, termed the "homeotic box," coding a domain of 60 amino acids which turned out to be highly conserved in many homeotic genes throughout eukaryotes (see Gehring, 1998; McGinnis, 1994). The homeoboxes and other such evolutionarily conserved motifs were found to act as master controllers of development (McGinnis, Garber, Wirz, Kuroiwa, and Gehring, 1984; and see Chapter 16). Genetic analysis now could often be reversed, and instead of looking for the molecular basis of phenomenological changes it became possible to predict phenomenological effects of molecular changes.

The sophisticated methods of genetic analysis and the large array of known mutations placed Drosophila again in center-stage of genetic control of differentiation. But it was the large-scale genetic screening for embryonic lethals in Drosophila by Christiane Nüsslein-Volhard and Eric Wieschaus, "attempting to identify all loci required for the establishment of [the metameric] pattern" (first those involved in early zygotic mutations, then also those acting as maternal mutations – those acting even before the genotype of the newly formed zygote exerts its impact) which generated the intensive analysis of the embryological and molecular foundations of pattern formation (Nüsslein-Volhard and Wieschaus, 1980). It was especially significant that the mutations in the fifteen gene loci of their analysis could be reduced to three types of pattern alteration: pattern duplication in each segment, pattern deletion in alternating segments, and deletion of a group of adjacent segments. Obviously, genetic analysis provided only the material for the embryological and molecular research, but in working out the details they were referred back to the genes and their interactions, and, perhaps most significantly, to the discovery of new systems of genetic control, such as those at the level of noncoding RNA regulation patterns (see, e.g., Chapter 18).

However, the growing crisis of reductionist molecular genetics in the late 1960s indicated that the biology of Drosophila might not provide the most appropriate experimental system for the new problems. Sydney Brenner conceived of two different sets of problems for "molecular biology":

> The first is to put the knowledge that we've got into a detailed physical and chemical basis . . .Thus one part of molecular biology is moving to become almost a branch of physics and chemistry. But then there is the second line, which is to bring other biological phenomena down to their molecular basis.
>
> Judson (1979, 204–205)

With an eye to the second line, Brenner referred to the work in molecular genetics as nothing more than to work out "Morgan's diversion" (Judson, 1979, 217): the ability to analyze development requires an appropriate organism. Sea urchins have long been a favorite organism of embryologists. But the difficulty of making genetic analysis excluded them. Geneticists favored Morgan's Drosophila, but its development is very complex and difficult to study.

> Although there are many theories suggesting how the extra DNA might be used for complex genetic regulation, the problem is still opaque. . . . These questions arise in particularly acute form in elaborate structures like nervous systems. . . . How is this complexity represented in the genetic program? Is it the outcome of a global dynamical system with a very large number of interactions? Or are there defined subprograms that different cells can get a hold of and execute for themselves? What controls the temporal sequences that we see in development?
>
> One experimental approach to these problems is to investigate the effects of mutations on nervous systems. . . . However, one surmises that genetical analysis alone would have provided only a very general picture of the organization of those processes. Only when genetics was coupled with methods of analyzing other properties of mutants, by assays of enzymes or *in vitro* assembly, did the full power of this approach develop.
>
> Brenner (1974, 71–72)

Genetic analysis was an essential but not a sufficient tool for the development of a research program that would resolve the problems of individual development and differentiation. Thus, somewhere in 1966 Brenner "decided that what was needed was an experimental organism which was suitable for genetical study and in which one could determine the complete structure of the nervous system." His choice eventually settled on the small nematode *Caenorhabditis elegans*. "It has a small and possibly fixed number of cells" (Brenner, 1974, 72). It reproduces parthenogenetically, but a sexual reproductive cycle can be induced. Being quite transparent, its anatomy can be followed under the microscope; and the fact that it can be grown in Petri dishes on a lawn of *E. coli* provides it with many of the advantages of bacteria (actually, bacteriophages) as an ideal experimental system, in spite (or just because?) of its very deterministic (preformationist) pattern of differentiation (Brenner, 1974).

In a sense, the opposite position has been taken by Nüsslein-Volhard, who extended her research from Drosophila to the genetic control of the embryogenesis of a vertebrate, the Zebra fish (*Danio rerio*), presumably a more complex organism than Drosophila and the nematode. However, being small, transparent, and easy to grow in large numbers made *Danio* an ideal organism for Nüsslein-Volhard's skills for screening experiments and she identified over a thousand genes. The combination of the genetic and molecular analyses of such factors significantly contributed to the prosperous discipline of evolutionary development.

With these developments genetic analysis completed its role as a separate discipline designed to elucidate the problem of inheritance. It became a central tool for the study of any problem of the life sciences. Yet, all these sciences contributed their share to better understanding the mechanism and evolution of inheritance and the systems' development that maintains life on earth.

Concluding comments

In the beginning of the 1950s, when I was a graduate student in Stockholm, my professor, Gert Bonnier, asked during one of our lunch breaks, what is the difference between a gene and a locus? I answered: "none," and failed. Mendel, like Kepler, strove to analyze the laws according to which (God's) world is run. He adopted the hybridist research tradition as the experimental method of analysis and followed the laws of segregation of factors for discernible trait variants. It was de Vries who imposed on genetics the notion that organisms are nothing but the consequences of the (physical chemical) properties of their components. Johannsen rejected such preformationism and defined *genes* as that something which represents hereditary transmission of trait variations. Finally, with Morgan's chromosomal theory of inheritance a *locus* was assigned to each gene. *Genes' properties in replication, function, and mutation were assigned to specific loci on the chromosomes.*

Although much of genetic analysis was performed without bothering about whether the entities that were discussed – the genes – were molecules of DNA or pieces of cardboard, the effort "to grind genes in a mortar and cook them in a beaker" (Muller, 1922) intensified. Genetic analysis became reductionist, not only in research methodology, but also increasingly in conception. Once Watson and Crick proposed the structure of DNA as the "genetic material," it was ascribed the role of the determinant genotype.

Benzer's genetic analysis of a gene down to (almost) the molecular level indicated that genes could be assigned to loci of functional consequences, and the semantic information of these was shown to be colinear with that of the amino acids of polypeptides. Sidney Brenner (2000) suggested using the term "genetic locus" to mean "the stretch of DNA that is characterized either by mapped mutations as in the old

287

genetics or by finding a complete open reading frame as in the new genomics." Crick's Central Dogma was the ultimate triumph of the reductionist, bottom-up conception of genetics: DNA sequences were the *genotype*, and its genes were deterministically transcribed and translated into a polypeptide *phenotype* that became part of the cell's structural and metabolic assembly, to eventually end up as green peas or blue eyes.

This dichotomy of the gene as that unit which is *identified* by a phenotypic "marker" but is *defined* as a DNA sequence has been developed to its extreme by Lenny Moss:

> The rhetoric of the gene as code and information . . . turns on . . . a conflation of two distinctly different meanings of the gene. When scientists and clinicians speak of genes for breast cancer, genes for cystic fibrosis, or genes for blue eyes, they are referring to a sense of the gene defined by its relationship to a phenotype . . . and not to a molecular sequence.
>
> It continues to be useful, in some contexts, to employ this usage of the word "gene." To speak of a gene for a phenotype is to speak as if, but *only* as if, it directly determines the phenotype. It is a form of preformationism but one deployed for the sake of instrumental utility. I call this sense of the gene – Gene-P, with the P for preformationist. . . .
>
> Quite unlike Gene-P, *Gene-D is defined by its molecular sequence*. A Gene-D is a developmental resource . . . which in itself is *indeterminate* with respect to phenotype. To be a Gene-D is to be a transcriptional unit on a chromosome within which are contained molecular template resources.
>
> Moss (2003, 44–46)

This notion culminated in recent decades with researchers talking about "reverse genetics" to indicate a change in direction: from analyzing the phenomena in order to discover their causes to analyzing causes in order to understand the phenomena. This may appear to be a shift from attempting to find out the laws of how the world is run to accepting that the most we can do is describe the history of the constraints that contributed to what we happen to encounter today. It ignores, however, consequences such as those of the "bewildering gene" (Falk, 1986). Once it has been shown that the sequences of DNA that are transcribed to RNA resemble only vaguely the RNA that is translated to a polypeptide, due to transcriptional and post-transcriptional processes, such a definite duality or any other discontinuity in the evolution of the gene concept seems unwarranted. Clearly there is no one DNA sequence that *on its own*, independently of the context of the product's function, can be unequivocally defined as a gene (Falk, 2000b, 2004). Furthermore, not all the cells' DNA is transcribed and much of the transcribed DNA is not translated. Cellular processes, by widely intervening in what is transcribed and how it is transcribed, and by intervening as well in what is translated, refute the determinist bottom-up conception of DNA sequences as the genotype that (together with the cellular systems) shapes phenotypes. As I hope to have shown, the deduction of the properties of genes from traits and the induction of properties from those of genes or DNA sequences are just two inferences on the continuum that is a dialog between research traditions and the consequent research methodologies. Genes and genotypes take on again, as in Johannsen, roles of intervening variables, generic terms for the consequences of cellular or organismic heredity that involve loci of DNA sequences (genome, on the other hand, is the technical term for the phenotype of the DNA – sometimes RNA – of the cell).

What's more, the duality of the gene-*P* and the gene-*D* yields to the vernacular use of the term "gene" as an icon of genetic determinism in daily language (see, e.g., Nelkin and Lindee, 1995). Towards the end of the twentieth century genocentricity, the conceptual deterministic reduction of all aspects of life to "genes," became a problem that greatly contaminated all ways of life (Barkow, Cosmides, and Tooby, 1992; Dawkins, 1976; Wilson, 1975). The image of the "gene" replaced the symbolic role of "blood" in former times for claims of biological hereditary relatedness. Genes meant "Nature," as opposed to "Nurture," even in contexts that described explicitly non-biological characteristics, such as "cultural genes." At the beginning of the twenty-first century even the term "gene" seems to have lost its popularity to the term "DNA." Moss *de facto* identifies the man-in-the-street's term with a "preformationist" notion of the gene-*P* and ascribes to it the determinate properties of the gene-*D*, inadvertently acknowledging the deceptive notion of the "gene for." Even if allowance is made for the practicing researcher to use the term "the gene for X" as an instrumental short cut for the complex relationships between the genotypic and phenotypic levels, how dangerous the inversion of the gene-*P* to a gene-*D* is may be learned from declarations such as Watson's that instead of the *belief* that our fate was in our stars, now we *know* that our fate is in our genes.

For a century the science of genetics managed to construct out of the dialectic tension of the reductive methodologies and the determinist conceptions, or out of the notions of intervening variables and hypothetical construct, a theory of inheritance that has wide operative

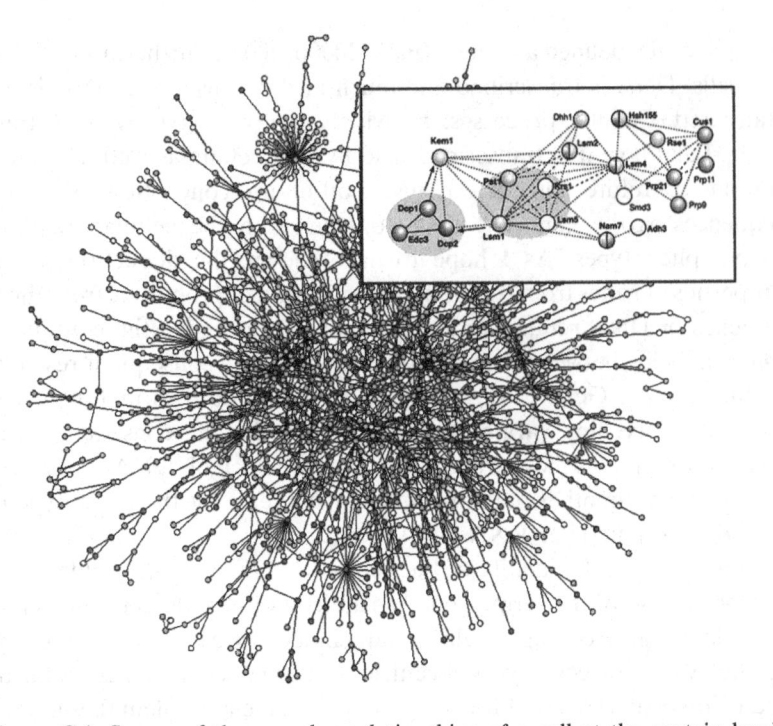

Figure C.1. Image of the complex relationships of a cell at the protein level. Insert shows in detail an (enlarged) minute fraction for discerning and studying single (inter)-action.

consequences. The theory of inheritance not only affects agriculture and medicine, but also influences the ways we look at the theory of organic evolution as well as how we understand organismic development and the acquisition of cognitive properties. The bewilderment of facing the complexities of so many processes – whether at the level of DNA sequences and the processes that depend on cellular functions, or at the level of the evolutionary processes that depend on cellular and organismic historic constraints, or at the level of the facts of systems for the evolutionary coordination of developmental processes – encouraged the establishment of Developmental System Theories (DST) (Oyama, 2000; Oyama *et al.*, 2001) (Figure C.1). It is, however, up to Oyama, Griffiths, and their colleagues to resolve the conflict of maintaining the concept of a multi-interactive complex system theory without giving up the tools of reductionist scientific methodology.

Genetic research today is no longer limited to the problems of inheritance in the sense of the transmission of biological capacities and structures from organisms of one generation to those of the following generations. Also the involvement with the molecules of genetic research, DNA (and sometimes RNA) is not restricted to departments of genetics: practically any biological research project now starts with or involves DNA, or at least "genes." So, what is left of genetic analysis? Reading the current papers in journals that carry "genetics" in the title, or attending a seminar in a department of genetics, leaves little doubt that what denotes these as "genetic" is their application of *genetic methodologies*. Mendel's reductionist hybridist research *methodology* has prevailed, even if our conception of the truth that we wish to establish has greatly changed.

Bibliography

Allen, G. E. (1966). Thomas Hunt Morgan and the problem of sex determination, 1903–1910. *Proceedings of the American Philosophical Society*, 110, 48–57.

(1975 [1978]). *Life Science in the Twentieth Century*. Cambridge: Cambridge University Press.

(1978). *Thomas Hunt Morgan: The Man and His Science*. Princeton, NJ: Princeton University Press.

(1979). Naturalists and experimentalists: The genotype and the phenotype. In W. Coleman and C. Limoges (eds.), *Studies in the History of Biology* (Vol. III, pp. 179–209). Baltimore, MD: Johns Hopkins University Press.

(1984). Thomas Hunt Morgan: Materialism and experimentalism in the development of modern genetics. *Social Research*, 51(3), 709–738.

Altenburg, E. (1933). The production of mutations by ultra-violet light. *Science*, 78, 587.

Altenburg, E., and Muller, H. J. (1920). The genetic basis of truncate wing – an inconstant and modifiable character in Drosophila. *Genetics*, 5, 1–59.

Amundson, R. (2006). EvoDevo as cognitive psychology. *Biological Theory*, 1 (1), 10–11.

Anfinsen, C. B., and Redfield, R. R. (1956). Protein structure in relation to function and biosynthesis. *Advances in Protein Chemistry*, 11, 1–100.

Ankeny, R. A. (2000). Marvelling at the marvel: The supposed conversion of A. D. Darbishire to Mendelism. *Journal of the History of Biology*, 33(2), 315–347.

Artavanis-Tsakonas, S., Rand, M. D., and Lake, R. J. (1999). Notch signaling: Cell fate control and signal integration in development. *Science*, 284(5415), 770–776.

Auerbach, C. (1967). The chemical production of mutations. *Science*, 158, 1141–1147.

Auerbach, C., and Robson, J. M. (1947). The production of mutations by chemical substances. *Proceedings of the Royal Society of Edinburgh, series B*, 62, 271–283.

Auerbach, C., Robson, J. M., and Carr, J. G. (1947). The chemical production of mutations. *Science*, 105, 243–247.

Bibliography

Avery, O. T., MacLeod, C. M., and McCarty, M. (1944). Studies on the chemical nature of the substance inducing transformation of Pneumococcal types. *Journal of Experimental Medicine*, 79, 137–158.

Baltimore, D. (1970). RNA-dependent polymerase in virions of RNA tumor viruses. *Nature and System*, 226, 1209–1211.

Barkow, J. H., Cosmides, L., and Tooby, J. (1992). *The Adapted Mind: Evolutionary Psychology and the Generation of Culture*. New York: Oxford University Press.

Bacq, Z. M., and Alexander, P. (1955). *Fundamentals of Radiobiology*. London: Butterworths Scientific Publications.

Bateson, W. (1894). *Materials for the Study of Variation*. London: Macmillan.
  (1905 [1928]). A suggestion as to the nature of the "walnut" comb in fowls. In R. C. Punnett (ed.), *Scientific Papers of William Bateson* (Vol. II, pp. 135–138). Cambridge: Cambridge University Press.
  (1907). Facts limiting the theory of heredity. *Science N.S.*, 26(672), 649–660.
  (1913 [1979]). *Problems of Genetics*. New Haven: Yale University Press.
  (1926 [1928]). Segregation. In R. C. Punnett (ed.), *Scientific Papers of William Bateson* (Vol. II, pp. 405–440). Cambridge: Cambridge University Press.

Bateson, W., and Punnett, R. C. (1911). On gametic series involving reduplication of certain terms. *Journal of Genetics*, 1(4), 293–302.

Bateson, W., and Saunders, E. R. (1902). *Reports to the Evolution Committee of the Royal Society* (pp. 1–160) (Report I. – Experiments undertaken by W. Bateson, F.R.S. and Miss E. R. Saunders). London: Royal Society.

Baur, E. (1912). Ein Fall von geschlechtsbegrenzter Vererbung bei *Melandrium album*. *Zeitschrift für induktive Abstammungs- und Vererbungslehre*, 8, 335–336.

Beadle, G. W. (1945). Biochemical genetics. *Chemical Review*, 37, 15–96.

Beadle, G. W., and Ephrussi, B. (1936). The differentiation of eye pigments in Drosophila as studied by transplantation. *Genetics*, 21(3), 225–247.

Beadle, G. W., and Tatum, E. L. (1941a). Experimental control of development and differentiation: Genetic control of developmental reactions. *The American Naturalist*, 75, 107–116.
  (1941b). Genetic control of biochemical reaction in Neurospora. *Proceedings of the National Academy of Sciences of the USA*, 27, 499–506.

Belling, J. (1928). A working hypothesis for segmental interchange between homologous chromosomes in flowering plants. *University of California Publications in Botany*, 14(8), 283–291.

Ben-Chetrit, E., and Levy, M. (1998). Familial Mediterranean fever. *Lancet*, 351, 659–664.

Bender, W., Akam, M., Karch, F., Beachy, P., Pfeifer, M., Spierer, P., et al. (1983). Molecular genetics of the bithorax complex in *Drosophila melanogaster*. *Science*, 221, 23–29.

Benzer, S. (1957). The elementary units of heredity. In B. Glass and W. D. McElroy (eds.), *The Chemical Basis of Heredity* (pp. 70–93). Baltimore, MD: Johns Hopkins Press.
  (1973). Genetic dissection of behavior. *Scientific American*, 229(6), 24–37.

Bibliography

Benzer, S., and Champe, S. P. (1961). Ambivalent rII mutants of phage T4. *Proceedings of the National Academy of Sciences of the USA*, 47, 1025–1038. (1962). A change from nonsense to sense in the genetic code. *Proceedings of the National Academy of Sciences of the USA* 48, 1114–1121.

Benzer, S., and Freese, E. (1958). Induction of specific mutations with 5-bromouracil. *Proceedings of the National Academy of Sciences of the USA*, 44(2), 112–119.

Berry, A. J., Ajioka, J. W., and Kreitman, M. (1991). Lack of polymorphism on the Drosophila fourth chromosome resulting from selection. *Genetics*, 129, 1111–1117.

Bishop, B. E. (1996). Mendel's opposition to evolution and to Darwin. *Journal of Heredity*, 87(3), 205–213.

Blakeslee, A. F. (1922). Variations in Datura due to changes in chromosome number. *The American Naturalist*, 56, 16–31.

Blixt, S. (1975). Why didn't Mendel find linkage? *Nature*, 256, 206.

Bolton, E. T., and McCarthy, B. J. (1962). A general method for the isolation of RNA complementary to DNA. *Proceedings of the National Academy of Sciences of the USA*, 48(8), 1390–1397.

Boveri, T. (1902). Über mehrpolige Mitosen als Mittel zur Analyse des Zellkerns. *Verhandlungen der physikalische-medizinische Gesellschaft, Würzburg, N.F.*, 35, 67–90.

Boyce, R. P., and Howard-Flanders, P. (1964). Release of ultraviolet light-induced thymine dimers from DNA in *E. coli* K12. *Proceedings of the National Academy of Sciences of the USA*, 51, 293–300.

Brannigan, A. (1979). The reification of Mendel. *Social Studies of Science*, 9, 423–454.

Brenner, S. (1957). On the impossibility of all overlapping triplet codes in information transfer from nucleic acid to proteins. *Proceedings of the National Academy of Sciences of the USA*, 43(8), 687–694. (1974). The genetics of *Caenorhabditis elegans*. *Genetics*, 77, 71–94. (2000). The end of the beginning. *Science*, 287(5461), 2173–2174.

Brenner, S., Dove, W., Herskowitz, I., and Thomas, R. (1990). Genes and development: Molecular and logical themes. *Genetics*, 126(3), 479–486.

Bridges, C. B. (1913). Non-disjunction of the sex chromosomes of Drosophila. *Journal of Experimental Zoology*, 15, 587–605. (1914). Direct proof through non-disjunction that the sex-linked genes of Drosophila are borne by the X-chromosome. *Science*, 40, 107–109. (1916). Non-disjunction as proof of the chromosome theory of heredity. *Genetics*, 1, 1–52 and 107–163. (1917). Deficiency. *Genetics*, 2, 445–465. (1925). Sex in relation to chromosomes and genes. *The American Naturalist*, 59, 127–137. (1935). Salivary chromosome maps. *The Journal of Heredity*, 26, 60–64. (1938). A revised map of the salivary gland X-chromosome. *The Journal of Heredity*, 29, 11–13.

Bridges, C. B., and Anderson, E. G. (1925). Crossing over in the X chromosome of triploid females of *Drosophila melanogaster*. *Genetics*, 10, 418–441.

Bibliography

Bridges, C. B., and Bridges, P. N. (1939). A new map of the second chromosome. *The Journal of Heredity*, 30, 475–477.

Britten, R. J., and Davidson, E. H. (1969). Gene regulation for higher cells: A theory. *Science*, 165, 349–357.

Britten, R. J., and Kohne, D. E. (1968). Repeated sequences in DNA. *Science*, 161(3841), 529–540.

Brown, S. W., and Zohary, D. (1955). The relationship of chiasmata and crossing over in *Lilium formosanum*. *Genetics*, 40, 850–873.

Bruce, A. B. (1910). The Mendelian theory of heredity and the augmentation of vigor. *Science*, 32, 627–628.

Bull, A. (1966). Bicaudal, a genetic factor which affects the polarity of the embryo in *Drosophila melanogaster*. *Journal of Experimental Zoology*, 161, 221–241.

Cairns, J. (1963). The bacterial chromosome and its manner of replication as seen in autoradiography. *Journal of Molecular Biology*, 6(3), 208–213.

Cairns, J., Overbaugh, J., and Miller, S. (1988). The origin of mutants. *Nature*, 335, 142–145.

Campbell, M. (1982). Mendel's theory: Its context and plausibility. *Centaurus*, 26, 38–69.

Carlson, E. A. (1959). Allelism, pseudoallelism, and complementation at the dumpy locus in *D. melanogaster*. *Genetics*, 44, 347–373.

(1966 [1989]). *The Gene: A Critical History*. Philadelphia: W. B. Saunders; Ames, IA: Iowa State University Press.

(2004). *Mendel's Legacy: The Origin of Classical Genetics*. Cold Spring Harbor, NY: Cold Spring Harbor Laboratory Press.

Carothers, E. E. (1913). The Mendelian ratio in relation to certain Orthopteran chromosomes. *Journal of Morphology*, 24, 487–511.

Casperson, T., and Schultz, J. (1938). Nucleic acid metabolism of the chromosomes in relation to gene reproduction. *Nature*, 142, 294–295.

Castle, W. E. (1906). Yellow mice and gametic purity. *Science N.S.*, 24, 275–281.

(1913a). *Heredity*. New York: D. Appleton and Comp.

(1913b). Simplification of Mendelian formulae. *The American Naturalist*, 47 (555), 170–182.

(1915). Mr. Muller on the constancy of Mendelian factors. *The American Naturalist*, 49, 37–42.

(1919a). Is the arrangement of the genes in the chromosome linear? *Proceedings of the National Academy of Sciences of the USA*, 5(2), 25–32.

(1919b). Piebald rats and selection. *The American Naturalist*, 53, 370–376.

(1919c). Piebald rats and the theory of genes. *Proceedings of the National Academy of Sciences of the USA*, 5, 126–130.

Chovnick, A. (1961). The garnet locus in *Drosophila melanogaster*. I. Pseudoallelism. *Genetics*, 46, 493–507.

(1989). Intragenic recombination in Drosophila: The rosy locus. *Genetics*, 123 (4), 621–624.

Chovnick, A., Schalet, A., Kernaghan, R. P., and Kraus, M. (1964). The rosy cistron in *Drosophila melanogaster* genetic fine structure analysis. *Genetics*, 50, 1254–1259.

Bibliography

Churchill, F. B. (1974). William Johannsen and the genotype concept. *Journal of the History of Biology*, 7, 5–30.

(1987). From heredity theory to *Vererbung*. The transmission problem, 1850–1915. *Isis*, 78(3), 337–364.

Cohen, J. (2007). DNA duplications and deletions help determine health. *Science*, 317(5843), 1315–1317.

Coleman, W. (1970). Bateson and chromosomes: Conservative thought in science. *Centaurus*, 15(3–4), 228–314.

Cooper, K. W. (1948). A new theory of secondary non-disjunction in female *Drosophila melanogaster*. *Proceedings of the National Academy of Sciences of the USA*, 34, 179–187.

Correns, C. (1902 [1924]). Scheinbare Ausnahmen von der Mendel'schen Spaltungregel für Bastarde. In *Carl Correns: Gesammelte Abhandlungen zur Vererbungswissenschaft aus periodischen Schrifren, 1899–1924* (pp. 287–299). Berlin: Julius Springer.

(1903 [1924a]). Über die dominierenden Merkmale der Bastarde. In *Carl Correns: Gesammelte Abhandlungen zur Vererbungswissenschaft aus periodischen Schrifren, 1899–1924* (pp. 329–343). Berlin: Julius Springer.

(1903 [1924b]). Weitere beiträge zur Kenntnis der dominierenden Merkmale und der Mosaikbildung der Bastarde. In *Carl Correns: Gesammelte Abhandlungen zur Vererbungswissenschaft aus periodischen Schrifren, 1899–1924* (pp. 342–349). Berlin: Julius Springer.

Creager, A. N. H. (2004). Mapping genes in microorganisms. In J.-P. Gaudillière and H.-J. Rheinberger (eds.), *From Molecular Genetics to Genomics: The Mapping Cultures of Twentieth-Century Genetics* (Vol. XXII, pp. 9–41). London: Routledge.

Creighton, H. B., and McClintock, B. (1931). A correlation of cytological and genetical crossing-over in *Zea mays*. *Proceedings of the National Academy of Sciences of the USA*, 17, 492–497.

Crick, F. H. C. (1958). On protein synthesis. In *Symposium of the Society for Experimental Biology. The Biological Replication of Macromolecules* (Vol. 12, pp. 138–163). Cambridge: Cambridge University Press.

(1966). *On Molecules and Men*. Seattle, WA: University of Washington Press.

(1970). Central dogma of molecular biology. *Nature*, 227(5258), 561–563.

Crick, F. H. C., Barnett, L., Brenner, S., and Watts-Tobin, R. J. (1961). General nature of the genetic code for proteins. *Nature*, 192(4809), 1227–1232.

Crick, F. H. C., and Lawrence, P. A. (1975). Compartments and polyclones in insect development. *Science*, 189, 340–347.

Crow, J. F., and Kimura, M. (1970). *An Introduction to Population Genetics Theory*. New York: Harper and Row.

Crow, J. F., and Temin, R. G. (1964). Evidence for the partial dominance of recessive lethal genes in natural populations of Drosophila. *The American Naturalist*, 98, 21–33.

Danchin, A., and He'enaut, A. (1997). The map of the cell is in the chromosome. *Current Opinions in Genetics and Development*, 7(6), 852–854.

Darbishire, A. D. (1902). Note on the results of crossing Japanese Waltzing mice with European Albino races. *Biometrika*, 2, 101–104.

Bibliography

Darden, L. (1991). *Theory Change in Science: Strategies from Mendelian Genetics*. New York: Oxford University Press.

Darlington, C. D. (1931). Meiosis (precocity theory). *Biological Review*, 6, 221–264.

—— (1937). *Recent Advances in Cytology* (2nd edn.). London: J and A Churchill.

Darnell, J. E., Jr. (1999). E.B. Wilson Lecture, 1998. Eukaryotic RNAs: Once more from the beginning. *Molecular Biology of the Cell*, 10(6), 1685–1692.

Davis, R. H., and Perkins, D. D. (2002). *Neurospora*: a model of model microbes. *Nature Reviews: Genetics*, 3(5), 397–403.

Davis, R. L. (2000). Neurofibromin progress on the fly. *Nature*, 403(6772), 846–847.

Darwin, C. (1868). *The Variation of Animals and Plants under Domestication* (Vol. I and II). London: Murray.

—— (1981 [1871]). *The Descent of Man and Selection in Relation to Sex*. New York: D. Appleton.

Dawkins, R. (1976). *The Selfish Gene*. New York: Oxford University Press.

de Vries, H. (1889). *Intracelluläre Pangenesis*. Jena: Gustav Fischer.

—— (1889 [1910]). *Intracellular Pangenesis*. Chicago, IL: Open Court.

—— (1902–3). *Die Mutationstheorie: Versuche und Beobachtungen Über die Entstehung von Arten im Pflanzenreich*. Leipzig: Von Veit.

—— (1912 [c. 1904]). *Species and Varieties: Their Origin by Mutation* (3rd, corrected and revised edn.). Chicago, IL: The Open Court Publishing Company.

—— (1950 [1900]). Concerning the law of segregation of hybrids. *Genetics*, 35 Supplement (*The birth of Genetics*) (5, part 2), 30–32.

Deichmann, U. (2004). Early responses to Avery et al.'s paper on DNA as hereditary material. *Historical Studies in the Physical and Biological Sciences*, 34(2), 207–232.

Demerec, M., and Hoover, M. E. (1936). Three related X-chromosome deficiencies in Drosophila. *The Journal of Heredity*, 27, 206–212.

Dennis, C. (2002a). Gene regulation: The brave new world of RNA. *Nature*, 418 (6894), 122–124.

—— (2002b). Small RNAs: The genome's guiding hand? *Nature*, 420(6917), 732.

Depew, D. J., and Weber, B. H. (1995). *Darwinism Evolving. Systems Dynamics and the Genealogy of Natural Selection*. Cambridge, MA: The MIT Press.

Di Trocchio, F. (1991). Mendel's experiments: a reinterpretation. *Journal of the History of Biology*, 24, 485–519.

Dobzhansky, T. (1929). Genetical and cytological proof of translocations involving the third and the fourth chromosomes of *Drosophila melanogaster*. *Biologisches Zentralblat*, 49, 408–419.

—— (1931). Translocations involving the second and the fourth chromosomes of *Drosophila melanogaster*. *Genetics*, 16, 629–658.

—— (1937). *Genetics and the Origin of Species* (1st edn.). New York: Columbia University Press.

—— (1970). *Genetics of the Evolutionary Process*. New York: Columbia University Press.

—— (1973). Nothing in biology makes sense except in the light of evolution. *The American Biology Teacher*, 35, 125–129.

Bibliography

Duhem, P. (1914). *The Aim and Structure of Physical Theory.* Princeton, NJ: Princeton University Press.

Dunn, L. C. (1965). *A Short History of Genetics.* New York: McGraw Hill.

East, E. M. (1910). A Mendelian interpretation of variation that is apparently continuous. *The American Naturalist,* 44(518), 65–82.

(1911). The genotype hypothesis and hybridization. *The American Naturalist,* 45, 160–174.

(1912). The Mendelian notation as a description of physiological facts. *The American Naturalist,* 46, 633–655.

(1923). Mendel and his contemporaries. *The Scientific Monthly,* 6, 225–237.

Eldredge, N., and Gould, S. J. (1972). Punctuated equilibrium: An alternative to phyletic gradualism. In T. J. M. Schopf (ed.), *Models of Paleobiology* (pp. 82–115). San Francisco, CA: Freeman, Cooper and Co.

Emerson, R. A. (1914). The inheritance of a recurring somatic variation in variegated ears of maize. *The American Naturalist,* 48, 87–115.

Ereshefsky, M. (2001). *The Poverty of the Linnaean Hierarchy: A Philosophical Study of Biological Taxonomy.* New York: Cambridge University Press.

Falk, R. (1955). Some preliminary observations on crossing-over and non-disjunction. *Hereditas,* 41, 376–383.

(1961). Are induced mutations in Drosophila overdominant? II. Experimental results. *Genetics,* 46(7), 737–757.

(1967). Fitness of heterozygotes for irradiated chromosomes in Drosophila. *Mutation Research,* 4, 805–819.

(1986). What is a gene? *Studies in the History and Philosophy of Science,* 17(2), 133–173.

(1988). Species as individuals. *Biology and Philosophy,* 3(4), 455–462.

(1991). The dominance of traits in genetic analysis. *Journal of the History of Biology,* 24(3), 457–484.

(1995). The struggle of genetics for independence. *Journal of the History of Biology,* 28(2), 219–246.

(2000a). Can the norm of reaction save the gene concept? In R. S. Singh, C. B. Krimbas, D. B. Paul and J. Beatty (eds.), *Thinking About Evolution: Historical, Philosophical and Political Perspectives* (Vol. II, pp. 119–140). Cambridge: Cambridge University Press.

(2000b). The gene – a concept in tension. In P. J. Beurton, R. Falk and H.-J. Rheinberger (eds.), *The Concept of the Gene in Development and Evolution: Historical and Epistemological Perspectives* (pp. 317–348). Cambridge: Cambridge University Press.

(2001a). Mendel's hypothesis. In G. E. Allen and R. M. MacLeod (eds.), *Science, History and Social Activism: A Tribute to Everett Mendelsohn* (pp. 77–86). Dordrecht: Kluwer.

(2001b). The rise and fall of dominance. *Biology and Philosophy,* 16(3), 285–323.

(2003). Linkage: From particulate to interactive genetics. *Journal of the History of Biology,* 36(1), 87–117.

(2004). Long Live the Genome! So should the gene. *History and Philosophy of the Life Sciences,* 26, 105–121.

Bibliography

(2006). Mendel's impact. *Science in Context*, 19(2), 215–236.

(2007). Genetic analysis. In M. Matthen and C. Stephens (eds.), *Philosophy of Biology* (Vol. III, pp. 249–308). Dordrecht: Elsevier.

(2008). Wilhelm Johannsen: A rebel or a diehard? In O. Harman and M. R. Dietrich (eds.), *Rebels, Mavericks, and Heretics in Biology* (pp. 66–83). New Haven, CT: Yale University Press.

(in press). Genetic regulation before the 1960s. In M. Laubichler, H.-J. Rheinberger and P. Hammerstein (eds.), *Regulation: Historical and Current Themes in Theoretical Biology*.

Falk, R., and Sarkar, S. (1991). The real objective of Mendel's paper. *Biology and Philosophy*, 6, 447–451.

(2006). Genetics. In S. Sarkar and J. Pfeifer (eds.), *The Philosophy of Science: An Encyclopedia* (Vol. I, pp. 330–339). New York: Routledge.

Falk, R., and Schwartz, S. (1993). Morgan's hypothesis of the genetic control of development. *Genetics*, 134(3), 671–674.

Farber, P. L. (1976). The type-concept in zoology during the first half of the nineteenth century. *Journal of the History of Biology*, 9(1), 93–119.

Feldman, M. (2001). The origin of cultivated wheat. In A. Bonjean and W. Angus (eds.), *The World of Wheat Book* (pp. 3–56). Paris: Lavoisier Tech. and Doc.

Fincham, J. R. S. (1998). Fungal genetics – past and present. *Journal of Genetics*, 77(2 and 3), 55–63.

Fincham, J. R. S., Day, P. R., and Radford, A. (1979). *Fungal Genetics* (4th edn.). London: Blackwell.

Fisher, R. A. (1918). The correlation between relatives on the supposition of Mendelian inheritance. *Transactions of the Royal Society, Edinburgh*, 52, 399–433.

(1922). On the dominance ratio. *Proceedings of the Royal Society of Edinburgh, series B*, 42, 321–341.

(1928). The possible modification of the response of the wild type to recurrent mutation. *The American Naturalist*, 62, 115–126.

(1930). *The Genetical Theory of Natural Selection*. Oxford: Clarendon.

(1936). Has Mendel's work been rediscovered? *Annals of Science*, 1, 115–137.

Ford, E. B. (1964). *Ecological Genetics*. London/New York: Methuen/John Wiley.

Franklin, I., and Lewontin, R. C. (1970). Is the gene the unit of selection? *Genetics*, 65(4), 707–734.

Freese, E. (1957). The correlation effect for a histidine locus of *Neurospora crassa*. *Genetics*, 42, 671–684.

(1959). The specific mutagenic effect of base analogues. *Journal of Molecular Biology*, 1, 87–105.

Fuerst, J. A. (1982). The role of reductionism in the development of molecular biology: Peripheral or central? *Social Studies of Science*, 12, 241–278.

Fuller, S. (2000). *Thomas Kuhn: A philosophical history for our times*. Chicago, IL: University of Chicago Press.

Galton, F. (1875). A theory of heredity (Revised version). *Journal of the Anthropological Institute*, 5, 329–348.

300

Bibliography

(1908). *Memories of My Life* (2nd edn.). London: Methuen.

Gamow, G. (1954). Possible relation between deoxyribonucleic acid and protein structure. *Nature*, 173, 318.

Garcia-Bellido, A. (1998). The engrailed story. *Genetics*, 148(2), 539–544.

Garcia-Bellido, A., Lawrence, P. A., and Morata, G. (1979). Compartments in animal development. *Scientific American*, 241(1), 102–110.

Garcia-Bellido, A., and Merriam, J. R. (1969). Cell lineage of the imaginal disc in Drosophila gynandromorphs. *Journal of Experimental Zoology*, 170, 61–76.

(1971). Parameters of the wing imaginal disc development of *Drosophila melanogaster*. *Developmental Biology*, 24, 61–87.

Garcia-Bellido, A., Rippoll, P., and Morata, G. (1973). Developmental compartmentalisation of the wing disc of Drosophila. *Nature: New Biology*, 245, 251–253.

Garcia-Bellido, A., and Santamaria, P. (1972). Developmental analysis of the wing disc in the mutant engrailed of *Drosophila melanogaster*. *Genetics*, 72, 87–104.

Garen, A. (1992). Looking for the homunculus in Drosophila. *Genetics*, 131(1), 5–7.

Garrod, A. E. (1908). *Inborn Errors of Metabolism* (2nd edn. 1923). Oxford: Oxford University Press.

Gasper, P. (1992). Reduction and instrumentalism in genetics. *Philosophy of Science*, 59, 655–670.

Gayon, J. (2000). From measurement to organization: A philosophical scheme for the history of the concept of heredity. In P. J. Beurton, R. Falk and H.-J. Rheinberger (eds.), *The Concept of the Gene in Development and Evolution: Historical and Epistemological Perspectives* (pp. 69–90). Cambridge: Cambridge University Press.

Gehring, W. J. (1967). Clonal analysis of determination dynamics in cultures of imaginal discs of *Drosophila melanogaster*. *Developmental Biology*, 16, 438–456.

(1998). *Master Control Genes in Development and Evolution: The Homeobox Story*. New Haven, CT: Yale University Press.

Gehring, W. J., and Nöthiger, R. (1973). The imaginal discs of *Drosophila*. In S. J. Counce and C. H. Waddington (eds.), *Developmental Systems* (Vol. II, pp. 211–290). London: Academic Press.

Gerstein, M. B., Lan, N., and Jansen, R. (2002). Integrating interactomes. *Science*, 295(5553), 284–287.

Gerstein, M. B., Bruce, C., Rozowsky, J. S., Zheng, D., Du, J., Korbel, J. O., *et al.* (2007). What is a gene, post-ENCODE? History and updated definition. *Genome Research*, 17(6), 669–681.

Ghiselin, M. T. (1971). The individual in the Darwinian revolution. *New Literary History*, 3, 113–134.

(1987). Species concepts, individuality, and objectivity. *Biology and Philosophy*, 2(2), 127–143.

Gibbs, W. W. (2003). The unseen genome: Gems among the junk. *Scientific American*, 289(5), 46–53.

Bibliography

Gilbert, S. F. (1978). The embryological origins of the gene theory. *Journal of the History of Biology*, 11(2), 307–351.

(1991a). *Developmental Biology* (3rd edn.). Sunderland, MA: Sinauer.

(1991b). Induction and the origins of developmental genetics. In S. Gilbert (ed.), *Developmental Biology. A Comprehensive Synthesis. A Conceptual History of Modern Embryology* (pp. 181–206). Baltimore, MD: Johns Hopkins University Press.

(1998). Bearing crosses: A historiography of genetics and embryology. *American Journal of Medical Genetics*, 76(2), 168–182.

Gilbert, S. F., Opitz, J. M., and Raff, R. A. (1996). Resynthesizing evolutionary and developmental biology. *Developmental Biology*, 173, 357–372.

Gingeras, T. R. (2007). Origin of phenotypes: Genes and transcripts. *Genome Research*, 17(6), 682–690.

Glass, B. (1963). The establishment of modern genetical theory as an example of the interaction of different models, techniques, and inferences. In A. C. Crombie (ed.), *Scientific Change: Historical studies in the intellectual, social and technical conditions for scientific discovery and technical invention, from antiquity to the present* (pp. 521–541). London: Heinemann.

Gliboff, S. (1999). Gregor Mendel and the law of evolution. *History of Science*, 37, 217–235.

Golding, B. (1994). *Non-Neutral Evolution: Theories and Molecular Data*. New York: Chapman and Hall.

Goldschmidt, R. (1917). Crossing over Ohne Chiasmatypie? *Genetics*, 2(1), 82–95.

(1938a). The theory of the gene. *Scientific Monthly*, 45, 268–273.

(1938b). *Physiological Genetics*. New York: McGraw-Hill.

(1950). Fifty years of Genetics. *The American Naturalist*, 84, 313–340.

(1954). Different philosophies of genetics. *Science*, 119, 703–710.

(1955). *Theoretical Genetics*. Berkeley, CA: University of California Press.

Gould, S. J. (1977). *Ontogeny and Phylogeny*. Cambridge, MA: Harvard University Press.

Gould, S. J., and Lewontin, R. C. (1979). The spandrels of San Marco and the Panglossian paradigm: A critique of the adaptationist programme. *Proceedings of the Royal Society of London, series B*, 205, 581–598.

Gowen, J. W. (1952). *Heterosis*. Ames, IA: Iowa State College Press.

Gray, J. (1969). *Near Eastern Mythology, Mesopotamia, Syria, Palestine*. London: The Hamlyn Publishing Group.

Green, M. M. (1963). Interallelic complementation and recombination at the rudimentary wing locus in *Drosophila melanogaster*. *Genetica*, 34, 242–253.

Greider, C. W., and Blackburn, E. H. (1985). Identification of a specific telomere terminal transferase activity in Tetrahymena extracts. *Cell*, 43, 405–413.

Grell, R. F. (1976). Distributive pairing. In M. Ashburner and E. Novitski (eds.), *The Genetics and Biology of Drosophila* (Vol. Ia, pp. 435–486). London: Academic Press.

Griesemer, J. R. (2000). Reproduction and reduction of genetics. In P. J. Beurton, R. Falk and H.-J. Rheinberger (eds.), *The Concept of the Gene in Development and Evolution: Historical and Epistemological Perspectives* (pp. 240–285). Cambridge: Cambridge University Press.

Griffiths, A. J. F., Gelbart, W. M., Miller, J. H., and Lewontin, R. C. (1999). *Modern Genetic Analysis*. New York: Freeman.

Griffiths, P. E. (2006). The fearless vampire conservator: Philip Kitcher, genetic determinism, and the informational gene. In E. M. Neumann-Held and C. Rehmann-Sutter (eds.), *Genes in Development: Re-reading the Molecular Paradigm* (pp. 175–198). Durham: Duke University Press.

Griffiths, P. E., and Stotz, K. (2006). Genes in the postgenomic era. *Theoretical Medicine and Bioethics*, 27(6), 499–521.

Gruneberg, H. (1938). An analysis of the "pleiotropic" effects of a new lethal mutation in the rat (*Mus norvegicus*). *Proceedings of the Royal Society of London, series B*, 125, 123–144.

Hadorn, E. (1961). *Developmental Genetics and Lethal Factors*. New York: John Wiley.

(1968). Transdetermination in cells. *Scientific American*, 219 (5), 110–118.

Haldane, J. B. S. (1930). A note on Fisher's theory of the origin of dominance, and a correlation between dominance and linkage. *The American Naturalist*, 64, 87–90.

(1954). *The Biochemistry of Genetics*. London: George Allen and Unwin.

(1957). The cost of natural selection. *Journal of Genetics*, 55, 511–524.

Hamilton, W. D. (1963). The evolution of altruistic behavior. *The American Naturalist*, 97, 354–356.

(1964a). The genetical evolution of social behavior. I. *Journal of Theoretical Biology*, 7, 1–16.

(1964b). The genetical evolution of social behavior. II. *Journal of Theoretical Biology*, 7, 17–52.

Hardy, G. H. (1908). Mendelian proportions in a mixed population. *Science*, 28, 49–50.

(1940). *A Mathematician's Apology*. Cambridge: Cambridge University Press.

Harman, O. S. (2004). *The Man Who Invented the Chromosome: A Life of Cyril Darlington*. Cambridge, MA: Harvard University Press.

Harper, L., Golubovskaya, I., and Cande, W. Z. (2004). A bouquet of chromosomes. *Journal of Cell Science*, 117(18), 4025–4032.

Harris, H. (1966). Enzyme polymorphism in man. *Proceedings of the Royal Society of London, series B*, 164, 298–310.

Hartman, P. E., Loper, J. C., and Serman, D. (1960). Fine structure mapping by complete transduction between histidine-requiring Salmonella mutants. *Journal of General Microbiology*, 22, 323–353.

Harwood, J. (1987). National style in science: Genetics in Germany and the United Stated between the world wars. *Isis*, 78, 390–414.

(1993). *Styles of Scientific Thought. The German Genetics Community 1900–1933*. Chicago, IL: The University of Chicago Press.

(1996). Weimar culture and biological theory: A study of Richard Woltereck (1877–1944). *History of Science*, 34, 347–377.

Hastings, P. J., and Whitehouse, H. L. K. (1964). A polaron model of genetic recombination by the formation of hybrid deoxyribonucleic acid. *Nature*, 201, 1052–1054.

Bibliography

Hayes, W. (1953). The mechanism of genetic recombination in *E. coli. Cold Spring Harbor Symposia on Quantitative Biology*, 18, 75–93.

(1964). *The Genetics of Bacteria and Their Viruses: Studies in Basic Genetics and Molecular Biology*. New York: John Wiley.

Heitz, E. (1929). Heterochromatin, Chromozentern, Chromomeren. *Berichte der deutschen botanischen Gesellschaft*, 47, 274–284.

Heitz, E., and Bauer, H. (1933). Beweise für die Chromosomennatur der Kernschleifen in den Knäuelkernen von *Bibio hortulanus. Zeitschrift für Zellforschung und mikroskopische Anatomie*, 17, 76–82.

Helinski, D. R., and Yanofsky, C. (1962). Correspondence between genetic data and the position of amino acid alteration in a protein. *Proceedings of the National Academy of Sciences of the USA*, 48, 173–183.

Hershey, A. D., and Chase, M. (1952). Independent functions of viral protein and nucleic acid in growth of bacteriophage. *Journal of General Physiology*, 36 (1), 39–56.

Holliday, R. (1964). A mechanism for gene conversion in fungi. *Genetical Research*, 5, 282–304.

Holmes, F. L. (2000). Seymour Benzer and the definition of the gene. In P. J. Beurton, R. Falk and H.-J. Rheinberger (eds.), *The Concept of the Gene in Development and Evolution* (pp. 115–155). Cambridge: Cambridge University Press.

(2001). *Meselson, Stahl, and the Replication of DNA: A History of "The Most Beautiful Experiment in Biology"*. New Haven, CT: Yale University Press.

Hornstein, E., and Shomron, N. (2006). Canalization of development by micro-RNAs. *Nature Genetics*, 38 (Supplement), S20–S24.

Horowitz, N. H. (1991). Fifty years ago: The Neurospora revolution. *Genetics*, 127(4), 631–635.

Hotta, Y., and Benzer, S. (1972). Mapping of behavior in *Drosophila melanogaster. Nature*, 240, 527–535.

Hotta, Y., Ito, M., and Stern, H. (1966). Synthesis of DNA during meiosis. *Proceedings of the National Academy of Sciences of the USA*, 56, 1184–1191.

Howard, A., and Pelc, S. R. (1951). Nuclear incorporation of P32 as demonstrated by autoradiographs. *Experimental Cell Research*, 2, 178–197.

Hoyer, B. H., McCarthy, B. J., and Bolton, E. T. (1964). A molecular approach in the systematics of higher organisms. *Science*, 144, 959–967.

Hubby, J. L., and Lewontin, R. C. (1966). A molecular approach to the study of genic heterozygosity in natural populations. I. The number of alleles at different loci in *Drosophila pseudoobscura. Genetics*, 54, 577–594.

Huberman, J. A., and Riggs, D. A. (1968). On the mechanism of DNA replication in mammalian chromosomes. *Journal of Molecular Biology*, 32, 327–341.

Hull, D. L. (1973). *Darwin and His Critics*. Chicago, IL: University of Chicago Press.

(1974). *Philosophy of Biological Science*. Englewood Cliffs, NJ: Prentice-Hall.

(1976). Are species really individuals? *Systematic Zoology*, 25, 174–191.

Bibliography

Hurst, C. C. (1906). Mendelian characters in plants and animals. In *Report of Conference on Genetics* (pp. 114–129). London: Royal Horticultural Society.

Hutchinson, G. E., and Rachootin, S. (1979). Historical introduction. In *William Bateson: Problems of Genetics* (pp. vii–xx). New Haven, CT: Yale University Press.

Huxley, J. (1943). *Evolution: The Modern Synthesis*. New York: Harper and Row.

Ingram, V. M. (1957). Gene mutations in human haemoglobin: the chemical difference between normal and sickle cell haemoglobin. *Nature*, 180, 326–328.

—— (1963). *The Hemoglobin in Genetics and Evolution*. New York: Columbia University Press.

Jacob, F., and Monod, J. (1961). Genetic regulatory mechanisms in the synthesis of proteins. *Journal of Molecular Biology*, 3, 318–356.

Jacob, F., and Wollman, E. L. (1961). *Sexuality and the Genetics of Bacteria*. New York: Academic Press.

Jahn, I. (1957/58). Zur Geschichte der Wiederentdeckung der Mendelschen Gesetze. *Wissenschaftliche Zeitschrift der Friedrich-Schiller Universität Jena, Mathematisch-naturwissenschaftliche Reihe*, 7(2/3), 215–227.

Janssens, F. A. (1909). La Théorie de la chiasmatypie. Nouvelle interprétation des cinèses de maturation. *La Cellule*, 25, 389–411.

Johannsen, W. (1903). *Über Erblichkeit in Populationen und reinen Linien*. Jena: Gustav Fischer.

—— (1903 [1955]). Concerning heredity in populations (Über Erblichkeit in Populationen und in reinen Linien) (H. Gall and E. Putschar, Trans.). By The Staff of Natural Science 3 (ed.), *Selected Readings in Biology for Natural Sciences* (Vol. III, pp. 172–215). Chicago, IL: University of Chicago Press.

—— (1909). *Elemente der exakten Erblichkeitslehre*. Jena: Gustav Fischer.

—— (1911). The genotype conception of heredity. *The American Naturalist*, 45 (531), 129–159.

—— (1923). Some remarks about units in heredity. *Hereditas*, 4, 133–141.

—— (1926). *Elemente der exakten Erblichkeitslehre* (3rd edn.). Jena: Gustav Fischer.

Judson, H. F. (1979). *The Eighth Day of Creation: Makers of Revolution in Biology*. New York: Simon and Schuster.

Kacser, H. (1987). Dominance not inevitable but very likely. *Journal of theoretical Biology*, 126, 505–506.

Kacser, H., and Burns, J. A. (1981). The molecular basis of dominance. *Genetics*, 97, 639–666.

Kauffman, S. A. (1973). Control circuits for determination and transdetermination. *Science*, 181, 310–318.

Kavenoff, R., and Zimm, B. H. (1973). Chromosome-sized DNA molecules from Drosophila. *Chromosoma*, 41, 1–27.

Kay, L. E. (2000). *Who Wrote the Book of Life? A History of the Genetic Code*. Stanford, CA: Stanford University Press.

Bibliography

Keeble, F., and Pellew, C. (1910). The mode of inheritance of stature and of time of flowering in peas (*Pisum sativum*). *Journal of Genetics*, 1, 47–56.

Keightley, P. D. (1996). A metabolic basis for dominance and recessivity. *Genetics*, 143(2), 621–625.

Keller, E. F. (1995). *Refiguring Life: Metaphors of Twentieth-Century Biology*. New York: Columbia University Press.

(2000). *The Century of the Gene*. Cambridge, MA: Harvard University Press.

(2002). *Making Sense of Life: Explaining Biological Development with Models, Metaphors, and Machines*. Cambridge, MA: Harvard University Press.

Kelner, A. (1949). Effect of visible light on the recovery of *Streptomyces griseus* conidia from ultraviolet irradiation injury. *Proceedings of the National Academy of Sciences of the USA*, 35, 73–79.

Kettlewell, H. B. D. (1955). Selection experiments on industrial melanism in the Lepidoptera. *Heredity*, 9, 323–342.

(1956). Further selection experiments on industrial melanism in the Lepidoptera. *Heredity*, 10, 287–301.

Kimura, M. (1968). Evolutionary rate at the molecular level. *Nature*, 217, 624–626.

King, J. L., and Jukes, T. H. (1969). Non-Darwinian evolution. *Science*, 164, 788–798.

Kitcher, P. (1984). 1953 and all that. A tale of two sciences. *The Philosophical Review*, 93(3), 335–373.

Kohler, R. E. (1994). *Lords of the Fly. Drosophila Genetics and Experimental Practice*. Chicago, IL: University of Chicago Press.

Kostoff, D. (1930). Discoid structure of the spireme. *The Journal of Heredity*, 21 (7), 323–324.

Kottler, M. J. (1979). Hugo de Vries and the rediscovery of Mendel's laws. *Annals of Science*, 36, 517–538.

Kreitman, M. (1983). Nucleotide polymorphism at the alcohol dehydrogenase locus of *Drosophila melanogaster*. *Nature*, 304, 412–417.

Kreitman, M., and Antezana, M. (2000). The population and evolutionary genetics of codon bias. In R. S. Singh, R. C. Lewontin, and C. B. Krimbas (eds.), *Evolutionary Genetics: From Molecules to Morphology* (Vol. I, pp. 82–101). Cambridge: Cambridge University Press.

Kühn, A. (1941). Über eine Gen-Wirkkette der Pigmentbildung bei Insekten. *Nachrichten der Akademie der Wissenschaften in Göttingen, Mathematisch-Physkalische Klasse*, 1941, 231–261.

Laubichler, M., and Rheinberger, H.-J. (2004). Alfred Kühn (1885–1968) and developmental evolution. *Journal of Experimental Zoology*, 302B, 103–110.

Lea, D. E. (1962). *Actions of Radiations on Living Cells*. Cambridge: Cambridge University Press.

Lederberg, J. (1947). Gene recombination and linked segregation in *Escherichia coli*. *Genetics*, 32, 505–525.

(ed.). (1990). *The Excitement and Fascination of Science: Reflections by Eminent Scientists* (Vol. III, part 1). Palo Alto, CA: Annual Reviews.

(1994). The transformation of genetics by DNA: An anniversary celebration of Avery, MacLeod, and McCarty (1944). *Genetics*, 136(2), 423–426.

Bibliography

(1996). Genetic recombination in *Escherichia coli*: Disputation at Cold Spring Harbor, 1946–1996. *Genetics*, 144(2), 439–443.

Lederberg, J., and Lederberg, E. M. (1952). Replica plating and indirect selection of bacterial mutants. *Journal of Bacteriology*, 63, 399–406.

Lederberg, J., Lederberg, E. M., Zinder, N. D., and Lively, E. R. (1951). Recombination analysis of bacterial heredity. *Cold Spring Harbor Symposia on Quantitative Biology*, 16, 413–443.

Lederberg, J., and Tatum, E. L. (1946). Gene recombination in *Escherichia coli*. *Nature*, 158, 558.

Lenoir, T. (1982). *The Strategy of Life*. Chicago, IL: University of Chicago Press.

Lerner, I. M. (1954). *Genetic Homeostasis*. New York: John Wiley.

Levine, M. (1988). Molecular analysis of dorsal-ventral polarity in Drosophila. *Cell*, 52, 785–786.

Lewis, E. B. (1951). Pseudoallelism and gene evolution. *Cold Spring Harbor Symposia on Quantitative Biology*, 16, 159–172.

(1963). Genes and developmental pathways. *American Zoologist*, 3, 33–56.

(1967). Genes and gene complexes. In R. A. Brink (ed.), *Heritage from Mendel* (pp. 17–47). Madison, WI: University of Wisconsin Press.

(1978). A gene complex controlling segmentation in Drosophila. *Nature*, 276, 565–570.

(1998). Antonio Garcia-Bellido in Caltech. *International Journal of Developmental Biology*, 42, 523–524.

Lewontin, R. C. (1974). *The Genetic Basis of Evolutionary Change*. New York: Columbia University Press.

(1991). Twenty-five years ago in Genetics: Electrophoresis in the development of evolutionary genetics: Milestone or millstone? *Genetics*, 128(4), 657–662.

(1992). Genotype and phenotype. In E. F. Keller and E. A. Lloyd (eds.), *Keywords in Evolutionary Biology* (pp. 137–144). Cambridge, MA: Harvard University Press.

Lewontin, R. C., and Berlan, J.-P. (1986). Technology, research, and the penetration of capital: The case of U.S. agriculture. *Monthly Review*, 38, 21–34.

(1990). The political economy of agricultural research: The case of hybrid corn. In C. R. Carroll, J. H. Vandermer and P. Rosset (eds.), *Agroecology* (pp. 613–628). New York: McGraw-Hill.

Lewontin, R. C., and Hubby, J. L. (1966). A molecular approach to the study of genic heterozygosity in natural populations. II. Amount of variation and degree of heterozygosity in natural populations of *Drosophila pseudoobscura*. *Genetics*, 54, 595–609.

Lewontin, R. C., and Kojima, K.-I. (1960). The evolutionary dynamics of complex polymorphism. *Evolution*, 14(4), 458–472.

Lewontin, R. C., Moore, J. A., Provine, W. B., and Wallace, B. (1981). *Dobzhansky's Genetics of Natural Populations I-XLIII*. New York: Columbia University Press.

Lewontin, R. C., Paul, D., Beatty, J., and Krimbas, C. B. (2000). Interview of R. C. Lewontin. In R. S. Singh, C. B. Krimbas, D. B. Paul, and J. Beatty (eds.), *Thinking About Evolution: Historical, Philosophical and Political Perspectives* (Vol. II, pp. 22–61). Cambridge: Cambridge University Press.

Bibliography

Liebe, B., Alsheimer, M., Höög, C., Benavente, R., and Scherthan, H. (2004). Telomere attachment, meiotic chromosome condensation, pairing, and bouquet stage duration are modified in spermatocytes lacking axial elements. *Molecular Biology of the Cell*, 15, 827–837.

Lifschytz, E., and Falk, R. (1969a). Fine structure analysis of a chromosome segment in *Drosophila melanogaster*. Analysis of ethyl methanesulphonate-induced lethals. *Mutation Research*, 8, 147–155.

(1969b). A genetic analysis of the Killer-prune (*K-pn*) locus of *Drosophila melanogaster*. *Genetics*, 62, 353–358.

Lindegren, C. C. (1953). Gene conversion in *Saccharomyces*. *Journal of Genetics*, 51, 625–637.

Lindsley, D. L., and Grell, E. H. (1968). *Genetic Variations of* Drosophila melanogaster. Washington, DC: Carnegie Institute of Washington Publication No. 627.

Loeb, J. (1912). *The Mechanistic Conception of Life: Biological Essays*. Chicago, IL: The University of Chicago Press.

Lovejoy, A. O. (1936 [1950]). *The Great Chain of Being*. Cambridge, MA: Harvard University Press.

Lüning, K. G. (1952a). Studies on the origin of apparent gene mutations in *Drosophila melanogaster*. *Acta Zoologica*, 33, 193–207.

(1952b). X-ray induced chromosome breaks in *Drosophila melanogaster*. *Hereditas*, 38, 321–338.

Luria, S. E., and Delbrück, M. (1943). Mutations of bacteria from virus sensitivity to virus resistance. *Genetics*, 28(6), 491–511.

MacCorquodale, K., and Meehl, P. E. (1948). On distinction between hypothetical constructs and intervening variables. *Psychological Review*, 55, 95–107.

Malling, H. V., and de Serres, F. J. (1968). Identification of genetic alterations induced by ethyl methanesulfonate in *Neurospora crassa*. *Mutation Research*, 6, 181–193.

Marmur, J., and Lane, D. (1960). Strand separation and specific recombination in deoxyribonucleic acids: biological studies. *Proceedings of the National Academy of Sciences of the USA*, 46(4), 453–461.

Mattick, J. S. (2003). Challenging the dogma: the hidden layer of non-protein-coding RNAs in complex organisms. *BioEssays*, 25(10), 930–939.

(2005). The functional genomics of noncoding RNA. *Science*, 309, 1527–1528.

Mayr, E. (1987). The ontological status of species: scientific progress and philosophical terminology. *Biology and Philosophy*, 2(2), 145–166.

McCarthy, B. J., and Bolton, E. T. (1963). An approach to the measurement of genetic relatedness among organisms. *Proceedings of the National Academy of Sciences of the USA*, 50(1), 156–164.

McCarthy, B. J., and Hoyer, B. H. (1964). Identity of DNA and diversity of messenger RNA molecules in normal mouse tissues. *Proceedings of the National Academy of Sciences of the USA*, 52(4), 915–922.

McClintock, B. (1938). The production of homozygous deficient tissues with mutant characteristics by means of the aberrant mitotic behavior of ring-shaped chromosomes. *Genetics*, 23, 315–376.

Bibliography

(1951). Chromosome organization and genic expression. *Cold Spring Harbor Symposia on Quantitative Biology*, 16, 13–47.

McGinnis, W. (1994). A century of homeosis; A decade of homeoboxes. *Genetics*, 137(3), 607–611.

McGinnis, W., Garber, R., Wirz, J., Kuroiwa, A., and Gehring, W. (1984). A homologous protein-coding sequence in Drosophila homeotic genes and its conservation in other metazoans. *Cell*, 37, 403–408.

McLaughlin, P. (2002). Naming Biology. *Journal of the History of Biology*, 35(1), 1–4.

Meijer, O.G. (1985). Hugo de Vries no Mendelian? *Annals of Science*, 42, 189–232.

Mendel, G. (1866). Versuche über Pflanzenhybriden. *Verhandlungen Naturforscher Verein, Brunn*, 4, 3–47.

Meselson, M., and Stahl, F.W. (1958a). The replication of DNA. *Cold Spring Harbor Symposia on Quantitative Biology*, 23, 9–12.

(1958b). The replication of DNA in *Escherichia coli*. *Proceedings of the National Academy of Sciences of the USA*, 44, 671–682.

Meselson, M.S., and Radding, C.M. (1975). A general model for genetic recombination. *Proceedings of the National Academy of Sciences of the USA*, 72, 358–361.

Metz, C.W. (1935). Structure of the salivary gland chromosomes in Sciara. *The Journal of Heredity*, 26, 177–188.

Mitchell, M.B. (1955). Aberrant recombination of pyridoxine mutants of Neurospora. *Proceedings of the National Academy of Sciencess of the USA*, 41, 215–220.

Moberg, C.L. (2005). *René Dubos: Friend of the Good Earth. Microbiologist, Medical Scientist, Environmentalist*. Washington, DC: ASM Press.

Mohr, O.L. (1923). Das Defizienz-Phänomen bei *Drosophila melanogaster*. *Zeitschrift für induktive Abstammungs- und Vererbungslehre*, 30, 279–283.

(1924). A genetic and cytological analysis of a section deficiency involving four units of the X-chromosome in *Drosophila melanogaster*. *Zeitschrift für induktive Abstammungs- und Vererbungslehre*, 32, 108–232.

Monaghan, F.V., and Corcos, A.F. (1990). The real objective of Mendel's paper. *Biology and Philosophy*, 5, 267–292.

Monod, J. (1972). *Chance and Necessity*. New York: Vintage Books.

Morange, M. (1994). *Histoire de la biologie moléculaire*. Paris: La Découverte.

(1998). *A History of Molecular Biology* (M. Cobb, Trans.). Cambridge, MA: Harvard University Press.

Morgan, L.V. (1922). Non-criss-cross inheritance in *Drosophila melanogaster*. *Biological Bulletin, Woods Hole*, 42, 267–274.

Morgan, T.H. (1910a). Chromosomes and heredity. *The American Naturalist*, 44, 449–498.

(1910b). Sex limited inheritance in Drosophila. *Science*, 32, 120–122.

(1911). Random segregation versus coupling in Mendelian inheritance. *Science*, 34(873), 384.

(1912). A modification of the sex ratio, and of other ratios, in Drosophila through linkage. *Zeitschrift für induktive Abstammungs- und Vererbungslehre*, 7, 323–345.

(1913a). Factors and unit characters in Mendelian heredity. *The American Naturalist*, 47(553), 5–16.

(1913b). *Heredity and Sex*. New York: Columbia University Press.

(1913c). Simplicity versus adequacy in Mendelian formulae. *The American Naturalist*, 47(558), 372–374.

(1919). *The Physical Basis of Heredity*. Philadelphia and London: Lippincott.

(1934a). *Embryology and Genetics*. New York: Columbia University Press.

(1934b). The relation of genetics to physiology and medicine. In *Nobel Lectures, Physiology or Medicine 1922–1941*. Amsterdam: Elsevier, 1965.

Morgan, T. H., Sturtevant, A. H., Muller, H., and Bridges, C. B. (1915). *The Mechanism of Mendelian Heredity*. New York: Henry Holt and Company.

Moss, L. (2003). *What Genes Can't Do*. Cambridge, MA: The MIT Press.

Muller, H. J. (1914a). The bearing of the selection experiments of Castle and Philips on the variability of Genes. *The American Naturalist*, 48, 567–576.

(1914b). A gene for the fourth chromosome of Drosophila. *The Journal of Experimental Zoology*, 17, 325–336.

(1916). The mechanism of crossing over. *The American Naturalist*, 50(592–595), 193–221, 284–305, 350–366, 421–434.

(1918). Genetic variability, twin hybrids and constant hybrids, in a case of balanced lethal factors. *Genetics*, 3, 422–499.

(1920). Are the factors of heredity arranged in a line? *The American Naturalist*, 54(631), 97–121.

(1922). Variation due to change in the individual gene. *The American Naturalist*, 56, 32–50.

(1927a). Artificial transmutation of the gene. *Science*, 66, 84–87.

(1927b). Quantitative methods in gene research. *The American Naturalist*, 61, 407–419.

(1928). The measurement of gene mutation rate in Drosophila, its high variability, and its dependence upon temperature. *Genetics*, 13, 279–357.

(1929). The gene as the basis of life. *Proceedings of the International Congress of Plant Sciences, Ithaca 1926*, 1, 897–921.

(1932). Further studies on the nature and causes of gene mutations. *Proceedings of the 6th International Congress of Genetics, Ithaca*, 1, 213–255.

(1940). An analysis of the process of structural change in chromosomes of Drosophila. *Journal of Genetics*, 40, 1–66.

(1947). The gene. *Proceedings of the Royal Society of London, series B*, 134, 1–37.

(1950a). Evidence of the precision of genetic adaptation. *The Harvey Lectures*, 43, 165–229.

(1950b). Our load of mutations. *American Journal of Human Genetics*, 2(2), 111–176.

(1954). The manner of production of mutations by radiation. In A. Hollaender (ed.), *Radiation Biology* (Vol. I, pp. 475–626). New York: McGraw Hill.

Bibliography

(1956). On the relation between chromosome changes and gene mutations. *Brookhaven Symposia in Biology*, 8 (591), 126–147.

Muller, H. J., and Falk, R. (1961). Are induced mutations in Drosophila over-dominant? I. Experimental design. *Genetics*, 46(7), 727–735.

Muller, H. J., and Herskowitz, I. H. (1954). Concerning the healing of chromosome ends produced by breakage in Drosophila melanogaster. *The American Naturalist*, 88, 177–208.

Muller, H. J., and Painter, T. S. (1929). The cytological expression of changes in gene alignment produced by X-rays in Drosophila. *The American Naturalist*, 63(686), 193–200.

Müller-Hill, B. (1996). *The lac Operon: A Short History of a Genetic Paradigm*. Berlin, New York: Walter Gruyer.

Müller-Wille, S. (1998). "Reducing varieties to their species". The Linnaean research program and its significance for modern biology. In E. M. Ruis and M. de P. P. Corrales (eds.), *Carl Linnaeus and Enlightened Science in Spain* (pp. 113–126). Madrid: Fundacion Berndt Wistedt (Spain and Sweden: Encounters throughout History).

Müller-Wille, S., and Orel, V. (2007). From Linnaean species to Mendelian factors: Elements of hybridism, 1751–1870. *Annals of Science*, 64(2), 171–215.

Neel, J. V., and Schull, W. J. (1954). *Human Heredity*. Chicago, IL: The University of Chicago Press.

Nelkin, D., and Lindee, M. S. (1995). *The DNA Mystique. The Gene as a Cultural Icon*. New York: Freeman.

Neumann-Held, E. M., and Rehmann-Sutter, C. (eds.). (2006). *Genes in Development: Re-reading the Molecular Paradigm*. Durham: Duke University Press.

Newcombe, H. B. (1949). The origin of bacterial variants. *Nature*, 164, 150–151.

Nicklas, R. B. (1997). How cells get the right chromosomes. *Science*, 275(5300), 632–637.

Nicklas, R. B., and Koch, C. A. (1969). Chromosome micromanipulation: 3. Spindle Fiber Tension and the Reorientation of Mal-Oriented Chromosomes. *Journal of Cell Biology*, 43(1), 40–50.

Nöthiger, R. (2002). Ernst Hadorn, a Pioneer of Developmental Genetics. *International Journal of Developmental Biology*, 46, 23–27.

Novitski, C. E. (2004). Revision of Fisher's analysis of Mendel's garden peas experiments. *Genetics*, 166(3), 1139–1140.

Novitski, E. (2004). On Fisher's criticism of Mendel's results with the garden pea. *Genetics*, 166(3), 1133–1136.

Novitski, E., and Blixt, S. (1978). Mendel, linkage and synteny. *BioScience*, 28(1), 34–35.

Nüsslein-Volhard, C. (1977). Genetic analysis of pattern-form organization in the embryo of *Drosophila melanogaster*. Characterization of the maternal effect mutant bicaudal. *Roux's Archiv für Entwicklungsmechanik*, 183, 249–268.

Nüsslein-Volhard, C., and Wieschaus, E. (1980). Mutations affecting segment number and polarity in Drosophila. *Nature*, 287, 795–801.

Ohno, S. (1970). *Evolution by Gene Duplication*. Berlin: Springer.

Bibliography

Okazaki, R., Okazaki, T., Sakabe, K., Sugimoto, K., and Sugino, A. (1968). Mechanism of DNA chain growth. I. Possible discontinuity and unusual secondary structure of newly synthesized chains. *Proceedings of the National Academy of Sciences of the USA*, 59, 598–605.

Olby, R. C. (1974). *The Path to the Double Helix: The Discovery of DNA*. London: Macmillan.

(1979). Mendel no Mendelian? *History of Science*, 17, 53–72.

(1985). *Origins of Mendelism* (2nd edn.). Chicago, IL: University of Chicago Press.

(1987). William Bateson's introduction of Mendelism to England: A reassessment. *British Journal for the History Science*, 20, 399–420.

(1990). The molecular revolution in biology. In R. C. Olby, G. N. Cantor, J. R. R. Christie and M. J. S. Hodge (eds.), *Companion to the History of Modern Science* (pp. 503–520). London: Routledge.

Oliver, B., and Leblanc, B. (2003). How many genes in a genome? *Genome Biology*, 5(1), 204.1–3.

Orel, V. (1996). *Gregor Mendel: The First Geneticist* (S. Finn, Trans.). Oxford: Oxford University Press.

(2005). Contested memory: debates over the nature of Mendel's paradigm. *Hereditas*, 142, 98–102.

(personal communication a). Science studies and nature of Mendel's paradigm.

(unpubl. ms. b). The useful research question of heredity in the background of the origin of heredity and evolution.

Orel, V., and Hartl, D. (1994). Controversies in the interpretation of Mendel's discovery. *History and Philosophy of the Life Sciences*, 16, 423–464.

Orel, V., and Wood, R. J. (1998). Empirical genetic laws published in Brno before Mendel was born. *Journal of Heredity*, 89(1), 79–82.

Orr-Weaver, T. L., Szostak, J. W., and Rothstein, R. J. (1981). Yeast transformation: A model system for the study of recombination. *Proceedings of the National Academy of Sciences of the USA*, 78, 6354–6358.

Orr, H. A. (1991). A test of Fisher's theory of dominance. *Proceedings of the National Academy of Sciences of the USA*, 88, 11413–11415.

Oyama, S. (2000). *The Ontogeny of Information: Development Systems and Evolution* (2nd edn.). Durham: Duke University Press.

Oyama, S., Griffiths, P. E., and Gray, R. D. (eds.). (2001). *Cycles of Contingency: Developmental Systems and Evolution*. Cambridge, MA: MIT Press.

Painter, T. S. (1934a). A new method for the study of chromosome aberrations and the plotting of chromosome maps in *Drosophila melanogaster*. *Genetics*, 19, 175–188.

(1934b). Salivary chromosomes and the attack on the gene. *The Journal of Heredity*, 25, 465–476.

Panshin, I. B. (1936). A demonstration of the specific nature of position effect. *Compts Rendus (Doklady) Academie Science U.R.S.S., N.S.*, 1(10), 83–86.

(1938). The cytogenetic nature of the position effect of the genes white (mottled) and cubitus interuptus (translated from Russian). *Biologicheskii zhurnal*, 7(4), 837–865.

312

Bibliography

Pardee, A. B., Jacob, F., and Monod, J. (1959). The genetic control and cytoplasmic expression of inducibility in the synthesis of β-galactosidase by *E. coli. Journal of Molecular Biology*, 1, 165–178.

Pardue, M. L., and Gall, J. G. (1970). Chromosomal localization of mouse satellite DNA. *Science*, 168, 1356–1358.

Parnas, O. S. (2006). On the shoulders of generations: The new epistemology of heredity in the nineteenth century. In S. Müller-Wille and H.-J. Rheinberger (eds.), *Heredity Produced: At the Crossroads of Biology, Politics, and Culture 1500–1870* (Vol. I, pp. 315–346). Cambridge, MA: MIT Press.

Paul, D. (1992). Heterosis. In E. Fox Keller and E. A. Lloyd (eds.), *Keywords in Evolutionary Biology* (pp. 167–169). Cambridge, MA: Harvard University Press.

Pearson, H. (2006). What is a gene? *Nature*, 441, 399–401.

Pearson, K. (1900). *The Grammar of Science* (2nd edn.). London: Macmillan.

Peaslee, M. H., and Orel, V. (2007). The evolutionary ideas of F. M. (Ladimir) Klácel, teacher of Gregor Mendel. *Biomedical Papers of the Medical Faculty, University Palacky Olomouc Czech Republic*, 151(1), 151–156.

Pick, D. (1989). *Faces of Degeneration. A European Disorder, c. 1848 – c. 1918.* Cambridge: Cambridge University Press.

Plunkett, C. R. (1932). Temperature as a tool of research in phenogenetics: Methods and results. *Proceedings of the 6th International Congress of Genetics, Ithaca*, 2, 158–169.

Polanyi, M. (1968). Life's irreducible structure. *Science*, 160, 1308–1312.

Pontecorvo, G. (1952). Genetic formulation of gene structure and gene action. *Advances in Enzymology*, 13, 121–149.

  (1958). *Trends in Genetic Analysis*. New York: Columbia University Press.

Portin, P. (1993). The concept of the gene: Short history and the present status. *The Quarterly Review of Biology*, 68(2), 173–223.

Pritchard, R. H. (1955). The linear arrangement of a series of alleles of *Aspergillus nidulans. Heredity*, 9, 343–371.

Provine, W. B. (1971). *The Origins of Theoretical Population Genetics*. Chicago, IL: University of Chicago Press.

Punnett, R. C. (1909). *Mendelism* (American Edition of 2nd edn., 1907). New York: Wilshire.

  (1911). *Mendelism* (3rd edn.). London: Macmillan.

  (1913). Reduplication series in sweet peas. *Journal of Genetics*, 3(2), 77–103.

  (1950). Early days of genetics. *Heredity*, 4(1), 1–10.

Raff, R. A., and Kaufman, T. C. (1983). *Embryos, Genes, and Evolution: The Developmental-Genetic Basis of Evolutionary Change*. New York: Macmillan.

Raffel, D., and Muller, H. J. (1940). Position effect and gene divisibility considered in connection with three strikingly similar scute mutations. *Genetics*, 25, 541–583.

Rheinberger, H.-J. (1997). *Toward a History of Epistemic Things: Synthesizing Proteins in the Test Tube*. Stanford, CA: Stanford University Press.

  (2000). Ephestia: The experimental design of Alfred Kühn's physiological developmental genetics. *Journal of the History of Biology*, 33(3), 535–576.

Bibliography

(2006). *Epistemologie des Konkreten. Studies zur Geschichte der modernen Biologie.* Frankfurt am Mein: Suhrkamp.

Rieger, R., Michaelis, A., and Green, M. M. (1976). *Glossary of Genetics and Cytogenetics* (4th edn.). Berlin: Springer.

Roll-Hansen, N. (1978). The genotype theory of Wilhelm Johannsen and its relation to plant breeding and the study of evolution. *Centaurus,* 22, 201–235.

Rosenberg, A. (1979). From reductionism to instrumentalism? In M. Ruse (ed.), *What the Philosophy of Biology Is: Essays Dedicated to David Hull* (pp. 245–262). Dordrecht: Kluwer.

(1985). *The Structure of Biological Science.* Cambridge: Cambridge University Press.

Roux, W. (1894). The problems, methods, and scope of developmental mechanics. In J. Maienschein (ed.), *Defending Biology: Lectures from the 1890s* (pp. 104–148). Cambridge, MA: Harvard University Press.

Sandler, I., and Sandler, L. (1985). A conceptual ambiguity that contributed to the neglect of Mendel's paper. *History and Philosophy of the Life Sciences,* 7, 3–70.

Sapp, J. (1983). The struggle for authority in the field of heredity, 1900–1932: New perspectives on the rise of genetics. *Journal of the History of Biology,* 16, 311–342.

(1990). *Where the Truth Lies: Franz Moewus and the Origins of Molecular Biology.* Cambridge: Cambridge University Press.

Sarkar, S. (1998). *Genetics and Reductionism.* Cambridge: Cambridge University Press.

(1999). From the *Reaktionsnorm* to the adaptive norm: The norm of reaction, 1909–1960. *Biology and Philosophy,* 14(2), 235–252.

Schaffner, K. F. (1969). The Watson-Crick model and reductionism. *British Journal for the Philosophy of Science,* 20, 325–348.

(1976). Reduction in biology: prospects and problems. In R. S. Cohen, C. A. Hooker, G. Pearce, A. C. Michalos and J. W. van Evra (eds.), *Proceedings of the 1974 Biennial Meeting of the Philosophy of Science Association* (pp. 613–632). Dordrecht: Reidel.

(1993). *Discovery and Explanation in Biology and Medicine.* Chicago, IL: University of Chicago Press.

Scherthan, H. (2001). A bouquet makes ends meet. *Nature Review: Molecular Cell Biology,* 2, 621–627.

(2007). Telomere attachment and clustering during meiosis. *Cellular and Molecular Life Sciences,* 64(2), 117–124.

Schmitt, J., Benavente, R., Hodzic, D., Höög, C., Stewart, C. L., and Alsheimer, M. (2007). Transmembrane protein Sun2 is involved in tethering mammalian meiotic telomeres to the nuclear envelope. *Proceedings of the National Academy of Sciences of the USA,* 104(18), 7426–7431.

Schrödinger, E. (1962 [1944]). *What Is Life? The Physical Aspect of the Living Cell.* Cambridge: Cambridge University Press.

Schwartz, S. (1998). *The significance of the trait in genetics, 1900–1945.* Unpublished Ph.D. Dissertation, The Hebrew University of Jerusalem.

(2000). The differential concept of the gene. In P. J. Beurton, R. Falk and H.-J. Rheinberger (eds.), *The Concept of the Gene in Development and Evolution: Historical and Epistemological Perspectives* (pp. 26–39). Cambridge: Cambridge University Press.

(2002). Characters as units and the case of the presence and absence hypothesis. *Biology and Philosophy*, 17(3), 369–388.

Selzer, J. (ed.). (1993). *Understanding Scientific Prose*. Madison, WI: University of Wisconsin Press.

Setlow, R. B., and Carrier, W. L. (1964). The disappearance of thymine dimers from DNA: An error-correcting mechanism. *Proceedings of the National Academy of Sciences of the USA*, 51, 226–231.

Shull, G. H. (1909). The "presence and absence" hypothesis. *The American Naturalist*, 43, 410–419.

Simpson, G. G. (1964). *This View of Life*. New York: Harcourt, Brace and World.

Sinnott, E. W., Dunn, L. C., and Dobzhansky, T. (1950). *Principles of Genetics* (4th edn.). New York: McGraw-Hill.

(1958). *Principles of Genetics* (5th edn.). New York: McGraw-Hill.

Sirks, M. J., and Zirkle, C. (1964). *The Evolution of Biology*. New York: The Ronald Press Company.

Sloan, P. R. (1976). The Buffon-Linnaeus controversy. *Isis*, 67(3), 358–375.

Snyder, M., and Gerstein, M. (2003). Defining genes in the genomics era. *Science*, 300(5617), 258–260.

Sober, E. (1984). *The Nature of Selection*. Cambridge, MA: The MIT Press.

Sober, E., and Wilson, D. S. (1998). *Unto Others: The Evolution and Psychology of Unselfish Behavior*. Cambridge, MA: Harvard University Press.

Srb, A. M., and Owen, R. D. (1953). *General Genetics*. San Francisco: Freeman.

Stadler, D. (1997). Ultraviolet-induced mutation and the chemical nature of the gene. *Genetics*, 145(4), 863–865.

Stadler, L. J. (1928a). Genetic effects of X-rays in maize. *Proceedings of the National Academy of Sciences of the USA*, 14, 69–75.

(1928b). Mutations in barley induced by X-rays and radium. *Science*, 68, 186–187.

(1954). The gene. *Science*, 120, 811–819.

Stadler, L. J., and Sprague, G. F. (1937). Contrasts in the genetic effects of ultra-violet treatment. *Science*, 85, 57–58.

Stahl, F. W. (1961). A chain model for chromosomes. *Journal de Chemie Physique*, 58, 1072–1077.

(1979). *Genetic Recombination: Thinking About It in Phage and Fungi*. San Francisco: Freeman.

(1988). A unicorn in the garden. *Nature*, 335, 112–113.

(1994). The Holliday junction on its thirtieth anniversary. *Genetics*, 138(2), 241–246.

Stamhuis, I. H., Meijer, O. G., and Zevenhuizen, E. J. A. (1999). Hugo de Vries on heredity, 1889–1903: Statistics, Mendelian laws, pangenes, mutations. *Isis*, 90(2), 238–267.

Stent, G. S. (1968). That was the molecular biology that was. *Science*, 160, 390–395.

Bibliography

(1969). *The Coming of the Golden Age: A View of the End of Progress*. New York: Natural History Press.

(1971). *Molecular Genetics: An Introductory Narrative*. San Francisco: Freeman.

Sterelny, K., and Griffiths, P. E. (1999). *Sex and Death: An Introduction to Philosophy of Biology*. Chicago, IL: Chicago University Press.

Stern, C. (1929). Über die additive Wirkung multipler allele. *Biologisches Zentralblat*, 49, 261–290.

(1931). Zytologisch-genetische Untersuchungen als Beweise fur die Morgansche Theorie des Faktorenaustauschs. *Biologisches Zentralblat*, 51 (10), 547–587.

(1936). Somatic crossing-over and segregation in *Drosophila melanogaster*. *Genetics*, 21, 625–730.

(1943). The Hardy-Weinberg law. *Science*, 97(2510), 137–138.

(1955). Two or three bristles. *Science in Progress*, 9, 41–84; 327–328.

(1959). Use of the term "superfemale". *Lancet*, 12, 1088.

(1960a). Dosage compensation – development of a concept and new facts. *Canadian Journal of Genetics and Cytology*, 2(2), 105–118.

(1960b). *Principles of Human Genetics* (2nd edn.). San Francisco, CA: Freeman.

Stern, C., and Sherwood, E. (1966). *The Origin of Genetics: A Mendel Sourcebook*. San Francisco, CA: Freeman.

Stern, H., and Hotta, Y. (1973). Biochemical control of meiosis. *Annual Review of Genetics*, 7, 37–66.

Streisinger, G., Okada, Y., Emrich, J., Newton, J., Tsugita, A., Terzaghi, E., *et al.* (1967). Frameshift mutations and the genetic code. *Cold Spring Harbor Symposia on Quantitative Biology*, 31, 77–84.

Strickberger, M. W. (1976). *Genetics* (2nd edn.). New York: Macmillan.

Sturtevant, A. H. (1913a). The Himalayan rabbit case, with some considerations on multiple allelomorphs. *The American Naturalist*, 47, 234–238.

(1913b). The linear arrangement of six sex-linked factors in Drosophila, as shown by their mode of association. *Journal of Experimental Zoology*, 14, 43–59.

(1914). The reduplication hypothesis as applied to Drosophila. *The American Naturalist*, 48, 535–549.

(1923). Inheritance of the direction of coiling in Limnaea. *Science*, 58, 269–270.

(1925). The effects of unequal crossing over at the Bar locus in Drosophila. *Genetics*, 10, 117–147.

(1929). The claret mutant type of *Drosophila simulans*: a study of chromosome elimination and cell-lineage. *Zeitschrift für Wissenschaftliche Zoologie*, 135, 323–356.

(1932). The use of mosaics in the study of the developmental effects of genes. In *Proceedings of the 6th International Congress of Genetics, Ithaca*, 1, 304–307.

(1965). *A Short History of Genetics*. New York: Harper and Row.

Bibliography

Sturtevant, A. H., and Beadle, G. W. (1936). The relations of inversions in the X chromosome of *Drosophila melanogaster* to crossing-over and disjunction. *Genetics*, 21, 554–604.

(1962 [1939]). *An Introduction to Genetics*. New York: Dover.

Sturtevant, A. H., and Dobzhansky, T. (1936). Inversions in the third chromosome of wild races of *Drosophila pseudoobscura*, and their use in the study of the history of the species. *Proceedings of the National Academy of Sciences of the USA*, 22, 448–450.

Sutton, W. (1903). The chromosomes in heredity. *Biological Bulletin*, 4, 231–251.

Swift, H. H. (1950). The desoxyribose nucleic acid content of animal nuclei. *Physiological Zoology*, 23, 169–198.

Swinburne, R. G. (1962). The presence-and-absence theory. *Annals of Science*, 18(3), 131–145.

Tabery, J. G. (2004). The "evolutionary synthesis" of George Udny Yule. *Journal of the History of Biology*, 37(1), 73–101.

Taylor, J. H. (1953). Autoradiographic detection of incorporation of P^{32} into chromosomes during meiosis and mitosis. *Experimental Cell Research*, 4, 164–173.

(1959). The organization and duplication of genetic material. *Proceedings of the Tenth International congress of Genetics*, Montreal 1958, 1, 63–78.

Taylor, J. H., Woods, P. S., and Hughes, W. I. (1957). The organization and duplication of chromosomes as revealed by autoradiographic studies using tritium-labeled thymidine. *Proceedings of the National Academy of Sciences of the USA*, 43, 122–128.

Temin, H. M., and Mizutani, S. (1970). RNA-dependent DNA polymerase in virions of Rous sarcoma virus. *Nature*, 226, 1211–1213.

Theunissen, B. (1994). Closing the door on Hugo de Vries' Mendelism. *Annals of Science*, 51, 225–248.

Thoday, J. M., and Read, J. (1947). Effect of oxygen on the frequency of chromosome aberrations produced by X-rays. *Nature*, 160, 608.

Timoféeff-Ressovsky, H. A., and Timoféeff-Ressovsky, N. W. (1926). Über das phaenotypische Manifestieren des Genotyps. II Über idio-somatische Variationsgruppen bei *Drosophila funebris*. *Wilhelm Roux's Archives of Developmental Biology*, 108, 146–170.

Timoféeff-Ressovsky, N. W., Zimmer, E. G., and Delbrück, M. (1935). Über die Natur der Genmutation und der Genstruktur. *Nachrichten von der Gesellschaft der Wissenschaften zu Göttingen*, 1, 189–245.

Tjio, J. H., and Levan, A. (1956). The chromosome number of man. *Hereditas*, 42, 1–6.

Toulmin, S. (1972). *Human Understanding. The Collective Use and Evolution of Concepts*. Princeton, NJ: Princeton University Press.

Troland, L. T. (1917). Biological enigmas and the theory of enzyme action. *The American Naturalist*, 51(606), 321–350.

Tschermak, E. (1950 [1900]). Concerning artificial crossing in *Pisum sativum*. *Genetics*, 35 Supplement (*The birth of Genetics*)(5, part 2), 42–47.

Vega, J. M., and Feldman, M. (1998). Effect of the pairing gene Ph1 on centromere misdivision in common wheat. *Genetics*, 148, 1285–1294.

Verdun, R. E., and Karlseder, J. (2007). Replication and protection of telomeres. *Nature*, 447, 924–931.

Vogelstein, B., and Kinzler, K. W. (1992). P53 function and disfunction. *Cell*, 70, 523–526.

Vogt, O. (1926). Psychiatrisch wichtige Tatsachen der zoologisch-botanischen Systematik. *Zeitschrift für die gesamte Neurologie und Psychiatrie*, 101, 805–832.

Volkin, E., and Astrachan, L. (1957). RNA metabolism in T2-infected *Escherichia coli*. In B. Glass and W. D. McElroy (eds.), *The Chemical Basis of Heredity* (pp. 686–695). Baltimore, MD: Johns Hopkins Press.

Waddington, C. H. (1942). Canalization of development and the inheritance of acquired characters. *Nature*, 150, 563–565.

(1957). *The Strategy of the Genes. A Discussion of Some Aspects of Theoretical Biology*. London: George Allen and Unwin.

Wallace, B. (1958). The average effect of radiation-induced mutations on viability in *Drosophila melanogaster*. *Evolution*, 12, 532–556.

Watson, J. D. (1968). *The Double Helix: A Personal Account of the Discovery of the Structure of DNA*. New York: Atheneum.

Watson, J. D., and Crick, F. H. C. (1953a). Genetical implications of the structure of deoxyribose nucleic acid. *Nature*, 171, 964–967.

(1953b). Molecular structure of nucleic acids. *Nature*, 171, 737–738.

(1953c). The structure of DNA. *Cold Spring Harbor Symposia on Quantitative Biology*, 18, 123–131.

Weaver, W. (1970). Molecular biology: Origin of the term. *Science*, 170(3958), 581–582.

Weinberg, R. A. (2007). *The Biology of Cancer*. New York: Garland.

Weiner, J. (1999). *Time, Love, Memory: A Great Biologist and His Quest for the Origins of Behavior*. New York: Alfred A. Knopf.

Weinstein, A. (1918). Coincidence of crossing over in *Drosophila melanogaster* (*ampelophila*). *Genetics*, 3, 135–172.

(1977). How unknown was Mendel's paper? *Journal of the History of Biology*, 10(2), 341–364.

Weinstock, G. M. (2007). ENCODE: More genomic improvement. *Genome Research*, 17, 667–668.

Westergaard, M. (1958). The mechanism of sex determination in dioecious plants. *Advances in Genetics*, 9, 217–281.

White, M. J. D. (1954). *Animal Cytology and Evolution* (2nd edn.). Cambridge: Cambridge University Press.

(1973). *The Chromosomes* (6th edn.). London: Chapman and Hall.

Whitehouse, H. L. K. (1965). *Towards an Understanding of the Mechanism of Heredity*. London: Edward Arnold.

Wilkie, A. O. M. (1994). The molecular basis of genetic dominance. *Journal of Medical Genetics*, 31, 89–98.

Wilkins, A. S. (1997). Canalization: a molecular genetic perspective. *BioEssays*, 19(3), 257–262.

(2002). *The Evolution of Developmental Pathways*. Sunderland, MA: Sinauer.

Bibliography

Williams, G. C. (1974 [1966]). *Adaptation and Natural Selection: A Critique of Some Current Evolutionary Thought*. Princeton, NJ: Princeton University Press.

Wilson, D. S. (1983). The group selection controversy: History and current status. *Annual Review of Ecology and Systematics*, 14, 159–187.

(1992). Group selection. In E. F. Keller and E. A. Lloyd (eds.), *Keywords in Evolutionary Biology* (pp. 145–148). Cambridge, MA: Harvard University Press.

Wilson, D. S., and Sober, E. (1989). Reviving the superorganism. *Journal of Theoretical Biology*, 136, 337–356.

Wilson, E. B. (1893 [1986]). The mosaic theory of development. In J. Maienschein (ed.), *Defending Biology: Lecture from the 1890s* (pp. 67–80). Cambridge, MA: Harvard University Press.

(1896). *The Cell in Development and Inheritance*. New York: Macmillan.

(1924). *The Cell in Development and Inheritance* (3rd edn.). New York: Macmillan.

Wilson, E. O. (1975). *Sociobiology: The New Synthesis*. Cambridge, MA: Harvard University Press.

Winkler, H. (1930). *Die Konversion der Gene*. Jena: Fischer.

Wollman, E. L., Jacob, F., and Hayes, W. (1956). Conjugation and genetic recombination in *Escherichia coli* K-12. *Cold Spring Harbor Symposia on Quantitative Biology*, 21, 141–162.

Woltereck, R. (1909). Weitere experimentelle Untersuchungen über Artveränderung, speziell über des Wesen quantitativer Artunterschiede bei Daphnien. *Verhandlungen der Deutschen Zoologischen Gesellschaft*, 19, 110–173.

Wood, R. J., and Orel, V. (2005). Scientific breeding in central Europe during the early nineteenth century: Background to Mendel's later work. *Journal of the History of Biology*, 38, 239–272.

Woodger, J. H. (1967). *Biological Principles*. London: Routledge and Kegan Paul.

Wright, S. (1929a). The evolution of dominance: Comment on Dr. Fisher's reply. *The American Naturalist*, 63, 556–561.

(1929b). Fisher's theory of dominance. *The American Naturalist*, 63, 274–279.

(1931). Evolution in Mendelian populations. *Genetics*, 16, 97–159.

(1945). Tempo and mode in evolution: A critical review. *Ecology*, 26, 415–419.

(1958). *Systems of Mating and Other Papers*. Ames, IA: The Iowa State College Press.

Wynne-Edwards, V. C. (1962). *Animal Dispersion in Relation to Social Behaviour*. Edinburgh: Oliver and Boyd.

Yanofsky, C., Carlton, B. C., Guest, J. R., Helinski, D. R., and Henning, U. (1964). On the colinearity of gene structure and protein structure. *Proceedings of the National Academy of Sciences of the USA*, 51(2), 266–272.

Yanofsky, C., Drapeau, G. R., Guest, J. R., and Carlton, B. C. (1967). The complete amino acid sequence of the tryptophan synthetase A protein (a subunit) and its colinear relationship with the genetic map of the A gene. *Proceedings of the National Academy of Sciences of the USA*, 57(2), 296–298.

Yanofsky, C., and Lennox, E. S. (1959). Transduction and recombination study of linkage relationships among the genes controlling tryptophan synthesis in *Escherichia coli. Virology*, 8, 425–447.

Yule, G. U. (1902). Mendel's laws and their probable relations to intra-racial heredity. *The New Phytologist*, 1(9 and 10), 193–207 and 222–238.

Yule, G. U. (1903). Professor Johannsen's experiments in heredity: A review. *The New Phytologist*, 2, 235–242.

Zacharias, H. (1995). Emil Heitz (1892–1965): Chloroplasts, heterochromatin and polytene chromosomes. *Genetics*, 141(1), 7–14.

Zevenhuizen, E. (1998). The hereditary statistics of Hugo de Vries. *Acta Botanica Neerlandica*, 47(4), 427–463.

Zirkle, C. (1935a). *The Beginnings of Plant Hybridization* (Morris Arboretum Monographs). Philadelphia, PA: University of Pennsylvania Press.

(1935b). The inheritance of acquired characters and the provisional hypothesis of pangenesis. *The American Naturalist*, 69, 417–445.

(1945). The early history of the idea of the inheritance of acquired characters and of pangenesis. *Transactions of the American Philosophical Society*, 35 (part II), 91–150.

Zuckerman, H., and Lederberg, J. (1986). Postmature scientific discovery? *Nature*, 324(6098), 629–631.

Index

aberrations, *see* chromosome
    aberration
allele, 48, 55, 60, 133, 135, 153, 213,
    271, 273
allelomorph, *see* allele
altruistic trait, 279, 280
Aspergillus nidulans, 96, 103, 154,
    155
atom of heredity, 73, 98, 125, 129,
    195
autocatalysis, 132
auxotroph, 96, 154, 179, 184, 186
Avery, Oswald T. (1877–1955), 2,
    185, 192–3

bacteria, 3, 105, 139, 142, 155, 156,
    172–3, 178, 180, 184–8, 192, 198,
    202, 205, 223, 228, 247, 250, 259,
    271, 284
bacteriophage, 127, 183, 188, 194, 271,
    see also phage
    T4, 154, 155
Baltimore, David (1938–), 210
Bateson, William (1861–1926), 6, 24,
    41–2, 46–50, 54–7, 70, 80, 87–9, 94,
    158, 178, 209, 212, 216, 263, 268,
    272, 273
Beadle, George W. (1903–1989), 5, 72,
    94, 112, 148, 172, 178–81, 234, 237
Belling, John (1866–1933), 92, 202
Benzer, Seymour (1921–2007), 154–6,
    188–90, 236, 239, 246, 263, 271,
    287

biochemical analysis, 175, 178–80, 185,
    193, 205, 208, 210, 214, 219, 223,
    225, 227, 260, 276
Biogenetic Law, *see* Haeckel, Ernst
    (1834–1919)
Biston betularia, 160
bithorax complex, 153, 242, 281, 282
bottom-up, 6, 8, 15, 16, 21, 50, 52, 61,
    65, 131, 134, 149, 151, 209, 219,
    233, 235, 247, 254, 260, 274, 288,
    289, *see also* top-down
Boveri, Theodor (1862–1915), 74, 78,
    108, 120
breeders, 4, 13, 19, 25, 26, 41, 60, 61,
    109, 161
Brenner, Sydney (1927–), 225, 246,
    283, 284, 287
Bridges, Calvin B. (1889–1938), 74,
    83–6, 98, 100, 101, 108, 109, 110
Britten, Roy, 250–4, 254, 255, 256, 257
Brno (Brünn), 1, 25, 26, 27
Buffon, Georges-Louis Leclerc
    (1707–1788), 13, 18, 21, 22, 23, 276
Butenandt, Adolf (1903–1995), 179

Caenorhabditis elegans, 246, 262, 266,
    284
Cairns, John (1922–), 184, 185
canalization, 280
cancer, 119, 242, 273
    tumor-suppressor, 242
Carson, Hampton, 161

Caspari, Ernst, 179, 237
Casperson, Torbjörn, 192, 193
Castle, William E. (1867–1962), 59,
    60, 64, 65, 94, 97, 129, 159, 213,
    233, 234
cell nucleus, 23, 45, 54, 78, 112, 144,
    181, 191, 236
cell theory, 33, 34, 54, 83, 236
Central Dogma, 7, 23, 126, 168, 171,
    175, 210, 223, 226–30, 245, 260,
    261, 264, 288
centromere, 99, 111, 119, 120, 121,
    200
Chargaff, Erwin (1905–2002), 193
Chase, Martha C. (1927–2003), 193,
    194
Chetverikov, Sergei S. (1880–1959),
    159
chiasma, 91, 113–16, 122
chiasmatype theory, 74, 89–91, 93, 95,
    98, 99, 103, 114–16, 205
chromatin, 71, 135, 192, 264
chromosome, 23, 74, 77, 79, 123, 132,
    195, 198, 199, 200
    non-disjunction, 83, 85, 122, 236
    polytene, 74, 100, 118, 134, 161
    sex chromosome, 74, 78, 81, 109
chromosome aberration, 74, 93, 108,
    118, 141, 149, *see also*
    chromosome rearrangement
chromosome map, 74, 92, 106, *see also*
    linkage map
    circular map, 106, 188
    cytological map, 101, 117
    genetic map, 74, 92, 93, 96, 97, 99, 105
    molecular map, 102, 106
    somatic map, 118
chromosome mechanics, 74, 84, 113
chromosome rearrangement, 110, 111,
    133, 134, 140, 146, 147, 252
    attached-X, 98
    deficiency, 100, 150
    deletion, 188
    induced rearrangement, 141
    inversion, 112, 115, 161, 278
    mottled rearrangement, 147
    translocation, 138, 139

chromosome theory, 7, 63, 71, 72,
    74, 80, 83, 84, 90, 97, 102, 108,
    177, 287
cistron, 189, 223, 226, 263, 271
ClB chromosome, 112, 136, 137, 139,
    140, 180
colinearity, 7, 126, 156, 287
complementation, 84, 104, 132, 153,
    154, 188, 203, 204, 217, 222, 224,
    226, 249, 263, 271
conjugation, 90, 92, 105, 187
copy choice, 92, 202, 203
Correns, Carl (1864–1933), 40–2, 53,
    54, 57, 74, 87, 94, 213
C_ot, 251
coupling and repulsion, 48, 74, 80, 87,
    88, 90
Creighton, Harriet S. (1909–2004), 93
Crick, Francis H. C. (1916–2004), 3, 7,
    23, 73, 126, 173, 175, 195, 196, 210,
    220–1, 225–30, 245, 246, 249, 253,
    276, 287, 288
crossing over, 74, 90, 92–4, 98, 104, 113,
    114, 134, 176, 202, 204–6
    interference, 96, 103, 202
Crow, James F. (1916–), 164
Cuvier, George (1766–1832), 22
C-value paradox, 252, 253
cytogenetics, 74, 108, 113, 117, 146,
    147, 191, 252
cytological theory of heredity, 113
cytoplasm, 23, 45, 191, 193, 214, 224,
    228, 271, 272

Darlington, Cyril D. (1903–1981),
    113–17, 121
Darwin, Charles (1809–1882), 13, 16,
    22, 23, 27, 32, 33, 37, 40, 41, 44–6,
    49, 60, 66, 166, 275, 276, 279
Davidson, Eric H. (1937–), 254–7
de Vries, Hugo (1848–1935), 6, 24,
    39–42, 44–7, 52–4, 58, 61, 158, 209,
    211, 227, 254, 287
Delbrück, Max (1906–1981), 142, 172,
    182–3, 185, 196, 205
deoxyribonucleic acid, 195, 219, 221,
    225

design of experiments, 29, 31, 35
development, 24, 36, 46, 47, 54, 56, 59,
 108, 169, 178, 181, 232, 248, 282
 compartmentalization, 240–2
 developmental constraint, 275, 280
Developmental System Theories, 290
Dicer, 266
differentiation, 80, 181, 283, 284
disease, 12, 159, 178, 219, 272, 273
DNA, 2–4, 7, 23, 73, 92, 102, 104, 106,
 117–19, 126, 135, 155–7, 167, 168,
 173, 176, 177, 189–91, 193–201,
 204–8, 218, 220, 223, 225, 227,
 228, 230, 246, 247, 249–53, 255–7,
 259–64, 270, 272, 278, 287
 junk, 8, 166, 247, 250, 264, 267
 non-coding, 260, 264–6
 repair, 206
 repetitive sequences, 117, 247,
  250–3, 257
 sequence, 104, 156, 157, 196, 206,
  261, 265, 278, 281–3, 288,
  see also nucleotide sequence
 synthesis, 199, 200, 204, 206, 208
DNA-polymerase, 200
Dobzhansky, Theodosius
 (1900–1975), 3, 138, 159, 161,
 163, 164, 172, 276
dominance, 30, 41, 48, 54, 153–4, 158,
 164, 209, 211, 213, 242, 268–72,
 see also recessivity
double helix, 7, 104, 155, 195, 196,
 203, 222, 229, 246, 259, *see also*
 DNA
Driesch, Hans (1867–1941), 59
Drosophila, 5, 9, 57, 74, 82, 83–6, 88, 90,
 93, 96, 100, 103, 106, 108–12,
 117–19, 121, 122, 127, 134, 136–8,
 143–6, 154, 155, 161, 172, 177, 178,
 188, 200, 205, 214, 217, 218, 231,
 232, 234, 235, 237, 239, 242, 246,
 256, 262, 269, 281–3, 284
 D. ampelophila, 84
 D. melanogaster, 74, 81, 101, 136,
  141, 164, 241, 278
 D. pseudoobscura, 112, 138, 161,
  165

D. simulans, 236, 278
D. willistoni, 161
Dunn, Leslie C. (1893–1974), 1, 2, 5,
 29, 64

E. coli, 7, 106, 156, 184, 186–8, 190, 197,
 205, 206, 223, 228, 239, 245, 252,
 284
East, Edward M. (1879–1938), 64–6,
 125, 129, 130, 139, 162
editing, *see* RNA processing
Eldredge, Niles (1943–), 253, 281
embryogenesis, 172, 217, 233, 235–7,
 239, 256, 283
entelechy, *see* Driesch, Hans (1867–1941)
environment, 16, 46, 63, 73, *see also*
 heredity
 Nature *versus* Nurture, 16, 73, 289
enzyme, 142, 156, 161, 165, 180, 181,
 184, 191, 204–8, 212, 214, 218, 223,
 224, 228, 237, 266, 269, 270
 β-galactosidase, 223, 224
Ephestia kühniella, 179, 235, 237
Ephrussi, Boris (1901–1979), 178, 237
epigenesis, 15, 22, 67, 78, 238, 241, 266
epistasis, 216, 272
Escherichia coli, see E. coli
essentialism, 17
euchromatin, 117, 146, *see also*
 heterochromatin
eukaryote, 105, 120, 187, 190, 202, 250,
 252, 261, 262, 283
evo-devo, 248, 258, 277
evolution, 27, 28, 32, 41, 42, 44–8, 58,
 61, 72, 123, 126, 132, 134, 135, 158,
 159, 185, 212, 246, 248, 250, 252,
 254, 259, 267, 275
exchange, 91–4, 96, 98, 105, 114–16,
 121, 122, 202, 228
 exchange pairing, 203
 molecular, 104
 physical, 91, 115
exon, 261, 262

Faktoren, 7, 30, 33–6, 39, 42, 44, 53, 54,
 56–8, 60, 77, 79, 86, 125, 128, 131,
 171, 211, 263

fate map, 236, 239
Fisher, Ronald A. (1890–1962), 6, 28,
    32, 52, 72, 159, 160, 161, 268
    Fundamental Theorem of Natural
    Selection, 72
fitness, 163, 165, 278, 279
    inclusive fitness, 279
fluctuation, 45, 58, 65, 66
Ford, Edmund B. (1901–1988), 160

Galton, Francis (1829–1911), 16, 44,
    49, 51, 60, 61, 63, 158
    stirp theory, 50
Gamow, George (1904–1968), 225
Garcia-Bellido, Antonio (1936–), 240,
    241
Garrod, Archibald (1857–1936), 178,
    212
Gärtner, Carl Friedrich (1772–1850),
    20, 21, 28, 158
Gehring, Walter J. (1939–), 238, 239,
    282
gene, 3, 5–7, 63, 69–71, 96, 98, 102, 106,
    117, 125–36, 139, 141, 143–56, 159,
    165, 172, 176, 177, 180, 188, 189,
    192, 195, 206, 214, 218, 221–3,
    231–3, 247–8, 255, 259–61, 264–6,
    268, 271, 272, 277, 278, 282,
    287, 288
    conversion, 92, 103, 104, 176, 202–5,
    207
    expression, 147, 200, 248, 254
    frequency, 163, 167, 169
    function, 84, 165, 178, 212, 214–16,
    219, 222, 225, 234, 273
    "gene for", 69, 234
    gene-*D*, 3, 288, 289
    gene-*P*, 3, 288, 289
    jumping gene, 234, 266
    mosaic gene, 259
    mutation, 134, 140, 147–51, 238
    overlapping gene, 156
    regulation, 254, 267
    regulator gene, 224
    selector gene, 241
    structure, 180, 214, 245
    sub-gene, 147, 148

generation, 15, *see* reproduction and
    generation
genetic code, 167, 173, 190, 223, 225,
    277
    triplet, 190, 225, 226
genetic load, 163
genetic regulation, 255
genetic variability, 159
genetics, 42
    classical, 3, 4, 173, 174, 176, 177, 205
    molecular, 3, 4, 7, 35, 156, 157, 173,
    174, 176, 177, 190, 205, 247,
    282, 284
    phenomenological, 4, 7, 50, 82, 125,
    126, 133, 143, 155, 156, 165, 172,
    175, 176, 190, 270, 282
    population, 51, 52, 106, 152, 158–60,
    275
    gene pool, 276
    reverse genetics, 288
genocentricity, 7, 73, 210, 216, 234, 289
genome, 119, 123, 163, 177, 216, 232,
    246, 252, 254, 255, 264, 265, 278
genomics, 23, 246, 267, 288
genotype, 5, 6, 43, 54, 56, 62–4, 67–74,
    125, 128, 131, 153, 157, 167, 168,
    185, 195, 210, 212, 214, 216, 217,
    223, 232, 233, 236, 260, 261, 268,
    272, 277, 279, 287, 288, 289
Geoffroy Saint Hilaire, Etienne
    (1772–1844), 22
germ line, 23, 59, 71, 72, 77, 82, 83,
    see also phage
Gluecksohn-Waelsch, Salome
    (1907–2007), 235
Goldschmidt, Richard B. (1878–1958),
    92, 97, 125, 126, 129, 133–5, 142,
    146, 148, 149, 217
Gould, Stephen J. (1941–2002), 168,
    253, 280, 281
Great Chain of Being, 15
Griffith, Frederick (1879–1941), 192
Gruneberg, Hans (1907–1982), 234
gynandromorph, 236, 240

Hadorn, Ernst (1902–1976), 237–9,
    256, 282

Haeckel, Ernst (1834–1919), 276
Haldane, J. B. S. (1892–1964), 159, 178, 192, 214, 222, 279
Hamilton, William D. (1936–2000), 279
Hardy, Godfrey H. (1877–1947), 158
Hardy-Weinberg Law, 159
Hayes, William (1913–1994), 105, 271
Heitz, Emil (1892–1965), 117
hemizygote, 136, 242
heredity, 16, 21, 44, 51, 289, *see also* environment
Hershey, Alfred (1908–1997), 193, 194
heterochromatin, 117, 146, 147, 251, *see also* euchromatin
heteroduplex, 104, 204, 207
heterokaryon, 180, 271
heterosis, 161, 162, 278
heterozygote, 30, 31, 81, 103, 115, 119, 133, 136, 161, 164, 224, 242, 268, 269
    cis heterozygote, 153
    heterozygosity *per se*, 162, 163, 164
    trans heterozygote, 153
Hfr, 105, 188
Holliday junction, 205–8
Holliday, Robin, 203–5
homeobox, 237, 283
homeosis, 41, 47, 241, 281
homozygote, 30, 99, 162
Human Genome Project, 8, 73, 246, 264
Huxley, Thomas H. (1825–1895), 44, 49
hybrid, 19–22, 32–5, 39, 74, 109, 158, 229, 230
hybridist tradition, 6, 17, 21, 22, 24, 28, 32, 35, 36, 39–42, 60, 74, 130, *see also* morphogenist tradition
hybridization, 4, 13, 15, 17–22, 25–9, 32–5, 40, 47, 50, 51, 69, 72, 78, 117, 131, 133, 155, 158, 161, 171, 176, 185–7, 193, 195, 196, 198, 201, 208, 211, 227, 249, 250, 261, 276
    in situ hybridization, 251
    in vitro hybridization, 276
    molecular hybridization, 117, 203, 246, 249, 266
    RNA-sequence hybridization, 267

hypothetical construct, 6, 7, 69, 72, 125, 128, 129, 247, 264, 289, *see also* intervening variable

imaginal disc, 238, 239, 256
inborn errors of metabolism, 178, 212
inbreeding coefficient, 159
information, 227, 230, 247, 255, 260–2, 277, 287
information theory, 3, 175
instrumentalism, 17, 147, *see also* nominalism
interactome, 265
intervening variable, 6, 7, 69, 70, 72, 125, 128, 129, 169, 247, 264, 289, *see also* hypotherical construct
intron, 261

Jacob, François (1920–), 105, 223–5, 228, 245, 250, 271
Janssens, Frans A. (1863–1924), 74, 90, 114
Johannsen, Wilhelm (1857–1927), 5–7, 17, 43, 57, 60–4, 66–73, 125, 128, 129, 210, 287, 289
Jukes, Thomas H. (1905–1999), 167

Kimura, Motoo (1924–1994), 166, 168
kinetochore, *see* centromere
King, Jack L. (1934–1983), 167
knock-out, 280
Kohne, David, 250
Kölliker, Albert (1817–1905), 24
Kölreuter, Joseph Gottlieb (1733–1806), 20, 27, 28, 32, 158
Kostoff, Dontcho, 100
Kühn, Alfred (1885–1968), 179, 237

Lamarck, Jean Baptiste (1744–1829), 13, 22, 210
Law of Ancestral Heredity, 49–52, 158
Lederberg, Joshua (1925–2008), 105, 183, 185–7, 202
lethal, 100, 136, 137, 141, 164, 237, 281
Lewis, Edward B. (1918–2004), 153, 242, 281, 282

Lewontin, Richard C. (1929–), 162, 165, 168, 276, 277, 280
Lindegren, Carl, 179
linkage, 84, 90, 91, 94, 95, 105, 187
  autosomes, 84
  centi-Morgans, 96
  linear map, 95, 97
  sex-linked, 83
linkage disequilibrium, 166, 275, 277
linkage map, 96, 97, 186, 187, 236
Linnæus, Carlos (1707–1778), 4, 13, 17–21, 23, 28, 61, 276
locus, 100, 103, 133–5, 142, 147, 153, 154, 184, 224, 242, 260, 287
  complex locus, 133, 281
Luria, Salvador E. (1912–1991), 182, 183, 185, 190

maize, 93, 96, 108, 111, 118, 130, 140, 149, 161, 172, 234
marker, 7, 54, 69, 83, 84, 92, 93, 96, 103, 110, 118, 131, 156, 178, 179, 186, 210, 235, 236, 239, 241, 242, 261, 282
  biochemical, 4, 7, 35
  cytological, 93, 99
  molecular, 4
maternal inheritance, 235
Mayr, Ernst (1904–2005), 276
McClintock, Barbara (1902–1992), 93, 108, 118, 234, 256
meiosis, 74, 77, 79, 84–6, 90–2, 96, 98, 99, 104, 112, 114, 116, 117, 181, 202, 208
  meiotic prophase, 93, 103, 114, 120
Mendel, Johann Gregor (1822–1884), 1, 4, 6–7, 13, 22, 25–37, 40, 41–2, 44–7, 49, 51–4, 60, 61, 82, 89, 114, 129, 131, 158, 209, 211, 254, 259, 265, 276, 287, 291
Mendelian factor, 155
Mendelian law of segregation, 30, 103
Mendelian particulate hypothesis, 50, 159
Meselson, Matthew S. (1930–), 196, 198
messenger-RNA, *see* mRNA

metabolic pathway, 179, 180, 218, 255, 269
methodology, 18, 46, 47, 52, 57, 79, 82, 96, 175, 282, 291
  molecular, 3, 242, 243, 254
  reductionist methodology, 29, 49, 80, 173, 177, 181, 209, 248, 254, 264, 270, 275, 278, *see also* reductionism: methodological
Miescher, Friedrich (1844–1895), 24
mitosis, 116, 236
  mitotic recombination, 239, 242
Mohr, Otto L. (1866–1967), 100
molecular biology, 4, 73, 126, 168, 173–5
Monod, Jacques (1910–1976), 175, 223–5, 228, 245, 250, 271
Morgan, Thomas H. (1866–1945), 3, 7, 24, 59, 67, 71, 72, 74, 79–83, 89, 90, 92, 94–7, 106, 113, 114, 125, 128, 171–3, 177, 179, 188, 213, 231–5, 239, 265, 284, 287
morphogenist tradition, 6, 17, 22, 23, 36, 40, 42, 58, 74, 130, *see also* hybridist tradition
mouse, 143, 144, 172, 192
mRNA, 127, 228, 247, 260, 261, 266
Muller, Herman J. (1890–1967), 64, 74, 84, 92, 97, 98, 103, 111, 112, 118, 125, 126, 129, 131–4, 136, 139, 142–8, 151–3, 155, 162–4, 166, 172, 180, 181, 195, 214–16, 218, 221, 263, 273
mutability, 39, 46, 144
mutagenesis, 126, 136, 140, 143
  chemical, 144–6, 172
  directed mutagenesis, 185
mutation, 46, 61, 66, 122, 132, 134, 135, 137, 141, 143, 225
  adaptive, 184
  back-mutation, 135
  base addition/deletion, 226
  base-substitution, 226
  detrimental, 138, 162, 166
  frame shift, 226
  induced, 141, 149, 213
  lethal, 218, 239

mosaic, 145
neutral, 166
nonsense, 273
point, 133, 135, 141, 142, 145, 146,
    149, 188
pre-adaptive, 183, 184
rate, 159
reverse, 213
single-hit, 141, 142, 149
transition, 154, 190, 213
transversion, 190
muton, 189

Nägeli, Carl (1817–1891), 36
Napp, Frantisek Cyril (1792–1867),
    25–7, 29
natural population, 112, 138, 164, 167,
    278
Nestler, Johann Karl (1783–1841), 27
Neurospora crassa, 96, 99, 172, 179,
    180, 218, 271
neutrality theory, 275, 277, 278
New Synthesis, 159, 163, 166, 167, 274,
    280, 281
Newcombe, H. B., 184
Nilsson-Ehle, Herman, 130
nominalism, 17
Norm of Reaction, 59, 210, *see also*
    Woltereck, Richard (1877–1944)
nuclear membrane, 120
nucleic acid, 176, 185, 191, 192, 194
nucleosome, 117
nucleotide, 4, 35, 102, 104, 155, 156,
    166, 168, 189, 191, 193, 203, 226,
    229, 260
  highly repetitive sequence, 126
  nucleotide sequence, 135, 250, 253,
    261, 278, *see also* DNA:
    sequence
Nussbaum, Moritz (1850–1915), 23, 83
Nüsslein-Volhard, Christiane (1942–),
    235, 283, 284

Okazaki fragments, 200
omni cellula e cellula, 23
one gene – one enzyme, 5, 72, 148, 152,
    176, 180, 181, 234, 246

operating system, 265
operator, 224, 245
operon, 245, 271, 272
ORF, 126, 247, 260, 265
Origin of Species, 23, 172
orthologous sequences, 281
oxygenation, 143

pairing,
  exchange *versus* distributive, 121,
    122
pangenes, 42, 44–6, 48, 53, 55
pangenesis, 15, 44
  Intracelluläre Pangenesis, 39, 44, 46,
    59, 209
Papilio dardanus, 160
paralogous sequences, 281
Paramecium, 181
Pardee, Arthur B. (1921–), 228
path-coefficients, 160
pattern formation, 283
Pearson, Karl (1857–1936), 49–51, 66,
    70, 73, 158
phage, 188, 189, 194, 205
phenocopy, 216
phenotype, 5, 43, 54, 56, 63–4, 67–70,
    74, 78, 86, 118, 127, 128, 131, 137,
    138, 146, 153, 157, 167, 185, 188,
    210, 212–16, 226, 232, 233, 236,
    260, 261, 268–72, 277, 288
phylogeny, 277
plasmid, 105, 106, 188, 208
pleiotropy, 86, 216–18, 224, 234, 256,
    260, 263
ploidy,
  aneuploidy, 78, 108, 110
  diploidy, 78, 108, 110, 150, 268, 269
  euploidy, 108
  haploidy, 101, 110, 150, 234, 269, 271
  polyploidy, 109
  triploidy, 108, 109, 114
polyclone, 241
polymorphism, 160, 161, 165, 166, 275,
    278
  balanced, 160, 161, 164
  electrophoretic, 161, 165, 276, 277
  industrial melanism, 160

polypeptide, 5, 127, 156, 165, 168, 190, 191, 223, 227, 228, 247, 260, 261, 270, 288, *see also* protein
polypeptide sequence, 126, 226
Pontecorvo, Guido (1907–1999), 154, 155
position effect, 134, 135, 146–8, 189, 219
variegated, 146, 147
predetermination, 235
preformation, 15, 22, 45–7, 57, 58, 60, 67, 72, 78, 82, 231, 284, 287
presence-and-absence hypothesis, 48, 52, 54–6, 68, 212, 268, 273
prokaryote, 105, 190, 218, 251, 262
protein, 7, 120, 121, 127, 145, 166–8, 191–4, 196, 210, 220, 222, 225, 226, 245, 246, 249, 253, 266, 276
amino acid sequence, 156, 196, 223, 229, 249, 272, 276–8
protein synthesis, 228
prototroph, 96, 186
pseudoallele, 152–4, 188, 242, 271
punctuated equilibrium, 253, 281
Punnett, Reginald C. (1875–1967), 56, 81, 87–9, 94, 158, 159
pure line, 62, 66

quantitative characteristics, 60, 62

random drift, 167
realism, *see* essentialism
reannealing, 251
recessivity, 31, 53, 55, 162, 213, 270, 271, *see also* dominance
recombination, 92, 95, 96, 100, 103, 120, 136, 186, 187, 201, 202, 203
enzyme-driven, 104
resolving power, 154, 155
recon, 189
reductionism, 8, 13, 162, 171–6, 238, 270, 274
conceptual, 4, 67, 209, 281, 283
instrumental, 125, 129, 131
methodological, 4, 7, 73, 181, 225, *see also* methodology: reductionist methodology

reductionist determinism, 7, 177, 228, 245, 254, 289
regression to the mean, 62, 67
regulation, 126, 173, 223, 245, 253, 264, 266
epigenetic regulation, 262
replication, 173, 202, 221, 222
semi-conservative, 196–8, 202
repressor, 225, 245
reproduction, *see* reproduction and generation
reproduction and generation, 11, 12, 15, 18, 21, 39, 44, 49
research tradition, 32, 277
Rhoades, Marcus M. (1903–1991), 108
ribosome, 5, 247, 253
*r*II, 154, 155, 156, 188–90, 223, 225
RNA, 5, 7, 8, 106, 118, 127, 176, 191, 193, 210, 225, 228–30, 246, 247, 249, 252, 256, 260, 262–6, 272, 276, 288
micro-RNA, 266
non-coding, 266, 267, 283
RNA-primer, 200
small RNA, 266
RNA processing, 127, 261
Roberts, Richard J. (1943–), 259, 261

Saccharomyces cerevisiae, *see* yeast
Salmonella typhimurium, 156
satellite DNA, 251
Schrödinger, Erwin (1887–1961), 143, 172
Schultz, Jack (1904–1971), 192
segregation, 30, 35, 39, 116
independent segregation, 30, 47, 55, 74, 79, 82, 95
Mendelian segregation, 42, 48, 74
selection, 62, 64, 66, 68, 165, 166, 214, 215, 268
group selection, 274, 275, 279
individual selection, 280
kin selection, 279
natural selection, 3, 44, 46, 66, 166–8, 213, 215, 258, 269, 274, 275, 279
purifying selection, 278
selection coefficient, 159, 278

Sequence Hypothesis, 228, 229
sex determination, 79, 108–10, 185, 235
Sharp, Phillip A. (1944–), 259, 261
Shull, George H. (1874–1954), 56, 162, 214
Simpson, Gerald G. (1902–1984), 254
Singer, Fred (1924–), 191
SNP, 4, 156
soma line, 45, 77, 82, *see also* germ line
species, 18, 19, 20–3, 27, 28, 41, 45, 46, 58, 61, 275, 279
splicing, 5, 127, 156, 247, 261, 262
    alternative splicing, 247, 262
    trans-splicing, 247, 262
"sport", 48, 61, 67, 81
Stadler, Lewis J. (1896–1954), 108, 125, 129, 140, 145, 148, 149–52, 192
Stahl, Franklin (1929–), 196, 198
Stent, Gunther (1924–2008), 246
Stern, Curt (1902–1981), 93, 214, 239
Streisinger, George (1927–1984), 226
Sturt, 236
Sturtevant, Alfred H (1891–1970), 1, 74, 94–6, 106, 112, 134, 172, 178, 181, 188, 219, 231, 234–6, 239
superorganism, 280
synapsis, 101, 111, 113–15, 119, 120
synaptonemal complex, 120, 121
system, 254
    complex system, 185, 239, 243

Target Theory, 142–3, 149, 172
Tatum, Edward (1909–1975), 5, 72, 105, 148, 179, 180, 181, 185, 186, 234
taxonomy, 8, 12, 13, 16, 17, 19, 23, 43, 58, 180, 249, 277
Taylor, J. Herbert, 198
telomere, 118, 120
    telomerase, 118
Temin, Howard (1934–1994), 210
tetrad analysis, 96–9, 103, 179
Theory of Repetition of Parts, 48, 88
Timoféeff-Ressovsky, Nikolai V. (1900–1981), 142, 143, 159, 210
top-down, 6, 8, 15, 44, 50, 52, 62, 63, 65, 70, 134, 209, 219, 233, 235, 243,

247, 248, 254, 260, 263, 264, 275, 280, 281, 282, *see also* bottom-up
Tradescantia, 149
trans, 188
transcription, 2, 8, 35, 126, 127, 173, 177, 210, 223, 247, 254, 260–2, 264–7, 288, 289
    reverse transcription, 230
transdetermination, 176, 238, 241, 256
transformation, 191, 192, 193
translation, 8, 126, 127, 165, 168, 173, 177, 247, 260, 261, 273, 288, 289
transmission, 2, 11–13, 21, 23, 24, 34, 42, 44, 45, 47, 49, 54, 55, 59, 60, 63, 67, 77, 81, 82, 125, 127, 168, 175, 181, 195, 211, 247, 287
Tschermak-Seysenegg, Erich von (1871–1962), 40, 41

Unger, Franz (1800–1870), 27, 33
unit character, 7, 39, 42, 44–9, 53–60, 62, 64, 65, 87, 88, 125, 129, 209, 212, 263
UV, 139, 145, 149, 172, 206

variables, 4, 7, 27, 43, 69
    interaction, 159
variation, 46, 58, 72, 132
viability, 87, 138, 151, 162, 164
virus, 139, 155, 178, 180, 181, 184, 191, 194, 259
Vogt, Oskar (1870–1959), 210

Waddington, Conrad H. (1905–1975), 275, 280
Wallace, Bruce (1920–), 164
Watson, James D. (1928–), 3, 7, 73, 126, 173, 177, 195, 196, 220–1, 225, 287, 289
Watson-Crick, 154, 155, 196, 197, 199, 210, 219, 222, 246
Weaver, Warren (1894–1978), 173
Weinberg, Wilhelm (1852–1937), 159
Weismann, August (1834–1914), 23, 24, 58, 59, 63, 71, 72, 74, 77, 78, 83, 113, 227, 237

Weldon, W. F. Raphael (1860–1906), 50, 66
Whitehouse, Harold L. K., 104, 203
Williams, George C. (1926–), 274
Wilson, Edmund B. (1856–1939), 71, 78, 79, 113, 181
Winkler, Hans (1877–1945), 92
Wollman, Elie, 105
Woltereck, Richard (1877–1944), 58, 210
Wright, Sewall (1889–1988), 159, 160, 167, 214, 268, 269, 279
Wynne-Edwards, V. C., 275

X-chromosome, 79, 84, 99, 109, 122, 136, 137, 141, 236
X-rays, 108, 111, 139, 140, 141, 144, 149, 164, 172, 179, 239

Y-chromosome, 84, 109, 110, 122
yeast, 96, 121, 191, 208
Yule, Udny G. (1871–1951), 50–2

Zebra fish (*Danio rerio*), 284
Zimmer, Karl G. (1911–1988), 142